Grundlehren der mathematischen Wissenschaften 220

A Series of Comprehensive Studies in Mathematics

Editors

S. S. Chern   J. L. Doob   J. Douglas, jr.
A. Grothendieck   E. Heinz   F. Hirzebruch   E. Hopf
S. Mac Lane   W. Magnus   M. M. Postnikov
F. K. Schmidt   W. Schmidt   D. S. Scott
K. Stein   J. Tits   B. L. van der Waerden

Managing Editors

B. Eckmann   J. K. Moser

A. A. Kirillov

Elements of the Theory of Representations

Translated from the Russian by Edwin Hewitt

Springer-Verlag
Berlin Heidelberg New York 1976

Aleksandr A. Kirillov
Moscow State University

Edwin Hewitt
University of Washington

Title of the Russian Original Edition:
Elementy Teorii Predstavleniĭ
Publisher: Nauka, Moscow, 1972

AMS Subject Classifications (1970):
Primary 43-xx   Secondary 20Cxx, 22A25, 22Dxx, 22Exx, 81A17, 81A78

ISBN-13: 978-3-642-66245-4   e-ISBN-13: 978-3-642-66243-0
DOI: 10.1007/978-3-642-66243-0

Library of Congress Cataloging in Publication Data. Kirillov, Aleksandr Aleksandrovich, 1936—. Elements of the theory of representations (Grundlehren der mathematischen Wissenschaften; 220). Translation of Elementy teorii predstavleniĭ. Bibliography: p. Includes index. 1. Representations of groups. I. Title. II. Series: Die Grundlehren der mathematischen Wissenschaften in Einzeldarstellung; 220. QA171. K5213. 512'.22. 75-31705

This work is subject to copyright. All rights are reserved, whether the whole or part of the material is concerned, specifically those of translation, reprinting, re-use of illustrations, broadcasting, reproduction by photocopying machine or similar means, and storage in data banks. Under § 54 of the German Copyright Law where copies are made for other than private use, a fee is payable to the publisher, the amount of the fee to be determined by agreement with the publisher.
© by Springer-Verlag Berlin Heidelberg 1976
Softcover reprint of the hardcover 1st edition 1976
Typesetting and printing: Zechnersche Buchdruckerei, Speyer
Bookbinding: Konrad Triltsch, Würzburg

Translator's Preface

The translator of a mathematical work faces a task that is at once fascinating and frustrating. He has the opportunity of reading closely the work of a master mathematician. He has the duty of retaining as far as possible the flavor and spirit of the original, at the same time rendering it into a readable and idiomatic form of the language into which the translation is made. All of this is challenging. At the same time, the translator should never forget that he is not a creator, but only a mirror. His own viewpoints, his own preferences, should never lead him into altering the original, even with the best intentions. Only an occasional translator's note is permitted.

The undersigned is grateful for the opportunity of translating Professor Kirillov's fine book on group representations, and hopes that it will bring to the English-reading mathematical public as much instruction and interest as it has brought to the translator. Deviations from the Russian text have been rigorously avoided, except for a number of corrections kindly supplied by Professor Kirillov. Misprints and an occasional solecism have been tacitly taken care of. The translation is in all essential respects faithful to the original Russian.

The translator records his gratitude to Linda Sax, who typed the entire translation, to Laura Larsson, who prepared the bibliography (considerably modified from the original), and to Betty Underhill, who rendered essential assistance.

Seattle, June 1975   Edwin Hewitt

Preface

The author of this book has, over a number of years, given courses and directed a seminar at Moscow State University on the theory of group representations.

The majority of the students in these courses and participants in the seminar have been university students of the first two years (and also occasionally graduate students and gifted secondary school pupils).

The membership of the seminar has constantly renewed itself. There have been new participants, not overly burdened with knowledge, but ready to study things new to them and to solve a huge number of problems.

For each new group of participants, it was necessary to organize a "primer" on the parts of mathematics needed for the theory of representations and also on the foundations of the theory of representations.

The author quickly got the idea of replacing himself by a book which should be on the one hand not too thick (which always frightens the reader) but which on the other hand should contain all of the needed information.

Through various circumstances, the realization of this idea required far more time than was originally contemplated. Nevertheless, through the moral support of friends and of my teacher I. M. Gel'fand, the book has now finally been written. The author begs forgiveness of his readers for the facts that the book is thicker than one would like and also contains only a part of what it ought to.

The first part of the book (§§ 1–6) are not directly connected with the theory of representations. Here we give facts needed from other parts of mathematics, with emphasis on those that do not appear in the prescribed curricula of elementary university courses. A reader familiar with this material may begin at once with the second part (§§ 7–15). This part contains the principal concepts and methods of the theory of representations. In the third part (§§ 16–19), we illustrate the general constructions and theorems of the second part by concrete examples.

The historical sketch found at the end of the book reflects the author's view of the development of the theory of representations, and makes no claim to be a reference work on the history of mathematics. At the end of the essay, we describe the present state of the theory of representations and give references to the periodical literature.

A particular feature of the book, and one which has reduced its size enormously, is the large number of problems. These problems, and the remarks appended to them, are printed in separate paragraphs. Nevertheless, one must not ignore them, since they play an essential rôle in the main text. In particular, a majority of the proofs are given in the form of a cycle of mutually connected problems. Almost all problems are supplied with remarks, which as a rule enable one to reconstruct the solution without difficulty. All the same it is useful to try to solve a problem independently and to turn to the remark appended only in case of failure.

We point out certain peculiarities in our choice of subject matter. Very little mention is made in the book of finite-dimensional representations of semisimple Lie groups and Lie algebras. The fact is that there are already available in the Russian language a sufficient number of good expositions of this part of the theory of representations (see [47], [46], [57]), and the author did not wish to repeat them.

The rôle of the theory of group representations in the theory of special functions is completely ignored in the present book. The monograph of N. Ja. Vikenkin [53] may serve as a good introduction to this topic.

The author's task has also not included a description of the manifold applications of the theory of representations in mathematical physics. At the present time there is a wide literature dealing with these applications (see for example [3], [37], [40], [28]).

We have apportioned a large space to the method of orbits, which has up to now not made its way into textbooks, and which by its simplicity and perspicuity without doubt belongs to the fundamentals of the theory of representations. A certain incompleteness in § 15, which deals with the method of orbits, is explained by the current state of knowledge in this area. Many important theorems have been proved only in special cases, or indeed exist only as conjectures.

At the present time, many mathematicians are working in this field both in the Soviet Union and in other countries. Beyond any peradventure our knowledge of the connections between orbits and representations will be much greater within a few years than it is now. The author hopes that some of the readers of this book may bring their contributions to the development of the theory of orbits.

A. Kirillov

Table of Contents

First Part. Preliminary Facts

§ 1. Sets, Categories, Topology 1
    1.1. Sets 1
    1.2. Categories and Functors 3
    1.3. The Elements of Topology 6

§ 2. Groups and Homogeneous Spaces 11
    2.1. Transformation Groups and Abstract Groups 11
    2.2. Homogeneous Spaces 15
    2.3. Principal Types of Groups 16
    2.4. Extensions of Groups 18
    2.5. Cohomology of Groups 21
    2.6. Topological Groups and Homogeneous Spaces 22

§ 3. Rings and Modules 24
    3.1. Rings 25
    3.2. Skew Fields 27
    3.3. Modules over Rings 28
    3.4. Linear Spaces 31
    3.5. Algebras 32

§ 4. Elements of Functional Analysis 35
    4.1. Linear Topological Spaces 35
    4.2. Banach Algebras 44
    4.3. C^*-Algebras 48
    4.4. Commutative Operator Algebras 52
    4.5. Continuous Sums of Hilbert Spaces and von Neumann Algebras 57

§ 5. Analysis on Manifolds 62
    5.1. Manifolds 62
    5.2. Vector Fields 68
    5.3. Differential Forms 74
    5.4. Bundles 77

§ 6. Lie Groups and Lie Algebras 83
    6.1. Lie Groups 83
    6.2. Lie Algebras 86
    6.3. The Connection between Lie Groups and Lie Algebras . . . 95
    6.4. The Exponential Mapping 101

Second Part. Basic Concepts and Methods of the Theory of Representations

§ 7. Representations of Groups 108
    7.1. Linear Representations 108
    7.2. Representations of Topological Groups in Linear Topological Spaces 111
    7.3. Unitary Representations 112

§ 8. Decomposition of Representations 115
    8.1. Decomposition of Finite Representations 115
    8.2. Irreducible Representations 119
    8.3. Completely Reducible Representations 121
    8.4. Decomposition of Unitary Representations 125

§ 9. Invariant Integration 129
    9.1. Means and Invariant Measures 129
    9.2. Applications to Compact Groups 133
    9.3. Applications to Noncompact Groups 138

§ 10. Group Algebras 139
    10.1. The Group Ring of a Finite Group 139
    10.2. Group Algebras of Topological Groups 141
    10.3. Application of Group C^*-Algebras 145
    10.4. Group Algebras of Lie Groups 148
    10.5. Representations of Lie Groups and their Group Algebras . . 152

§ 11. Characters 156
    11.1. Characters of Finite-Dimensional Representations 156
    11.2. Characters of Infinite-Dimensional Representations 160
    11.3. Infinitesimal Characters 163

§ 12. Fourier Transforms and Duality 166
    12.1. Commutative Groups 167
    12.2. Compact Groups 173
    12.3. Ring Groups and Duality for Finite Groups 177
    12.4. Other Results 179

§ 13. Induced Representations 181
    13.1. Induced Representations of Finite Groups 182
    13.2. Unitary Induced Representations of Locally Compact Groups 187
    13.3. Representations of Group Extensions 195
    13.4. Induced Representations of Lie Groups and their Generalizations 199
    13.5. Intertwining Operators and Duality 205
    13.6. Characters of Induced Representations 209

§ 14. Projective Representations 215
    14.1. Projective Groups and Projective Representations 215
    14.2. Schur's Theory 220
    14.3. Projective Representations of Lie Groups 222

Table of Contents   XI

§ 15. The Method of Orbits 226
   15.1. The Co-Adjoint Representation of a Lie Group 226
   15.2. Homogeneous Symplectic Manifolds 231
   15.3. Construction of an Irreducible Unitary Representation by an
         Orbit 235
   15.4. The Method of Orbits and Quantization of Hamiltonian
         Mechanical Systems 241
   15.5. Functorial Properties of the Correspondence between Orbits
         and Representations 249
   15.6. The Universal Formula for Characters and Plancherel Measures 251
   15.7. Infinitesimal Characters and Orbits 256

Third Part. Various Examples

§ 16. Finite Groups 259
   16.1. Harmonic Analysis on the Three-Dimensional Cube 259
   16.2. Representations of the Symmetric Group 262
   16.3. Representations of the Group $SL(2, \mathbf{F}_q)$ 265
   16.4. Vector Fields on Spheres 268

§ 17. Compact Groups 271
   17.1. Harmonic Analysis on the Sphere 271
   17.2. Representations of the Classical Compact Lie Groups 274
   17.3. Spinor Representations of the Orthogonal Group 276

§ 18. Lie Groups and Lie Algebras 279
   18.1. Representations of a Simple Three-Dimensional Lie Algebra . . 279
   18.2. The Weyl Algebra and Decomposition of Tensor Products . . 281
   18.3. The Structure of the Enveloping Algebra $U(\mathfrak{g})$ for $\mathfrak{g} = \mathfrak{sl}(2, \mathbf{C})$. 283
   18.4. Spinor Representations of the Symplectic Group 287
   18.5. Representations of Triangular Matrix Groups 290

§ 19. Examples of Wild Lie Groups 292

A Short Historical Sketch and a Guide to the Literature 297

Bibliography 302

Subject Index 310

First Part. Preliminary Facts

§ 1. Sets, Categories, Topology

1.1. Sets

The background in the theory of sets needed to read this book is amply supplied by what is given in ordinary university courses. (See for example the first chapters of the textbooks of A.N. Kolmogorov and S.V. Fomin [38] and of G.E. Šilov [49].) One can find more penetrating treatments (including an exact definition of the concept of set) in the books of Fraenkel and Bar-Hillel [17] and P. Cohen [12].

For the reader's convenience we list here some of the notation that we shall use and also recall certain definitions.

$\emptyset$ denotes the void set;

$x \in X$ means that the element x belongs to the set X;

$x \notin X$ means that the element x does not belong to the set X;

$X \subset Y$ means that the set X is contained in the set Y (and possibly coincides with it);

$\bigcup_{\alpha \in A} X_\alpha$ denotes the union of the system of sets X_α, which are indexed by the set A; if A is finite, then we also use the notation $X \cup Y \cup \cdots \cup Z$;

$\bigcap_{\alpha \in A} X_\alpha$ denotes the intersection of the system of sets X_α;

$X \setminus Y$ denotes the complement of the set X in the set Y;

$\prod_{\alpha \in A} X_\alpha$ denotes the Cartesian product of the sets X_α, that is, the collection of all functions $\{x_\alpha\}_{\alpha \in A}$, where $x_\alpha \in X_\alpha$;

$\left. \begin{array}{l} f: X \to Y \\ \text{or} \\ X \xrightarrow{f} Y \end{array} \right\}$ denotes the mapping f of the set X into the set Y;

$\left. \begin{array}{l} f: x \mapsto y \\ \text{or} \\ x \xrightarrow{f} y \end{array} \right\}$ means that the mapping f carries the element x into the element y;

X^Y is the set of all mappings of the set Y into the set X;

card X is the cardinal number of the set X; for finite sets X we also use the notation $|X|$;

$\{x; A\}$ denotes the set of all x that satisfy the condition A.

We say that a *binary relation* is defined in a set X if we have specified a certain subset R of $X \times X$. Instead of the relation $(x, y) \in R$ we also write xRy and we say that x and y stand in the relation R or that x and y are connected by the relation R. The *inverse relation* R^{-1} is defined as the set of all pairs (x, y) for which $(y, x) \in R$. The *product* $R_1 \cdot R_2$ is the set of all pairs (x, y) for which there exists an element z such that $(x, z) \in R_1$ and $(z, y) \in R_2$.

A relation R is called *reflexive* if R contains the diagonal $\Delta = \{(x, x); x \in X\}$, *symmetric* if $R = R^{-1}$, and *transitive* if $R \cdot R \subset R$. A relation with all three of these properties is called an *equivalence relation*. In this case, instead of $(x, y) \in R$, we say "x and y are R-equivalent" (or simply equivalent, if it is clear what relation R we have in mind).

The set of elements equivalent to a given $x \in X$ is the *equivalence class* containing x. The set $X_{(R)}$ of equivalence classes is called the *factor set* of the set X by the relation R. Assigning each $x \in X$ to the class containing it, we obtain the *canonical mapping* $p: X \to X_{(R)}$.

It is clear that a mapping $f: X \to Y$ can be embedded in the commutative diagram[1]

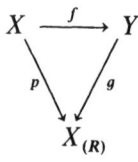

$$X \xrightarrow{f} Y$$
$$p \searrow \quad \swarrow g$$
$$X_{(R)}$$

if and only if f is constant on each equivalence class. In this case, g is defined uniquely from f and is called a *factor mapping*.

An *order relation* on a set X is a transitive binary relation R that is antisymmetric in the following sense: $R \cap R^{-1} \subset \Delta$. For an order relation R, we usually write $x \succ y$ instead of $(x, y) \in R$.

A set that is equipped with an order relation is said to be *ordered* (sometimes *partially ordered*). An ordered set is said to be *linearly ordered* if $R \cup R^{-1} = X \times X$ (i.e., if every pair of elements are comparable). The following assertions are equivalent:

1 (Zermelo's axiom of choice). *The product of an arbitrary family of nonvoid sets is nonvoid.*

2 (Zorn's lemma). *Suppose that X is an ordered set in which every linearly ordered subset Y is bounded* (that is, there is an element $x \in X$ such that $x \succ y$ for all $y \in Y$). *Then the set X contains at least one maximal element* (that is, an element x_0 such that if $x \succ x_0$, then $x = x_0$; maximality of x_0 does not imply that $x_0 \succ x$ for all $x \in X$).

Zorn's lemma is a generalization of the well-known principle of mathematical induction, and replaces this principle in situations where we are considering uncountable sets.

---

[1] See footnote 2 on page 4.

§ 1. Sets, Categories, Topology

A *directed set*[1] is a set A with an order relation R defined in it satisfying the following additional condition:

for arbitrary $\alpha, \beta \in A$, there exists an element $\gamma \in A$ such that $\alpha \prec \gamma$, $\beta \prec \gamma$.

Let (A, R) be a directed set and X an arbitrary set. A mapping of A into X is called a *net* or a *direction* in X. Clearly this notion is a generalization of the notion of a sequence in X (to which it reduces if A is the set of natural numbers with the ordinary order relation).

As a rule one considers in mathematics sets which are endowed with one or another *structure* (for example, ordered sets, groups, topological spaces, and so on). We can give an exact meaning to this notion.

A *tower of sets* over X is any set obtained from X and auxiliary sets $S, T, \ldots$ by the elementary operations listed above (see page 1). To define a structure on X is to fix an element of a certain tower of sets over X. (For the foregoing examples of structures, the corresponding towers have the form $(2)^{X \times X}$, $X^{X \times X}$, $(2)^{(2)^X}$, where (2) is an auxiliary set consisting of exactly 2 elements.)

1.2. Categories and Functors

The language of categories, which will be used in this book, is so simple and natural that it presents no difficulties even for a reader unfamiliar with it. Here we give only a few basic definitions. More information can be found, for example, in the book of A. Grothendieck [25, Ch. I] or in the appendix of D. Buchsbaum to the book [10]. See also the Appendix "The language of categories" to the lecture notes of Ju. I. Manin on algebraic geometry (Publishing House of Moscow State University, 1970), from which we borrow the first sentence.

"The language of categories embodies a "sociological" approach to a mathematical object: a group or a space is considered not as a set with an inherent structure by itself, but as a member of the society of objects similar to it."

We say that we are given a *category* K if
1) there is given a class $\text{Ob} K$ of *objects* of the category K;
2) for every pair A, B of objects of K there is given a set $\text{Mor}(A, B)$ of *morphisms* of the object A into the object B;
3) there is a *law of composition* defined for every triple A, B, C of objects in K, that is, a mapping

$$\text{Mor}(A, B) \times \text{Mor}(B, C) \to \text{Mor}(A, C).$$

The composition of the morphisms $f \in \text{Mor}(A, B)$ and $g \in \text{Mor}(B, C)$ is denoted by $g \circ f$ and satisfies the following conditions:

a) $f \circ (g \circ h) = (f \circ g) \circ h$ for arbitrary $f \in \text{Mor}(C, D)$, $g \in \text{Mor}(B, C)$, $h \in \text{Mor}(A, B)$;

b) for every $A \in \text{Ob} K$ there exists an element $1_A \in \text{Mor}(A, A)$ such that $1_A \circ f = f$, $g \circ 1_A = g$ for arbitrary $f \in \text{Mor}(B, A)$, $g \in \text{Mor}(A, B)$.

As an example, consider the category M where $\text{Ob} M$ is the class of all sets and $\text{Mor}(A, B) = B^A$. Many of the categories considered below are *subcategories*

---

[1] Bourbaki uses the expression *filtering to the right*.

of M, that is, the objects of these categories are sets, the morphisms are mappings of sets, and composition of morphisms is composition of mappings[1].

If a morphism $f \in \mathrm{Mor}(A, B)$ admits an *inverse* morphism f^{-1} (i.e., a morphism such that $f \circ f^{-1} = 1_B$, $f^{-1} \circ f = 1_A$), then it is called an *isomorphism*, and the objects A and B are called *isomorphic*.

For every category K, we can define the *dual* category K°. By definition, we have $\mathrm{Ob}\, K^\circ = \mathrm{Ob}\, K$, $\mathrm{Mor}(A, B)^\circ = \mathrm{Mor}(B, A)$. The composition of f and g in K° is defined as the composition of g and f in K.

An object X is called *universally repelling (universally attracting)* if for every $Y \in \mathrm{Ob}\, K$, the set $\mathrm{Mor}(X, Y)$ ($\mathrm{Mor}(Y, X)$) consists of exactly one element. From this definition, it follows that if there are several universal objects in a category K, then they are all canonically isomorphic.

It is clear that in going from a category to its dual, universally repelling objects turn into universally attracting objects, and conversely.

The concept of universal object permits us to consider from a single point of view a great number of constructions that are used in mathematics. In particular, we shall see below that tensor products, enveloping algebras, induced representations, and cohomology of groups can be defined as universal objects in appropriately chosen categories.

By way of an example, we now give the definition of the sum and product of objects of an arbitrary category.

Let $\{X_\alpha\}_{\alpha \in A}$ be a family of objects of a category K. We shall consider a new category K_A. The objects of K_A are the collections $(Y, \{f_\alpha\}_{\alpha \in A})$, where Y is an object in K and $f_\alpha \in \mathrm{Mor}(X_\alpha, Y)$.

A morphism from $(Y, \{f_\alpha\}_{\alpha \in A})$ to $(Z, \{g_\alpha\}_{\alpha \in A})$ is defined as a morphism $h: Y \to Z$ such that for all $\alpha \in A$, the following diagram is commutative[2]:

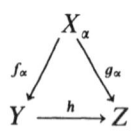

$$\begin{array}{ccc} & X_\alpha & \\ {\scriptstyle f_\alpha}\swarrow & & \searrow{\scriptstyle g_\alpha} \\ Y & \xrightarrow{h} & Z \end{array}$$

We shall suppose that there is a universally repelling object $\{X, i_\alpha\}$ in the category K_A (all such objects are canonically isomorphic, as we mentioned above). Then the object X is called the *sum* of the family $\{X_\alpha\}$ and the morphism i_α is called the *canonical embedding* of the summand X_α in the sum X.

The definition of the *product* P of the family $\{X_\alpha\}_{\alpha \in A}$ and of the *canonical projection* $p_\alpha \in \mathrm{Mor}(P, X_\alpha)$ is obtained from the definition of sum by *reversing arrows*, that is, by replacing K_A by $(K_A^\circ)^\circ$.

---

[1] The reader may wish to have an example of a category not of this type. Examples are the category of formal groups and the category of diagrams. The category K_A considered below is a special case of these.

[2] A diagram consisting of objects and morphisms of a category K is said to be *commutative* if the composition of morphisms along a path marked out by arrows of the diagram depends only upon the initial and terminal points of the path. In the following example, this means that $h \circ f_\alpha = g_\alpha$.

§ 1. Sets, Categories, Topology

Problem 1. Show that the sum and product exist for an arbitrary family of objects in the category M.

Hint. Consider the operations of disjoint union and ordinary product of sets.

A small change in the definition of sum and product leads to the concepts of inductive and projective limits. Suppose that the set of indices A is a directed set and, if $\alpha \prec \beta$, then there exists a morphism $f_{\alpha\beta} \in \mathrm{Mor}(X_\alpha, X_\beta)$ such that $f_{\alpha\gamma} = f_{\beta\gamma} \circ f_{\alpha\beta}$ for every triple $\alpha \prec \beta \prec \gamma$. We consider the category whose objects are collections $(Y, \{f_\alpha\}_{\alpha \in A})$, $f_\alpha \in \mathrm{Mor}(X_\alpha, Y)$, such that for all $\alpha \prec \beta$ the following diagrams commute:

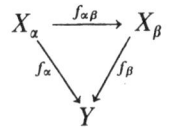

$$\begin{array}{ccc} X_\alpha & \xrightarrow{f_{\alpha\beta}} & X_\beta \\ & \searrow{\scriptstyle f_\alpha} \quad \swarrow{\scriptstyle f_\beta} & \\ & Y & \end{array}$$

The morphisms of $(Y, \{f_\alpha\})$ into $(Z, \{g_\alpha\})$ are those morphisms $h \in \mathrm{Mor}(Y, Z)$ such that for all $\alpha \in A$, the following diagrams commute:

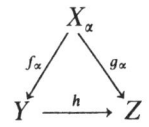

$$\begin{array}{ccc} & X_\alpha & \\ {\scriptstyle f_\alpha}\swarrow & & \searrow{\scriptstyle g_\alpha} \\ Y & \xrightarrow{h} & Z \end{array}$$

A universally repelling object in this category is called the *inductive limit* of the family $\{X_\alpha\}$. The definition of *projective limit* is obtained by reversing arrows.

Problem 2. Let A be the set of all natural numbers with the following definition of order: $m \prec n$ means that m divides n. Let X_n be the set of all integers, for every n. Let f_{mn} be the operation of multiplying by n/m. Prove that the inductive limit of the family $\{X_m\}$ can be identified in a natural way with the set of rational numbers, and the mapping f_m with division by m.

Let K_1 and K_2 be two categories. Suppose that to every object X in K_1, there corresponds an object $F(X)$ in K_2 and to every morphism $f \in \mathrm{Mor}(X, Y)$ there corresponds a morphism

$$F(f) \in \mathrm{Mor}(F(X), F(Y)).$$

Suppose further that the equalities

$$F(1_X) = 1_{F(X)}, \quad F(f \circ g) = F(f) \circ F(g),$$

hold. Then we say that F is a *covariant functor* from K_1 into K_2. We obtain the notion of a *contravariant functor* if we replace the last condition by $F(f \circ g) = F(g) \circ F(f)$. (This is equivalent to replacing one of the categories K_1 and K_2 by its dual category.)

Covariant functors from K_1 into K_2 are themselves a category. The morphisms of F into G are the so-called *functorial morphisms*. These assign to each object X in K_1 a morphism $\phi(X): F(X) \to G(X)$ such that the diagram

$$\begin{array}{ccc} F(X) & \xrightarrow{\phi(X)} & G(X) \\ {\scriptstyle F(f)}\downarrow & & \downarrow{\scriptstyle G(f)} \\ F(Y) & \xrightarrow{\phi(Y)} & G(Y) \end{array}$$

is commutative for all $f \in \mathrm{Mor}(X, Y)$.

The category of contravariant functors is defined analogously.

We can also define functors of several variables, covariant in some variables and contravariant in the others.

Problem 3. Let K be an arbitrary category. Show that the mapping $(X, Y) \to \mathrm{Mor}(X, Y)$ can be completed to a functor from $K \times K$ into M, contravariant in the first variable and covariant in the second.

Hint. For $f \in \mathrm{Mor}(X_1, X)$, $g \in \mathrm{Mor}(Y, Y_1)$, $\phi \in \mathrm{Mor}(X, Y)$, set $F(f, g): \phi \mapsto g \circ \phi \circ f$.

A covariant functor F from the category K into M is called *representable* if it is isomorphic to a functor $\mathrm{Mor}(X, \cdot)$, obtained from a bifunctor[1] Mor by fixing the first variable in an obvious way. The object X is called a *representing object* for the functor F.

Analogously, a contravariant functor has a representing object Y if F is isomorphic to $\mathrm{Mor}(\cdot, Y)$.

Many important functors are representable or become representable under suitable modification of the category.

1.3. The Elements of Topology

This section consists essentially of a list of terminology, concepts, and basic facts about topology. The fundamentals of topology can be found in the book [38, Ch. II]. A more detailed treatment is given in the book of J. L. Kelley [36].

A topological space is a set X in which we are given a family τ of subsets having the following properties:
1) the void set and X itself belong to τ;
2) the intersection of a finite number of elements of τ belongs to τ;
3) the union of an arbitrary family of elements of τ belongs to τ.

The system τ is called a *topology* on X.

A subsystem $\tau' \subset \tau$ is called a *base for the topology* τ if every element of τ is the union of a certain family of elements of τ'. Every system of subsets of X that satisfies 1) and 2) above is the base for a certain topology.

Sets belonging to the topology τ are called *open* relative to this topology. An open set containing a point $x \in X$ is called a neighborhood of that point. The complements of open sets are called *closed*. For every $Y \subset X$, there exists a smallest closed set containing Y. It is called the *closure* of Y. We say that a

---

[1] That is, a functor of two variables.

§ 1. Sets, Categories, Topology

subset Y is dense in X if the closure of Y coincides with X. The sets that are obtained from open and closed sets by the operations of countable unions, countable intersections, and complementation are called *Borel sets*.

One can define the general concept of a *limit* in a topological space in the following way. Let $\{x_\alpha\}_{\alpha \in A}$ be a direction in X (that is, a family of points of X, numbered by the elements of a directed set A; see para. 1.1). A point x is said to be a limit of the direction $\{x_\alpha\}$ if for every neighborhood U of the point x, there exists an element $\alpha \in A$ such that $x_\beta \in U$ for all $\beta > \alpha$. We write this as $x_\alpha \underset{A}{\to} x$ or $\lim_{\alpha \in A} x_\alpha = x$. The reference to the set A is often omitted.

A mapping of one topological space into another is called *continuous* if the inverse image of every open set is open, and is called a *Borel mapping* if the inverse image of every open set is a Borel set.

Problem 1. Prove that a mapping f is continuous if and only if the condition $x_\alpha \underset{A}{\to} x$ implies that $f(x_\alpha) \underset{A}{\to} f(x)$ (that is, f commutes with the operation of passing to the limit).

In spaces with a countable basis, the general directions of problem 1 may be replaced by ordinary sequences.

Topological spaces and continuous mappings form a category T, in which the sum and the product of an arbitrary family of objects is defined.

Problem 2. Prove that the product of a family of topological spaces $\{X_\alpha, \tau_\alpha\}_{\alpha \in A}$ is the set $\prod_{\alpha \in A} X_\alpha$ with the topology τ, a basis for which is formed by sets of the form $\prod_{\alpha \in A_1} U_\alpha \times \prod_{\alpha \in A \setminus A_1} X_\alpha$, where A_1 is a finite subset of A and the sets U_α belong to τ_α.

Hint. First consider the product of two spaces.

Every subset Y of a topological space (X, τ) is itself a topological space if we define the open subsets of Y to be the intersections with Y of open subsets of X. A subset Y with this topology is called a *subspace of* X.

Let R be an equivalence relation on X. The factor set $X_{(R)}$ will be a topological space if we define as open those subsets whose inverse images are open in X. The set $X_{(R)}$ with this topology is called a *factor space* of the space X.

A topological space is called *compact* if every covering of it by open sets admits a finite subcovering.

Problem 3. The product of an arbitrary family of compact spaces is compact. The image of a compact space under a continuous mapping is compact.

A topological space is called *separated* or *Hausdorff* if every pair of distinct points admit disjoint neighborhoods. It is called *semiseparated*, or a T_0 *space*, if one of each pair of distinct points admits a neighborhood not containing the other. A compact Hausdorff space is called a *compactum*.

In Hausdorff spaces, every direction can admit only one limit. This property accounts for the fact that the majority of topological spaces used in mathematics and in its applications are Hausdorff.

Nevertheless, there are important classes of topological spaces for which Hausdorff separation does not in general hold. An example is provided by factor spaces. (For example, spaces of orbits of transformation groups—see below, § 2—are in the most interesting cases T_0 but not Hausdorff.) A second example is spaces of algebraic geometry with the Zariski topology (in which the closed sets are the sets of simultaneous solutions of systems of algebraic equations).

A topological space is called *connected* if it cannot be represented as the union of two disjoint nonvoid sets, both of which are open and closed.

Problem 4. Every continuous mapping of a connected space into a discrete space (i.e., a space in which every subset is open) is necessarily constant.

Hint. Prove that the inverse image of every point must be either void or the entire space.

The following construction yields an important class of topological spaces.

We say that there is a *distance* or *metric* defined on a set X if there is given a nonnegative function ρ on $X \times X$ with the following properties:
1) $\rho(x,y) \geq 0$ and $\rho(x,y)=0$ if and only if $x=y$;
2) $\rho(x,y)=\rho(y,x)$;
3) $\rho(x,y)+\rho(y,z) \geq \rho(x,z)$.

A set X with a distance on it is called a *metric space*. Every metric space can be considered as a topological space if as a basis of open sets we take the family of open balls, that is, sets of the form $S_r(x) = \{y; \rho(x,y) < r\}$, $x \in X$, $r \in \mathbf{R}$, $r > 0$.

Plainly different metric spaces can lead to homeomorphic topological spaces. Furthermore, there exist topological spaces not obtainable by this construction: they are called *nonmetrizable*.

For spaces with a countable basis, metrizability is equivalent to each of the following conditions:

a) every pair of disjoint closed sets admit disjoint neighborhoods[1];

b) every continuous real-valued function defined on a closed subspace can be extended to a continuous function on the entire space.

There are two important concepts definable for metric spaces: uniform continuity and completeness. An analysis of these concepts leads to the following definition.

A set X is said to be a *uniform space* if there is given in X a system σ of reflexive symmetric binary relations having the following properties:

a) every relation in σ contains the square of a certain relation in σ;

b) the intersection of any two relations in σ contains a certain relation in σ.

For $R \in \sigma$, we say that x and y are *R-near* if $(x,y) \in R$.

A mapping f of a uniform space (X,σ) into a uniform space (X',σ') is called *uniformly continuous* if for every $R' \in \sigma'$, there exists an $R \in \sigma$ such that whenever x and y are R-near, then $f(x)$ and $f(y)$ are R'-near.

To every metric space (X,ρ) there corresponds a uniform space (X,σ) in which the system σ consists of all sets of the form $R_r = \{(x,y); \rho(x,y) < r\}$, $r \in \mathbf{R}$, $r > 0$.

To every uniform space (X,σ) there corresponds a topological space (X,τ) in which a basis for open sets is formed by the family of all sets of the form $R_x = \{y; (x,y) \in R\}$, $R \in \sigma$, $x \in X$.

---

[1] A *neighborhood* of a set X is any open set $U \supset X$.

§ 1. Sets, Categories, Topology

Problem 5. Prove that in the category of uniform spaces, the product of an arbitrary family is defined. Prove also that the operation of forming products commutes with the functor defined above from the category of uniform spaces into the category of topological spaces.

A direction $\{x_\alpha\}_{\alpha \in A}$ in a uniform space (X, σ) is called *fundamental* if for every $R \in \sigma$ there exists an index $\alpha \in A$ such that x_β and x_γ are R-close for all β and γ that follow α.

A uniform space is called *complete* if every fundamental direction has a limit.

Every uniform space X can be *completed*, that is, it can be embedded as a dense subspace in a certain complete space $\bar{X}$. For Hausdorff spaces, this embedding $i: X \to \bar{X}$ can be defined as a universal object in the category of uniformly continuous mappings into complete spaces. In this case, the completion is defined uniquely up to an isomorphism.

Problem 6. Let X be a complete Hausdorff space, and Y a subspace of X. Prove that the completion of Y coincides with its closure in X.

One of the most powerful tools in studying the category T of topological spaces and the categories connected with it is the construction and investigation of functors from T into other categories.

Let ΓT be the category of topological spaces, the morphisms of which are homotopy classes of continuous mappings. (Two mappings f_0 and f_1 of X into Y belong to the same *homotopy class* or are *homotopic* if there exists a family of mappings $f_t: X \to Y$, $t \in [0,1]$, such that the mapping

$$X \times [0,1] \to Y : (x,t) \mapsto f_t(x)$$

is continuous.)

The majority of the functors that we shall use have the form $F = F_1 \circ G$, where G is the natural functor from T into ΓT, which puts every object into correspondence with itself, and assigns to each morphism its homotopy class.

Example 1. $\pi_n(X, x)$, the n-th homotopy group of the space X with respect to the point $x \in X$. This is a covariant functor from the category T_1 of spaces with a designated point (the morphisms are the continuous mappings that carry the designated point into the designated point) into the category of sets for $n = 0$, into the category of groups for $n = 1$, and into the category of Abelian groups for $n > 1$.

We can formulate the usual definition of homotopy groups as follows. Let G_1 be the natural functor from T_1 to ΓT_1 (the category of spaces with a designated point and with morphisms the homotopy classes of mappings carrying the designated point into the designated point). Then we have $\pi_n = F_n \circ G_1$, where $F_n = \text{Mor}(S^n, \cdot)$. By S^n we denote the n-dimensional sphere (with a designated point).

The group $\pi_1(X)$ is called the *fundamental group* of the space X. If this group is trivial, the space is said to be *simply connected*.

Example 2. $H^n(X, \Pi)$, the n-dimensional cohomology group of the space X with coefficients in the group Π. There are many different definitions of this functor. These are the simplicial, the cellular, singular, Čech, de Rham, and others. For "sufficiently good" spaces (in particular, for smooth compact manifolds, see § 5), all of these definitions are equivalent. We give here a definition using the concept of a representable functor. Let $K(\Pi, n)$ be a topological space having the properties

$$\pi_m(K(\Pi, n), *) = \begin{cases} 0 & \text{for } m \neq n, \\ \Pi & \text{for } m = n. \end{cases}$$

(One can show that such a space exists and that all such spaces are isomorphic objects of the category ΓT defined above.)

Then we have $H^n(\cdot, \Pi) = F_\Pi^n \circ G$, where G is the natural functor carrying T to ΓT and $F_\Pi^n = \mathrm{Mor}(\cdot, K(\Pi, n))$.

Problem 7. If the space X can be contracted in itself to a point, then $H^n(X, \Pi) = 0$ for $n \geq 1$.

Hint. The object X is isomorphic to a point in the category ΓT.

The structure of the spaces $K(\Pi, n)$ is quite complicated. With the exception of certain spaces $K(\Pi, 1)$, all of them are infinite-dimensional.

Problem 8. Prove that if Z denotes the group of integers, then for the space $K(Z, 1)$, we can take the circle S^1.

Let us give yet another definition, more convenient for practical computations.

Let $\mathfrak{U} = \{U_\alpha\}_{\alpha \in A}$ be a covering of the space X by open sets. An n-dimensional *cochain* of the covering $\mathfrak{U}$ with coefficients in Π is a function c from $A \times A \times \ldots \times A$ ($n+1$ factors) into Π that has the following properties:

1) the domain of definition of c consists of the sequences $(\alpha_0, \alpha_1, \ldots, \alpha_n)$ for which the intersection $U_{\alpha_0} \cap U_{\alpha_1} \cap \cdots \cap U_{\alpha_n}$ is nonvoid.

2) The function c is skew-symmetric, that is, changes its sign under permutation of two arguments.

The set of all n-dimensional cochains forms a group $C^n(\mathfrak{U}, \Pi)$.

We introduce the *coboundary operator* d, which carries $C^n(\mathfrak{U}, \Pi)$ into $C^{n+1}(\mathfrak{U}, \Pi)$ by the following rule:

$$dc(\alpha_0, \ldots, \alpha_{n+1}) = \sum_{i=0}^{n+1} (-1)^i c(\alpha_0, \ldots, \hat{\alpha}_i, \ldots, \alpha_{n+1})$$

(the sign $\hat{}$ means that the corresponding argument is suppressed).

The cochain dc is called the *coboundary* of the the cochain c. The set of all n-dimensional coboundaries is denoted by the symbol $B^n(\mathfrak{U}, \Pi)$. If $dc = 0$, then the cochain c is called a *cocycle*. The set of all n-dimensional cocycles is denoted by the symbol $Z^n(\mathfrak{U}, \Pi)$.

As is easy to verify, the operator d has the property that $d^2 = 0$. It follows that

$$B^n(\mathfrak{U}, \Pi) \subset Z^n(\mathfrak{U}, \Pi).$$

The factor group

$$H^n(\mathfrak{U}, \Pi) = Z^n(\mathfrak{U}, \Pi)/B^n(\mathfrak{U}, \Pi)$$

is called the n-dimensional Čech cohomology group of the covering $\mathfrak{U}$. The group $H^n(X, \Pi)$ can be defined as the inductive limit of the groups $H^n(\mathfrak{U}, \Pi)$ by the directed set of all coverings. In point of fact, the computation of the limit can often be replaced by the following

Theorem of Leray. *If a covering* $\mathfrak{U} = \{U_\alpha\}$ *has the property that all of the sets* U_α *and their intersections have trivial cohomology in dimensions* $n \geq 1$, *then we have*

$$H^n(X, \Pi) = H^n(\mathfrak{U}, \Pi).$$

Problem 9' (D. A. Každan). Infer from Leray's theorem the following theorem of Helly.
If we are given $n+2$ *convex sets in the space* $\mathbf{R}^n$ *with the property that any* $n+1$ *of them have nonvoid intersection, then all* $n+2$ *of them have nonvoid intersection.*
Hint. Use problem 7 and the fact that $H^n(X, \mathbf{Z}) = 0$ for all subsets X of $\mathbf{R}^n$.

Corollary. *The hypothesis of convexity in Helly's theorem can be replaced by the following weaker hypothesis. All of the sets and their intersections have trivial cohomology groups in dimensions* $n \geq 1$.

Problem 10. Prove that

$$H^k(S^n, \Pi) = \begin{cases} \Pi & \text{for } k=0 \text{ and } k=n, \\ 0 & \text{in all other cases}. \end{cases}$$

Hint. Consider a covering of the sphere by $n+2$ open hemispheres.
For more details about the functors $\pi_n, H^n(\cdot, \Pi)$ and others, consult the books [2], [52], [31], [32].

§ 2. Groups and Homogeneous Spaces

2.1. Transformation Groups and Abstract Groups

A *transformation group* is a nonvoid set G of mappings of a certain set X onto itself with the following properties:
1) if $g_1 \in G$ and $g_2 \in G$, then $g_1 g_2 \in G$;
2) if $g \in G$, then $g^{-1} \in G$.

As examples of transformation groups, we have the group G_1 of all one-to-one mappings of the set X onto itself and the group G_0 consisting of the identity mapping alone. It is clear that every transformation group G on the space X is contained between these two: $G_0 \subset G \subset G_1$.

Problem 1. List all transformation groups on a set of three elements.
Suppose that we have some sort of structure on the set X. The set G of transformations g such that g and g^{-1} preserve the structure is evidently a group. This group is called the group of *automorphisms* of the given structure.

Examples. The group of homeomorphisms of a topological space, the group of isometries of a metric space, the group of invertible linear operators.

Indeed, the way just described for finding transformation groups has universal character: every transformation group is a group of automorphisms for an appropriate structure.

It is often convenient to forget that the elements of the group G are transformations of a certain set and to remember only the law of multiplication of these elements.

The concept arising in this way is called an *abstract group* (in contrast to a transformation group) or simply a *group*. More precisely, a *group* is a nonvoid set G on which there is defined a law of multiplication, that is to say, a mapping of $G \times G$ into G with certain properties, specified below. We will denote by $g_1 \circ g_2$ or simply $g_1 g_2$ the result of multiplying g_1 and g_2. The additional properties required are:

1) *associativity*: $g_1 \circ (g_2 \circ g_3) = (g_1 \circ g_2) \circ g_3$;
2) *the existence of an identity*: there exists an element e in the group G such that $e \circ g = g \circ e = g$ for all $g \in G$;[1]
3) *the existence of inverses*: for every $g \in G$, there exists an element $g^{-1} \in G$ such that $g \circ g^{-1} = g^{-1} \circ g = e$.

A group G is called *commutative* or *abelian* if in addition to the above properties one also has

4) *commutativity*: for all $g_1, g_2 \in G$, we have $g_1 \circ g_2 = g_2 \circ g_1$.

In commutative groups, the group operation is frequently called addition, $g_1 + g_2$ is written for $g_1 \circ g_2$, and the identity element is written as 0.

It is obvious that every transformation group is a group in the sense of the definition just given.

Also, groups that are quite different transformation groups may well be isomorphic as abstract groups. Consider for example the following four transformation groups:

  a) the group of isometric mappings of an equilateral triangle onto itself;
  b) the group of all permutations of a set of three elements;
  c) the group of invertible matrices of order 2 with coefficients in the field of integers modulo 2;
  d) the group of fractional linear transformations of the extended complex plane generated by the mappings $z \mapsto z^{-1}$ and $z \mapsto 1 - z$.

Each of these groups consists of six elements, and they multiply by one and the same law. (The reader should check this for himself.)

A mapping ϕ of a group G into a group H is called a *homomorphism* if $\phi(g_1 \circ g_2) = \phi(g_1) \circ \phi(g_2)$ for all $g_1, g_2 \in G$. Groups and homomorphisms of groups form the *category of groups*.

In case the group H is commutative, the homomorphisms of G into H themselves form a commutative group, the operation of addition being defined by $(\phi_1 + \phi_2)(g) = \phi_1(g) + \phi_2(g)$. This group is denoted by $\text{Hom}(G, H)$.

---

[1] Instead of e, one often writes 1.

§ 2. Groups and Homogeneous Spaces

Problem 2. Find the group $\text{Hom}(G,H)$ for all 16 pairs of groups (G,H), chosen from the following four:
  a) the group $\mathbf{Z}$ of integers under addition;
  b) the group $\mathbf{Z}_n$ of integers modulo n under addition;
  c) the group $\mathbf{Q}$ of rational numbers under addition;
  d) the group $\mathbf{Q}^*$ of nonzero rational numbers under multiplication.

Hint. The answer is given by the following table, where (m,n) denotes the greatest common divisor of the numbers m and n and $\oplus$ is the direct sum in the category of abelian groups:

G \ H	$\mathbf{Z}$	$\mathbf{Z}_m$	$\mathbf{Q}$	$\mathbf{Q}^*$
$\mathbf{Z}$	$\mathbf{Z}$	$\mathbf{Z}_m$	$\mathbf{Q}$	$\mathbf{Q}^*$
$\mathbf{Z}_n$	0	$\mathbf{Z}_{(m,n)}$	0	$\mathbf{Z}_{(n,2)}$
$\mathbf{Q}$	0	0	$\mathbf{Q}$	0
$\mathbf{Q}^*$	$\mathbf{Z}\oplus\mathbf{Z}\oplus\mathbf{Z}\oplus\cdots$	$\mathbf{Z}_{(m,2)}\oplus\mathbf{Z}_m\oplus\mathbf{Z}_m\oplus\mathbf{Z}_m\oplus\cdots$	$\mathbf{Q}\oplus\mathbf{Q}\oplus\mathbf{Q}\oplus\cdots$	$\mathbf{Z}_2\oplus\mathbf{Z}_2\oplus\cdots\oplus\mathbf{Z}\oplus\mathbf{Z}\oplus\cdots$

Suppose that we are given a homomorphism $\phi: G \to H$. Then we define
$\ker\phi = \{g \in G: \phi(g) = e\}$, the *kernel* of the homomorphism ϕ, and
$\text{im}\,\phi = \{\phi(g): g \in G\}$, the *image* of the homomorphism ϕ.
A sequence

$$\cdots \longrightarrow G_n \xrightarrow{\phi_n} G_{n+1} \longrightarrow \cdots$$

of groups and homomorphisms is called *exact* if the image of each homomorphism is the kernel of the following homomorphism. Consider for example the following sequences:

$$1 \longrightarrow G \xrightarrow{\phi} H, \quad G \xrightarrow{\phi} H \longrightarrow 1, \quad 1 \longrightarrow G \xrightarrow{\phi} H \longrightarrow 1.$$

Exactness of these sequences means, respectively, that ϕ is a *monomorphism* (i.e., $\ker\phi = 1$), an *epimorphism* (i.e., $\text{im}\,\phi = H$), and an *isomorphism*.

Let g_0 be a fixed element of the group G. The mapping $g \mapsto g_0^{-1} g g_0$ is an *automorphism* of the group G (that is, an isomorphism of the group onto itself). Automorphisms of this form are called *inner* automorphisms. We often write g^{g_0} instead of $g_0^{-1} g g_0$.

A homomorphism ϕ of an abstract group into a transformation group is called a *realization* or a *representation* of this group. A representation ϕ is called *faithful* if ϕ is a monomorphism.

A set X is called a *left G-space* if there exists a realization of the group G by transformations of the set X. We also say that the group G *acts on X on the left*.

We write the result of the action of $g \in G$ on the element $x \in X$ as gx. The equality $g_1(g_2 x) = (g_1 g_2)x$ holds.

We often encounter a situation in which elements g of a group G are mapped onto transformations $\phi(g)$ of a space X in such a way that $\phi(g_1 g_2) = \phi(g_2)\phi(g_1)$. In this case, ϕ is called an *antirepresentation* and the set X is called a *right G-space*. We also say that G *acts on X on the right*. We denote the element $\phi(g)x$ by the symbol xg. In this case we have the equality $(xg_1)g_2 = x(g_1 g_2)$.

Example 1. Consider the group $G = GL(n, K)$ of invertible matrices of order n over a field K. The set of all column vectors forms a left G-space, and the set of all row vectors forms a right G-space.

Example 2. For a given category and an object A of this category, let Aut A denote the group of automorphisms of the object A, that is, the group of invertible elements in Mor(A, A). Then the set Mor(A, B) is a left G_1-space with respect to $G_1 =$ Aut A, and it is a right G_2-space with respect to $G_2 =$ Aut B.

Left and right G-spaces form two categories, morphisms of which are the so-called *G-mappings*, that is, mappings that commute with the actions of the group. These two categories are canonically isomorphic, since every left G-space X becomes a right G-space if we define $xg = g^{-1}x$.

Let X and Y be two G-spaces. Consider the factor space $X \underset{G}{\times} Y$ of the product space $X \times Y$ with respect to the equivalence defined by the action of G on $X \times Y$. We call this factor space the *product of the spaces X and Y over G*. Suppose for example that X is a right G-space and that Y is a left G-space. Then two pairs (x, y) and (x', y') are equivalent if and only if $x' = xg$, $y' = g^{-1}y$ for some $g \in G$.

We obtain an important special case of the foregoing construction by taking as Y a group G_1 that contains G, with its obvious structure as a left G-space. In this case, we give $X \underset{G}{\times} G_1$ its natural structure as a right G_1-space.

The correspondence $X \mapsto X \underset{G}{\times} G_1$ generates a functor from the category K_G of all G-spaces into the category K_{G_1} of all G_1-spaces. This is easy to verify. The reader can obtain an intuitive grasp of this functor by considering the following example. Let G_1 be the group of rotations of 3-dimensional space around a fixed point 0, and let G be the subgroup of rotations about a fixed line l that passes through 0. Let X be a 2-dimensional plane orthogonal to l with the natural action of G on it. Then $X \underset{G}{\times} G_1$ is isomorphic as a G_1-space to the set of all tangent vectors to a sphere with center at 0.

This functor can also be obtained by another natural method.

Problem 3. Let X be an object of the category K_G. Consider the functor F from K_{G_1} to the category of sets:

$$F(Y) = \text{Mor}(X, \tilde{Y}),$$

where $\tilde{Y}$ denotes Y considered as an object from K_G (note that every G_1-space is also a G-space). Prove that this functor is representable and that $X \underset{G}{\times} G_1$ is a

§ 2. Groups and Homogeneous Spaces

representing object. In other words, prove that

$$\text{Mor}(X, \tilde{Y}) = \text{Mor}(X \underset{G}{\times} G_1, Y)$$

(as functors of Y).

2.2. Homogeneous Spaces

A group G is said *to act transitively* on X, or the G-space X is said to be *homogeneous*, if every point of X can be carried into every other point of X by some transformation in G. Every G-space is the union of homogeneous spaces. In fact, the *G-orbit* of a point $x \in X$ (that is, the set of all points of the form gx, $g \in G$) is a homogeneous G-space, and the entire set is the union of the orbits of its points.

In a certain sense, all homogeneous G-spaces can be listed. Let us fix a point $x \in X$. The set G_x of all elements of G that leave x fixed forms a subgroup of G. This subgroup is called a *stationary subgroup* or the *stabilizer* of the point x. Let g be an element of G. The set of elements of the form gh, where h runs through G_x, is called a *left coset* of the group G with respect to the subgroup G_x and is written as gG_x. It is not hard to verify that gG_x is the set of all elements of the group G that carry the point x into the point gx. Thus every homogeneous left G-space is isomorphic to the space of left cosets of G with respect to a certain subgroup. Conversely, if H is a subgroup of G, then the set G/H of left cosets of G with respect to H is a homogeneous G-space under the natural action of G:

$$g: g_1 H \mapsto g g_1 H.$$

Problem 1. Two homogeneous spaces G/H_1 and G/H_2 are isomorphic (as objects of the category K_G, see para. 2.1) if and only if the subgroups H_1 and H_2 are *conjugate*, that is, are carried into each other by inner automorphisms of the group G.

Everything said here can be carried over *verbatim* to right G-spaces and *right cosets*. The space of right cosets of G with respect to H will be denoted by the symbol $H \backslash G$.[1]

A subgroup H is called *invariant* (or *normal*, or a *normal divisor*) in G if all of the subgroups conjugate to H coincide with H. In this case, every right coset is a left coset and conversely. The homogeneous space G/H is itself a group under the operation $g_1 H \circ g_2 H = g_1 g_2 H$. Geometrically, this means that the product xy is obtained from y by the transformation that yields x from the initial point. The correctness of this definition is ensured by the fact that the stationary subgroups of all points coincide. (Therefore the transformation is defined by the result of its action on a single point.) The group obtained in this way is called the *factor group of G with respect to H.*

---

[1] Translator's note. This notation is not to be confused with the symbol $X \backslash Y$, which is used to to denote the complement of Y in the set X.

In the sequel we shall frequently need to know the general form of a G-mapping of a G-space X into itself or into another G-space Y. We shall now solve this problem for homogeneous G-spaces. Let $X = G/H$, $Y = G/K$, and suppose that $\phi: X \to Y$ is a G-mapping. Let us suppose that ϕ carries the class $H \in X$ to the class $gK \in Y$. Then we have $\phi(g_1 H) = g_1 \phi(H) = g_1 g K$, which is to say that the mapping ϕ is completely determined by the element $g \in G$. However, this element is not arbitrary. In fact, if $h \in H$, then we have $hH = H$. It follows that $\phi(hH) = \phi(H)$, that is, we have $hgK = gK$ for all $h \in H$, and from this it follows that $H^g = g^{-1}Hg \subset K$. Conversely, if the element $g \in G$ has this property, then the mapping $\phi: g_1 H \to g_1 g K$ will be a well-defined G-mapping. Elements g and g' generate one and the same G-mapping ϕ if and only if $g^{-1}g' \in K$.

We examine in more detail the case where $X = Y$ and the mapping ϕ is an automorphism. In this case we have $K = H$ and $H^g = H$, since ϕ^{-1} shares with ϕ the property of being an automorphism. The set of elements $g \in G$ for which $H^g = H$ is called the *normalizer* of the subgroup H in G and is denoted by the symbol $N_G(H)$ (or simply $N(H)$ if it is clear from the context what group G is under discussion). Therefore the group of all automorphisms of the G-space $X = G/H$ is isomorphic with the factorgroup $N(H)/H$.

Suppose that H and K are two subgroups of G, and that $g \in G$. The set of all elements of the form hgk, as h runs through H and k runs through K, is denoted by the symbol HgK and is called a *double coset* of the group G with respect to the subgroups H and K. The set of these double cosets is denoted by the symbol $H\backslash G/K$. We suggest that the reader verify that the set $H\backslash G/K$ is isomorphic to each of the following:
 a) the set of K-orbits in $X = H\backslash G$;
 b) the set of H-orbits in $Y = G/K$;
 c) the set of G-orbits in $X \times Y$;
 d) the set $X \underset{G}{\times} Y$.

2.3. Principal Types of Groups

The simplest of all groups to analyze are commutative groups. All such groups with a finite number of generators are easily classified.

Problem 1. Every group with a single generator (such groups are called *cyclic*) is isomorphic either to the group $\mathbf{Z}$ of integers or to the group $\mathbf{Z}_m$ of integers modulo m.

By induction on the number of generators, it is easy to prove that every abelian group G with a finite number of generators is isomorphic to the sum of cyclic groups. This is not true of an arbitrary abelian group, as is evident from the example $G = \mathbf{Q}$ ($\mathbf{Q}$ is the additive group of rational numbers).

Problem 2. Let M denote the least common multiple and d the greatest common divisor of the numbers m and n. Prove that the groups $\mathbf{Z}_m \oplus \mathbf{Z}_n$ and $\mathbf{Z}_M \oplus \mathbf{Z}_d$ are isomorphic.

§ 2. Groups and Homogeneous Spaces

From problem 2, it is easy to infer that every abelian group with a finite number of generators is isomorphic to exactly one group of the form

$$Z_{m_1} \oplus Z_{m_2} \oplus \cdots \oplus Z_{m_k} \oplus Z \oplus \cdots \oplus Z,$$

where each of the numbers m_i, $i = 1, 2, \ldots, k-1$, is a divisor of the number m_{i+1}.

Noncommutative groups can be classified according to the degree of their noncommutativity. Given two subsets X and Y of the group G, let $[X, Y]$ denote the set of all elements of the form $xyx^{-1}y^{-1}$, as x runs through X and y runs through Y. We define two sequences of subgroups of G. Let $G_0 = G$, and for positive integers n, let G_n be the subgroup of G generated by the set $[G_{n-1}, G_{n-1}]$. The subgroup G_n is called the n-th derived group of the group G. Define G^0 as G, and for positive integers n, let G^n be the subgroup generated by the set $[G, G^{n-1}]$. We obviously obtain the following inclusions:

$$G = G_0 \supset G_1 \supset G_2 \supset \cdots \supset G_n \supset \cdots,$$
$$G = G^0 \supset G^1 \supset G^2 \supset \cdots \supset G^n \supset \cdots.$$

For a commutative group, these sequences are trivial: we have $G_n = G^n = 1$ for all $n \geq 1$. The group G is called *solvable of class k* if $G_n = 1$ beginning with $n = k$. It is called *nilpotent of class k* if $G^n = 1$ beginning with $n = k$.

Problem 3. Prove that G is a solvable group of class k if and only if the following exist:

an abelian invariant subgroup $A_0 \subset G$;
an abelian invariant subgroup A_1 in $G_1 = G/A_0$;
an abelian invariant subgroup A_2 in $G_2 = G_1/A_1$,

and so on, where this chain breaks off at the k-th step, that is, $G_k = 1$.

Hint. For A_0, take the $(k-1)$-st derived group and use induction on k.

Let X be a subset of the group G. The set of elements of the group G that commute with all of the elements of X is called the *centralizer* of X in G and is denoted by the symbol $C_G(X)$. The centralizer of the entire group G is called the *center* of the group.

Problem 4. Prove that G is a nilpotent group of class k if and only if:

G has a nontrivial center C_0;
the factor group $G_1 = G/C_0$ has a nontrivial center C_1;
the factor group $G_2 = G_1/C_1$ has a nontrivial center C_2, and so on, and this chain breaks off at the k-th step, that is, $G_k = 1$.

Hint. The subgroup C_0 contains G^{k-1}.

A group that contains no nontrivial invariant subgroups (that is, different from the group itself and the subgroup consisting only of the identity element) is called *simple*.

The classification of all finite simple groups is one of the most fundamental and most difficult problems of the theory of groups.

For example, it was only recently that Burnside's famous conjecture, namely, that there are no simple (noncommutative) groups of odd order, was established. Recent years have seen remarkable progress in this field, brought about in part by the use of the theory of infinite groups (Lie groups and algebraic groups).

2.4. Extensions of Groups

Suppose that the group G is not simple, that is, contains a nontrivial normal subgroup G_0. Then we have the following exact sequence:

$$1 \to G_0 \to G \to G_1 \to 1.$$

This is called an *extension* of the group G_1 by the group G_0.

The study of G can in many respects be reduced to the study of two "smaller" groups, the subgroup G_0 and the factor group $G_1 = G/G_0$.

We note that in general the group G cannot be reconstructed merely from G_0 and G_1. In fact, every element $g \in G$ yields an inner automorphism of the group G, which carries G_0 onto itself. Thus we obtain a homomorphism of G into the group $\text{Aut } G_0$, the group of automorphisms of the group G_0. Under this homomorphism, elements of the group G_0 go into the subgroup $\text{Int } G_0$ of inner automorphisms. Thus we define a homomorphism of G/G_0 into the group $\text{Aut } G_0/\text{Int } G_0$. Therefore, to reconstruct G, it is necessary to know the homomorphism $\phi: G_1 \to \text{Aut } G_0/\text{Int } G_0$.

Supposing that we know this homomorphism ϕ, let us try to construct the group G. We note that the group G can be identified, as a set, with the direct product of the groups G_1 and G_0. In fact, if we choose a representative $\alpha(g_1) \in G$ from each of the cosets $g_1 \in G_1$, then every element $g \in G$ can be written uniquely in the form $g = \alpha(g_1) g_0$, $g_1 \in G_1$, $g_0 \in G_0$. This gives the desired "numeration" of the elements of G by pairs (g_1, g_0). In the sequel, it will be convenient for us to suppose that the representative chosen for the class G_0 is the identity element. Then, in the coordinates we have chosen, the elements of the subgroup G_0 have the form (e, g_0) and we have the following equality:

$$(g_1, g_0)(e, g_0') = (g_1, g_0 g_0').$$

Furthermore, the mapping

$$(e, g_0) \mapsto (g_1, e)(e, g_0)(g_1, e)^{-1}$$

is an automorphism of G_0, which we denote by $\psi(g_1)$. Note that the image $\psi(g_1)$ in $\text{Aut } G_0/\text{Int } G_0$ is already known and is in fact $\phi(g_1)$. Knowing not only ϕ but also ψ, we can now compute the product of an element (e, g_0) on the left with any element of G:

$$(e, g_0)(g_1, g_0') = (e, g_0)(g_1, e)(e, g_0') = (g_1, e)(e, \psi(g_1)^{-1} g_0)(e, g_0')$$

$$= (g_1, [\psi(g_1)^{-1} g_0] g_0').$$

§ 2. Groups and Homogeneous Spaces

The product of two elements of the form (g_1, e) must have the form

$$(g_1, e)(g_1', e) = (g_1 g_1', \chi(g_1, g_1')),$$

where $\chi: G_1 \times G_1 \to G_0$ is a certain mapping.

Knowing ψ and χ, we can compute the product of any two elements of G by the formula

$$(g_1, g_0)(g_1', g_0') = (g_1 g_1', \chi(g_1, g_1')[\psi(g_1')^{-1} g_0] g_0'). \tag{1}$$

One can verify easily that the functions $\psi: G_1 \to \operatorname{Aut} G_0$ and $\chi: G_1 \times G_1 \to G_0$ are connected by the identities

a) $\qquad \psi(g_1)\psi(g_2) = \psi(g_1 g_2) I(\chi(g_1, g_2)),$

where $I(g)$ is the inner automorphism of G_0 corresponding to the element g $(I(g): g_0 \mapsto g g_0 g^{-1})$, and also

b) $\qquad \chi(g_1 g_2, g_3) \psi(g_3)^{-1} \chi(g_1, g_2) = \chi(g_1, g_2 g_3) \chi(g_2, g_3).$

These identities follow from associativity of multiplication in G. Conversely, if we have any two functions ψ and χ satisfying conditions a) and b), then formula (1) defines a group operation on the set $G_1 \times G_0$.

It is possible for extensions constructed by functions ψ, χ and ψ', χ', to be *equivalent* in the sense that the following diagram is commutative:

$$\begin{array}{c} 1 \to G_0 \to G \to G_1 \to 1 \\ \| \quad \varepsilon \downarrow \quad \| \\ 1 \to G_0 \to G' \to G_1 \to 1. \end{array}$$

This happens if and only if there is a mapping $\xi: G_1 \to G_0$ with the following properties:

c) $\qquad \psi'(g_1) = \psi(g_1) I(\xi(g_1)^{-1});$

d) $\qquad \chi'(g_1, g_2) = \xi(g_1 g_2) \chi(g_1, g_2) [\psi(g_2^{-1})[\xi(g_1)^{-1}]] \xi(g_2)^{-1}.$

In this case, the mapping ε has the form

$$\varepsilon: (g_1, g_0) \mapsto (g_1, \xi(g_1) g_0).$$

It may well happen that there are no functions ψ and χ satisfying conditions a) and b). In fact, the automorphisms $\psi(g_1)\psi(g_2)$ and $\psi(g_1 g_2)$ lie in one and the same coset $\phi(g_1 g_2) \in \operatorname{Aut} G_0 / \operatorname{Int} G_0$ and therefore differ by an inner automorphism. Thus the equality a) is satisfied and defines the function $\chi(g_1, g_2)$ up to an element of the center of the group G_0 (since this center is the kernel of the homomorphism $I: G_0 \to \operatorname{Int} G_0$). Applying I to both sides of the equality b), we see without difficulty that these two sides differ by a certain element $\omega(g_1, g_2, g_3) \in C(G_0)$.

Now using the amount of choice left to us in defining $\chi(g_1,g_2)$, we can alter ω by an element of the form

$$\beta(g_1g_2,g_3)\beta(g_1,g_2)\beta(g_2,g_3)^{-1}\beta(g_1,g_2g_3)^{-1}, \qquad (2)$$

where β is a certain mapping of $G_1 \times G_1$ into $C(G_0)$.

On the other hand, one can show that the only restriction on the mapping ω from $G_1 \times G_1 \times G_1$ into $C(G_0)$ is the identity

$$\omega(g_1,g_2,g_3)^{-1}\omega(g_1,g_2,g_3g_4)\omega(g_1,g_2g_3,g_4)^{-1}$$
$$=\omega(g_2,g_3,g_4)\omega(g_1g_2,g_3,g_4)^{-1}. \qquad (3)$$

In general, such a mapping ω is not representable in the form (2) using a mapping β.

It thus turns out that the group G cannot always be constructed from given entities G_1, G_0, and $\phi: G_1 \to \mathrm{Aut}\,G_0/\mathrm{Int}\,G_0$. The obstruction to the construction of G is a certain mapping ω that satisfies (3), modulo a mapping of the form (2).[1]

The situation is greatly simplified if we are able to choose the mapping $\psi: G_1 \to \mathrm{Aut}\,G_0$ to be a homomorphism. Extensions for which such a choice of ψ is possible are called *central*. In this case, condition a) implies that $I(\chi(g_1,g_2)) \equiv e$ and consequently the values of the function χ belong to the center of the group G_0. Hence the obstruction ω automatically has the form (2), and therefore the desired functions χ exist. Furthermore, for a fixed function ψ, the set of functions χ that satisfy b) forms an abelian group under the natural definition of multiplication. The set of functions that are equivalent to the trivial function $\chi(g_1,g_2) \equiv e$ in the sense of condition d) form a subgroup.

In this way, the set of classes of equivalent extensions receives the structure of an abelian group, a fact which greatly simplifies its study.

As the identity element in this group, we have the extension associated with the function $\chi(g_1,g_2) \equiv e$. In this case the group G_1 is embedded in the group G as a subgroup, the extension is called *decomposable*, and the group G is called a *semi-direct product* of the groups G_1 and G_2.

Example 1. The group E_n of isometries of n-dimensional Euclidean space is a semidirect product of the subgroup of rotations and the subgroup of translations.

Example 2. Consider the group $GL(n,K)$ of all invertible matrices of order n with entries from a field K. This group is a direct product of K^* (the multiplicative group of the field K) and the group $SL(n,K)$ of unimodular matrices.

Example 3. The group $\mathbf{Z}_{mn}$ is a semidirect product of the groups $\mathbf{Z}_m$ and $\mathbf{Z}_n$ if and only if m and n are relatively prime.

The following problem is an excellent exercise to aid in mastering the concepts introduced above.

Problem 1. Describe all extensions of the form

$$1 \to \mathbf{Z}_m \to G \to \mathbf{Z}_n \to 1.$$

---

[1] For a formulation of this assertion in terms of cohomology of groups, see problem 2 in para. 2.5.

2.5. Cohomology of Groups

We have seen that the problem of extending groups leads to the study of functions of one, two, or three group elements, satisfying certain special relations. Many other problems in representation theory, algebraic geometry, and topology lead to exactly the same functions. A study of the situation that arose in this way has led to the evolution of the concept of cohomology of groups, which permits us to unite into a single scheme a large number of related theories. We shall here give only the principal definitions, referring the reader for details to the book of S. Mac Lane [41]. This monograph gives an accessible and reasonably complete exposition of homological algebra, a part of which is the theory of cohomology of groups.

Suppose that a group G acts on an abelian group M. A function $c(g_0, \ldots, g_n)$ on $G \times \cdots \times G$ with values in M is called an *n-dimensional cochain* if

$$c(gg_0, \ldots, gg_n) = g c(g_0, \ldots, g_n)$$

for all $g, g_i \in G$. The set of all n-dimensional cochains forms the group $C^n(G, M)$. We define an operator d carrying $C^n(G, M)$ into $C^{n+1}(G, M)$ by the following formula:

$$dc(g_0, \ldots, g_{n+1}) = \sum_{i=0}^{n+1} (-1)^i c(g_0, \ldots, \hat{g}_i, \ldots, g_{n+1}).$$

Here the sign $\hat{\ }$ means that the corresponding argument is suppressed. A cochain c is called a *coboundary of the cochain* b, if $c = db$. It is called a *cocycle* if $dc = 0$. The following fundamental property of the operator d is easily checked: $d \circ d = 0$. Therefore the group $Z^n(G, M)$ of all n-dimensional cocycles contains the group $B^n(G, M)$ of all coboundaries of $(n+1)$-dimensional cochains as a subgroup. The factor group

$$H^n(G, M) = Z^n(G, M) / B^n(G, M)$$

is called the *n-dimensional cohomology group* of the group G with coefficients in M. If M is a ring or an algebra, then $H^*(G, M) = \bigoplus_n H^n(G, M)$ can also be regarded as a (graded) ring or algebra.

For practical calculations, it is convenient to consider not the cochain $c(g_0, \ldots, g_n)$ but the function

$$\tilde{c}(h_1, \ldots, h_n) = c(e, h_1, h_1 h_2, \ldots, h_1 h_2 \cdots h_n).$$

This function defines the original cochain uniquely.

Problem 1. Prove the identity

$$\widetilde{dc}(h_1, \ldots, h_{n+1}) = h_1 \tilde{c}(h_2, \ldots, h_{n+1})$$
$$+ \sum_{i=1}^{n} (-1)^i \tilde{c}(h_1, \ldots, h_{i-1}, h_i h_{i+1}, h_{i+2}, \ldots, h_{n+1}) + (-1)^{n+1} \tilde{c}(h_1, \ldots, h_n).$$

In particular, for $n=0,1,2,3$, the operator d when written in terms of the functions $\tilde{c}$ assumes the following forms:

$$\widetilde{dc}(h) = h\tilde{c} - \tilde{c},$$
$$\widetilde{dc}(h_1,h_2) = h_1\tilde{c}(h_2) - \tilde{c}(h_1 h_2) + \tilde{c}(h_1),$$
$$\widetilde{dc}(h_1,h_2,h_3) = h_1\tilde{c}(h_2,h_3) - c(h_1 h_2, h_3) + c(h_1, h_2 h_3) - c(h_1, h_2),$$
$$\widetilde{dc}(h_1,h_2,h_3,h_4) = h_1\tilde{c}(h_2,h_3,h_4) - \tilde{c}(h_1 h_2, h_3, h_4) + \tilde{c}(h_1, h_2 h_3, h_4)$$
$$- \tilde{c}(h_1, h_2, h_3 h_4) + \tilde{c}(h_1, h_2, h_3).$$

Problem 2. Prove the following formulations of certain results obtained above.

An obstruction to the construction of an extension is a certain element of $H^3(G_1, C(G_0))$.

Classes of equivalence for central extensions are numbered by elements of $H^2(G_1, C(G_0))$.

The most effective way to study cohomology of groups is to consider $H^n(G, M)$ as a bifunctor contravariant in G and covariant in M.

There is also a connection between cohomology of groups and cohomology of topological spaces. In the simplest case, where G operates trivially on M, this connection assumes the form

$$H^n(G, M) = H^n(K(G, 1), M).$$

2.6. Topological Groups and Homogeneous Spaces

A *topological group* is a set G that is simultaneously a group and a topological space, in which the group and topological structures are connected by the following condition:

The mapping $G \times G \to G: (x,y) \mapsto xy^{-1}$ is continuous.

This requirement is equivalent to the following three conditions (which are more convenient for checking).

1. The mapping $(x,y) \mapsto xy$ is continuous in x and in y.
2. The mapping $x \mapsto x^{-1}$ is continuous at the point $x = e$.
3. The mapping $(x,y) \mapsto xy$ is continuous in both variables together at the point (e,e).

Every abstract group can be regarded as a topological group, if we give it the discrete topology (every subset is open).

We obtain more interesting examples by considering groups of continuous mappings of topological spaces onto themselves, preserving one or another additional property. For example, we obtain in this way the group of invertible matrices, the group of conformal mappings of the disk, the group of unitary operators on a Hilbert space, and others.

Topological groups form a category, in which the morphisms are continuous homomorphisms. (Sometimes one considers a narrower class of morphisms, adding the requirement that the mappings be open or perhaps closed.)

§ 2. Groups and Homogeneous Spaces

A set X is called a *(left) topological G-space* if it is a (left) G-space in the ordinary sense (see para. 2.1) and also is provided with a topology such that the mapping

$$G \times X \to X : (g, x) \mapsto gx$$

is continuous. If the space X is Hausdorff, then the stationary subgroup of each point is a closed subgroup of G. Conversely, if H is a closed subgroup of a Hausdorff group G, then the space G/H of left cosets is a homogeneous Hausdorff G-space, when given the usual factor space topology. If X is any other topological G-space with the same stationary subgroup, then the natural mapping $\phi: G/H \to X$ is one-to-one and continuous. If G and X are locally compact, then ϕ is a homeomorphism[1].

Thus, for locally compact groups and spaces, there is a one-to-one correspondence between homogeneous topological G-spaces and conjugacy classes of closed subgroups.

As an example of the relation between topological and group-theoretic properties, we cite the following fact.

Problem 1. Let G be a connected topological group and Γ a discrete normal divisor in G. Then Γ lies in the center of G.

Hint. Consider the mapping of G into Γ defined by $g \mapsto g\gamma g^{-1}$, for $\gamma \in \Gamma$.

Every topological group can be turned into a uniform space in two ways. First, one can define x and y as being V-near, where V is a neighborhood of the identity, if $xy^{-1} \in V$. Second, one can define x and y as being V-near if $x^{-1}y \in V$. Under the first definition, right translation is a uniformly continuous mapping, while left translation and mapping into the inverse are in general not uniformly continuous. For the second definition, left translation is uniformly continuous, while right translation and mapping into the inverse are in general not uniformly continuous.

It is natural to complete these two uniform spaces and to try give them topological group structures. Evidently this can be done only if the following condition is satisfied:

$$\left. \begin{array}{l} \text{for every fundamental net } \{x_\alpha\}_{\alpha \in A}, \\ \text{the net } \{x_\alpha^{-1}\}_{\alpha \in A} \text{ is also fundamental.} \end{array} \right\} \quad (1)$$

It turns out that if this condition is satisfied, then both completions are in fact topological groups and that these two groups are isomorphic.

In particular, condition (1) is satisfied for all commutative groups.

Example 1. Let $\mathbf{Q}$ be the additive group of rational numbers, with the topology defined by the usual metric:

$$\rho(x_1, x_2) = |x_1 - x_2|.$$

Then the completion of $\mathbf{Q}$ is isomorphic to $\mathbf{R}$.

---

[1] A topological space is said to be *locally compact* if every point admits a neighborhood with compact closure.

Example 2. Let us give a different topology in **Q**. Let p be a prime number. Every rational number can be written in the form $x = p^k m/n$, where the integers m and n are prime to p. We set $\rho_p(x, y) = p^{-k}$, if $x - y = p^k m/n$.

The completion of **Q** in this metric is called the group of *p-adic numbers* and is denoted by $\mathbf{Q}_p$. (As a matter of fact, $\mathbf{Q}_p$ is also a field; see § 3.)

The completion of the subgroup $\mathbf{Z} \subset \mathbf{Q}$ in this metric (which obviously coincides with the closure of **Z** in $\mathbf{Q}_p$) is called the group of *p-adic integers* and is denoted by $\mathbf{O}_p$.

Problem 2. Prove that the group $\mathbf{O}_p$ is compact and homeomorphic to Cantor's ternary set.

Hint. Show that every element of $\mathbf{O}_p$ can be written uniquely as the sum of a series $\sum_{k=0}^{\infty} a_k p^k$, where $0 \leq a_k \leq p-1$, and that convergence in $\mathbf{O}_p$ is just coordinatewise convergence of the sequences $\{a_k\}$.

The following is another important example in which we apply the notion of completion. Let G be an infinite group. As a basis of neighborhoods of the identity, take all of the subgroups of G that have finite index. We shall suppose that the topology obtained in this way satisfies Hausdorff's separation axiom. In this case, condition (1) holds, and the completion $\hat{G}$ turns out to be a compact topological group. Groups of this sort are called *profinite*. (They are projective limits of finite groups.)

Problem 3. If $G = \mathbf{Z}$, then $\hat{G} = \prod_p \mathbf{O}_p$.

Hint. If a sequence $\{n_k\}$ is fundamental in the topology under consideration, then it is fundamental in all of the metrics ρ_p that we have defined above, and conversely. From this it follows that $\hat{G}$ is embedded in $\prod_p \mathbf{O}_p$. The fact that this embedding is an epimorphism follows from the so-called "Chinese remainder theorem". This theorem asserts that there is an integer having any specified remainders modulo a finite set of moduli that are relatively prime in pairs.

Problem 4. Prove that the group $\hat{G}$ of the preceding problem can also be obtained as the projective limit of a family of groups G_n. Each G_n is isomorphic to **Z**, and the indices are ordered by the relation of divisibility (see problem 2 from para. 2.2), and the mapping of G_n onto G_m is multiplication by n/m.

For proofs of the facts stated in this § and for further information about topological groups, see the books of L.S. Pontrjagin [44] and N. Bourbaki [7, Chapter III].

§ 3. Rings and Modules

Proofs of the facts given in this section, along with more information, can be found in the textbook of S. Lang [39] and also in the treatise of N. Bourbaki [6].

3. Rings and Modules

3.1. Rings

A *ring* is a set K with two operations: addition, with respect to which K is an abelian group; and multiplication, connected with addition by the distributive law:

$$x(y+z)=xy+xz, \quad (x+y)z=xz+yz.$$

One ordinarily studies *associative* rings, that is, rings satisfying the condition $(xy)z=x(yz)$. In this book, the word "ring" will always mean "associative ring".

A ring K is called *commutative* if $xy=yx$ for all $x,y \in K$.

A subgroup $K' \subset K$ is called a *subring*, if $xy \in K'$ for all $x,y \in K'$. It is called a *left ideal (right ideal)* if $xy \in K'$ for all $x \in K$ and $y \in K'$ ($yx \in K'$). An ideal that is both a left and a right ideal is called a *two-sided ideal*.

If K' is a two-sided ideal, then the factor group K/K' can easily be given a multiplicative structure: the product of the cosets $x+K'$ and $y+K'$ is the coset $xy+K'$. The ring obtained in this way is called a *factor ring*.

We list several examples of rings, which we shall encounter frequently in the sequel.

The ring $\mathbf{Z}$ of integers.

The ring $\mathbf{Z}_n$ of residues modulo n (the factor ring of the ring $\mathbf{Z}$ with respect to the ideal of all multiples of n).

The ring $K[t]$ of polynomials in the variable t with coefficients in a ring K.

The ring $\mathrm{Mat}_n(K)$ of matrices of the n-th order with coefficients in a ring K.

Problem 1. Let K be a finite ring, consisting of p elements, where p is a prime number. Prove that K is isomorphic to the ring of residues modulo p or that the product of any two elements of K is zero.

The *unit* of a ring is an element such that multiplication on either right of left by this element is the identity mapping.

Every ring K can be embedded in a ring K_1 having a unit 1, in such a way that K_1 is generated by K and 1. The ring K_1 is defined by this property up to an isomorphism.

The operation of passing from K to K_1 is called *adjoining a unit*.

Problem 2. Prove that if K already possesses a unit, then the ring K_1 is isomorphic to $K \oplus \mathbf{Z}$.

Hint. Let e be the unit of K. Every element of K_1 can be written in the form $n(1-e)+x$, where $x \in X$ and $n \in \mathbf{Z}$.

An important part of a ring K is its *radical* $R(K)$. If K has a unit, then the radical $R(K)$ can be defined by any of the following equivalent statements:

1) $R(K)$ is the intersection of all maximal left ideals in K distinct from K itself;

2) $R(K)$ is the intersection of all maximal right ideals of K distinct from K itself;

3) $R(K)$ consists of all elements $x \in K$ such that the element $1+axb$ admits a multiplicative inverse for all $a,b \in K$.

One can prove that $R(K)$ is a two-sided ideal in K. If K is an arbitrary ring, one can define the radical $R(K)$ by

$$R(K) = R(K_1),$$

where K_1 is the ring obtained from K by adjoining a unit (as a matter of fact, $R(K_1)$ is always contained in K).

A ring K is called *semisimple* if $R(K) = \{0\}$, and a *radical ring* if $R(K) = K$.

Problem 3. The ring $K/R(K)$ is semisimple.

Problem 4. Suppose that K is a field. Then the ring $\text{Mat}_n(K)$ is semisimple.

Problem 5. Consider the subring $T_n(K) \subset \text{Mat}_n(K)$ consisting of all upper triangular matrices (i.e., of matrices $\|a_{ij}\|$ such that $a_{ij} = 0$ for $i > j$). Suppose again that K is a field. Prove that the radical of $T_n(K)$ is the ring $ST_n(K)$, which is defined as the set of strictly triangular matrices $\|a_{ij}\|$, i.e., $a_{ij} = 0$ for $i \geq j$.

Let A be an abelian group. A ring K is called A-*graded* if as an abelian group it is a direct sum of groups K^α, where α runs through A, with the property that $K^\alpha K^\beta \subset K^{\alpha+\beta}$, which is to say that the product of an element of K^α with an element of K^β belongs to $K^{\alpha+\beta}$. A $\mathbf{Z}$-graded ring is called *graded*, without any prefix.

Suppose that in a ring K there is given an increasing (decreasing) family of subrings K_i such that $K_i K_j \subset K_{i+j}$. Then we say that in K, there is given an *increasing (decreasing) filtration*.

Consider the category of rings with a filtration, in which the morphisms are homomorphisms of rings that preserve the filtration. Consider also the category of graded rings, in which the morphisms are homomorphisms of rings that preserve the grading. There is a natural functor from the first category into the second. We denote the functor by gr and construct it in the following way. Suppose for the sake of definiteness that $\{K_i\}$ is an increasing filtration. First we define

$$\text{gr}^i K = K_i / K_{i-1}$$

and

$$\text{gr}\, K = \bigoplus_i \text{gr}^i K \quad \text{(direct sum of abelian groups)}.$$

The structure of a graded ring in $\text{gr}\, K$ will be completely determined once we show the manner of multiplication of an element $x \in \text{gr}^i K$ by an element $y \in \text{gr}^j K$. Let $\tilde{x}$ be a representative of the class x in K_i and $\tilde{y}$ a representative of the class y in K_j. Then xy is defined as the class in $\text{gr}^{i+j} K$ that contains the element $\tilde{x}\tilde{y} \in K_{i+j}$.

It is not hard to verify that this definition is consistent, and we leave this verification to the reader.

Problem 6. Let $K = \mathbf{Z}$, the ring of integers, and let K_i be the subring of numbers that are multiples of m^i, m being some fixed number. Prove that $\text{gr}\, K$ is isomorphic to the ring $\mathbf{Z}_m[t]$, with the natural grading according to powers of t.

§ 3. Rings and Modules

Let K be a ring that is also a topological space. Suppose that K is a topological group under addition and also that multiplication is continuous in both variables. Then K is called a *topological ring*. Every topological ring admits a completion. The so-called I-adic completion is the most important example of this construction. Let I be a two-sided ideal in K. As a basis of neighborhoods of zero, we take the family $\{I^n\}$ of powers of the ideal I. If $\bigcap_n I^n = \{0\}$, this topology will satisfy Hausdorff's separation axiom. It is called the *I-adic topology*, and the completion of K in this topology is called the *I-adic completion*.

Example. Suppose that $K = \mathbf{Z}$ and that $I = m\mathbf{Z}$. Then the I-adic completion of $\mathbf{Z}$ is what is called the ring $\mathbf{Z}(m)$ of m-adic integers.

Problem 7. Let $p_1, \ldots, p_k$ be the prime divisors of m. Prove that the ring of m-adic integers is isomorphic with the direct sum of the rings of p-adic integers for $p = p_1, \ldots, p_k$. That is, prove that $\mathbf{Z}(m) = \bigoplus_i \mathbf{O}_{p_i}$.

Problem 8. Consider positive integers written to the base 10. Prove that there are exactly four idempotent "infinite integers":

$$\ldots 0\,000\,000, \quad \ldots 2\,890\,625, \quad \ldots 7\,109\,376 \quad \text{and} \quad \ldots 0\,000\,001.$$

This means that these four integers and no others have the following property. If the last k digits of a number N are the same as those of one of the four numbers listed above, then all powers of the number N end in exactly the same digits.

Hint. Use the result of Problem 7: $\mathbf{Z}(10) = \mathbf{O}_2 \oplus \mathbf{O}_5$. Use also the fact that there are no divisors of zero in the ring of p-adic integers.

3.2. Skew Fields

A set K is called a *skew field* (sometimes a *division algebra*) if all four arithmetic operations are defined in it in such a way that under addition and multiplication K is a ring, and such that the element x^{-1} is defined for all x different from zero. A skew field K that is also a topological space is called a *topological skew field* if:
1) K is a topological ring with respect to addition and multiplication;
2) the function $x \mapsto x^{-1}$ is continuous everywhere it is defined.

A skew field with commutative multiplication is called a *field*. It is known that every finite skew field is commutative. We shall suppose that the reader is thoroughly familiar with the following fields:
the field $\mathbf{Q}$ of rational numbers;
the field $\mathbf{R}$ of real numbers;
the field $\mathbf{C}$ of complex numbers.
Besides these basic fields, we shall define here and use throughout this book the following:
the skew field $\mathbf{H}$ of quaternions;
the field $\mathbf{Q}_p$ of p-adic numbers;
the field $\mathbf{F}_q$, consisting of a finite number q of elements;
the field $\mathbf{K}_q$ of formal power series with coefficients from $\mathbf{F}_q$.

The most convenient way to define **H** is as the skew field of all complex matrices of the form $\begin{pmatrix} x & y \\ -\bar{y} & \bar{x} \end{pmatrix}$ or real matrices of the form

$$\begin{pmatrix} a & b & c & d \\ -b & a & -d & c \\ -c & d & a & -b \\ -d & -c & b & a \end{pmatrix}.$$

Let **Q**′ be an algebraic extension of the field **Q** (such fields are called *algebraic number fields*). Let p be a prime ideal in the ring **Z**′ of integers in the field **Q**′. The field $\mathbf{Q}_\mathfrak{p}$ is defined as the completion of **Q**′ in its p-adic topology. The field $\mathbf{Q}_\mathfrak{p}$ can also be obtained as an algebraic extension of the field $\mathbf{Q}_p$ of p-adic numbers where p is the unique prime integer belonging to the ideal p.

The finite field $\mathbf{F}_q$ is completely determined by the number q of its elements. This number has to be a prime power: $q = p^n$. In the case where $n = 1$ and $q = p$, this is the field $\mathbf{Z}_p$ of residues modulo p. In the general case, $\mathbf{F}_q$ is an algebraic extension of $\mathbf{Z}_p$ and can be realized by matrices of the n-th order with elements from $\mathbf{Z}_p$. For example, $\mathbf{F}_4$ consists of the matrices

$$\begin{pmatrix} 1 & 0 \\ 0 & 1 \end{pmatrix}, \begin{pmatrix} 1 & 1 \\ 1 & 0 \end{pmatrix}, \begin{pmatrix} 0 & 1 \\ 1 & 1 \end{pmatrix}, \begin{pmatrix} 0 & 0 \\ 0 & 0 \end{pmatrix}$$

with coefficients in $\mathbf{Z}_2$.

The field $\mathbf{K}_q$ is made up of formal power series of the form $\sum_{k=k_0}^{\infty} a_k x^k$, $a_k \in \mathbf{F}_q$, with the usual algebraic operations.

More details about the fields described above can be found, for example, in the book of Z. I. Borevič and I. R. Šafarevič [5].

It is known that the skew fields $\mathbf{Q}_\mathfrak{p}$, $\mathbf{K}_q$, **R**, **C**, and **H** are all of the locally compact nondiscrete skew fields. For a proof of this fact, see, for example, the book of L. S. Pontrjagin [44].

The additive groups of all of these skew fields are self-dual in the sense of Pontrjagin's duality theorem (see § 12). This fact is important for the theory of representations.

3.3. Modules over Rings

An abelian group M is called a *left module* over a ring K or, for short, a *left K-module* if all elements of M can be multiplied on the left by elements of K to obtain again elements of M, and if in addition the following identities hold:

$$\lambda(x+y) = \lambda x + \lambda y, \quad (\lambda + \mu)x = \lambda x + \mu x, \quad (\lambda \mu)x = \lambda(\mu x)$$

for all $\lambda, \mu \in K$ and $x, y \in M$.

§ 3. Rings and Modules

If the ring K has a unit element 1, then one ordinarily also requires that $1 \cdot x = x$ hold for all $x \in M$.

One defines in like manner *right K-modules* and *two-sided K-modules*. If K is commutative, then every left K-module is automatically equipped with the structure of right and a two-sided K-module.

A subgroup $N \subset M$ that is invariant under multiplication by elements of K is called a *submodule*. In this case, the factor group M/N is also a K-module and is called a *factor module*. The collection of all left (right, two-sided) K-modules forms a category, the morphisms of which are group homomorphisms that commute with left (right, two-sided) multiplication by elements of K. We call these morphisms *linear mappings* or *K-mappings*. The set of all K-mappings of M into N is denoted by $\mathrm{Hom}_K(M,N)$. We also write $\mathrm{End}_K M$ for $\mathrm{Hom}_K(M,N)$. In the category of K-modules, sums, products, projective limits, and inductive limits all exist, for an arbitrary family of objects.

For a finite family of K-modules $M_1, \ldots, M_n$, the operations of sum and product lead to one and the same module M, consisting of all sequences $(x_1, \ldots, x_n)$ with $x_i \in M_i$. We leave it to the reader to give M its natural structure as a K-module and to construct mappings $i_k: M_k \to M$ and $p_k: M \to M_k$, in such fashion that $(M, \{i_k\})$ will be the sum and $(M, \{p_k\})$ the product of the modules $M_1, \ldots, M_n$.

We note that a ring K can be regarded as a left, right, or two-sided module over itself. The direct sum of n such modules is called the *free K-module of rank n*.

Problem 1. Let K be a ring with unit and M the free right K-module of rank n. Then the ring $\mathrm{End}_K M$ is isomorphic to $\mathrm{Mat}_n(K)$.

Hint. Consider first the case $n=1$.

A left, right, or two-sided K-module M is called *simple* if it admits no submodules except for $\{0\}$ and M.

Problem 2. Prove that an element $x \in K$ belongs to $R(K)$ if and only if $xM = \{0\}$ for every simple left module M.

Hint. If m is an element of a simple left module M, then the set of all $x \in K$ for which $xm = 0$ is a maximal left ideal in K.

The Jordan-Hölder Theorem. *Suppose that M is a module and that there is an increasing family of submodules $\{0\} = M_0 \subset M_1 \subset \cdots \subset M_n = M$, with the property that every factor module $N_i = M_i/M_{i-1}$, $i = 1, \ldots, n$, is simple. Then the equivalence classes of the modules $N_1, \ldots, N_n$ are determined completely (except for the order in which they are written) by the equivalence class of the module M and do not depend upon the choice of the family $\{M_i\}$.*

The proof of this theorem is a simple consequence of Schur's lemma (see § 7 *infra*).

Let M be a right K-module and N a left K-module. We can then define the *tensor product over K* of the modules M and N, denoted by $M \underset{K}{\otimes} N$. To do this, we first define the abelian group $M \square N$ generated by all symbols of the form $m \square n$, where m runs through M and n runs through N. Let $M \bigcirc N$ denote the subgroup of $M \square N$ generated by all elements of the form

a) $(m_1+m_2)\square n - m_1\square n - m_2\square n$,

b) $m\square(n_1+n_2) - m\square n_1 - m\square n_2$,

c) $m\lambda\square n - m\square \lambda n$,

where m, m_1, m_2 run through M, n, n_1, n_2 run through N, and λ runs through K. We define $M \underset{K}{\otimes} N = M\square N / M\bigcirc N$. The image of the element $m\square n \in M\square N$ in $M \underset{K}{\otimes} N$ is usually denoted by $m \underset{K}{\otimes} n$ or simply $m \otimes n$.

Note that the tensor product of two K-modules is an abelian group that carries no natural structure as a K-module. Nevertheless, if M (or N) has the structure of a left (or right) K-module, then this structure can be carried over to the tensor product. In particular, if K is a commutative ring, then M and N are two-sided K-modules and hence $M \underset{K}{\otimes} N$ is also a two-sided K-module.

Example. Every (abelian) group can be regarded as a **Z**-module, setting $nx = x + x + \cdots + x$ (n summands) for $n > 0$ and $nx = -(-nx)$ for $n < 0$.

Problem 3. Compute the tensor product of the groups $\mathbf{Z}_m$ and $\mathbf{Z}_n$ as **Z**-modules.

Answer. $\mathbf{Z}_{(m,n)}$, where (m, n) is the greatest common divisor of the numbers m and n.

The tensor product can also be defined starting with the notion of a universal object. Suppose for simplicity that the ring K is commutative and that $M_1, \ldots, M_n$ is a collection of K-modules. Consider the category whose objects are multilinear mappings (that is, linear in each argument) of the Cartesian product $M_1 \times \cdots \times M_n$ into an arbitrary K-module M (one for each object). The morphisms are the commutative diagrams

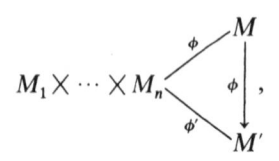

$$M_1 \times \cdots \times M_n \begin{array}{c} \phi \\ \nearrow \\ \\ \searrow \\ \phi' \end{array} \begin{array}{c} M \\ \\ \downarrow \psi \\ \\ M' \end{array},$$

where ϕ and ϕ' are multilinear mappings and ψ is a linear mapping of M into M'. The universal repelling object in this category is called the *tensor product* over K of the modules $M_1, \ldots, M_n$.

Problem 4. Show that the definition just given is, for the case $n = 2$, equivalent to the first definition of tensor product.

Hint. Consider the multilinear mapping of $M_1 \times M_2$ into $M_1 \underset{K}{\otimes} M_2$ defined by $(x_1, x_2) \mapsto x_1 \otimes x_2$.

Let K be a commutative ring and M a certain K-module. The set $M^* = \mathrm{Hom}_K(M, K)$ has a natural structure as a K-module. It is called the *dual module* to the module M.

The mapping $M \mapsto M^*$ is a contravariant functor in the category of K-modules.

§ 3. Rings and Modules 31

Problem 5. Let M and N be free K-modules of finite rank. Prove that there exist natural isomorphisms as follows:

$$(M^*)^* \approx M, \quad \mathrm{Hom}_K(M,N) \approx \mathrm{Hom}_K(N^*,M^*) = M^* \underset{K}{\otimes} N,$$
$$\left(M \underset{K}{\otimes} N\right)^* \approx M^* \underset{K}{\otimes} N^*.$$

3.4. Linear Spaces

Suppose that K is a field. A K-module M in this case is called a *linear* or a *vector space* over K and elements of M are called *vectors*.

In the case where K is a noncommutative skew field, one sometimes uses the terms "left (right) linear space over K" instead of "left (right) K-module".

Every module over a skew field K is free. In other words, every linear space *admits a basis*.

One can also show that all bases of a space M have the same cardinal number, which is called the *dimension of* M and is written $\dim_K M$ or simply $\dim M$, if it is clear what K is under consideration.

Suppose that M and N are linear spaces over K. Elements of the space $\mathrm{Hom}_K(M,N)$ are called *linear operators* from M into N. Let $\{x_\alpha\}$ and $\{y_\beta\}$ be bases in M and N respectively. Every linear operator A from M into N is uniquely determined by its matrix $\|a_{\alpha\beta}\|$:

$$A x_\alpha = \sum_\beta a_{\alpha\beta} y_\beta. \tag{1}$$

Conversely, every matrix $\|a_{\alpha\beta}\|$ in which every row contains only a finite number of nonzero entries defines, by formula (1), a linear operator from M into N.

Problem 1. Show that $\dim \mathrm{Hom}_K(M,N) = \dim M \dim N$ if both factors on the right side are finite.

Let A be a linear operator on a finite-dimensional space M over a field K. We define the *trace* of A as the element $\mathrm{tr} A \in K$ that is the image of A under composition of the mappings

$$\mathrm{Hom}_K(M,M) \xrightarrow{\phi} M^* \underset{K}{\otimes} M \xrightarrow{\psi} K.$$

Here ϕ is the natural isomorphism and ψ is the linear operator that corresponds to the bilinear mapping of $M^* \times M$ into K defined by $(m^*, m) \mapsto m^*(m)$.

Problem 2. Prove that $\mathrm{tr} A$ is equal to the sum of the diagonal elements of the matrix of the operator A in an arbitrary basis.

Problem 3. If $A: M \to N$ and $B: N \to M$ are linear operators on finite dimensional spaces, then $\mathrm{tr} AB = \mathrm{tr} BA$.

The operations of direct sum and tensor product of two spaces lead to the concepts of *direct sum* and *tensor product* of two operators. Suppose that

$A_1 \in \operatorname{Hom}_K(M_1, N_1)$, and $A_2 \in \operatorname{Hom}_K(M_2, N_2)$. Then $A_1 \oplus A_2$ denotes the operator from $M_1 \oplus M_2$ to $N_1 \oplus N_2$ given by

$$A_1 \oplus A_2 : (x_1, x_2) \mapsto (A_1 x_1, A_2 x_2).$$

We define $A_1 \otimes A_2$ as the operator from $M_1 \otimes M_2$ to $N_1 \otimes N_2$ given by the formula

$$A_1 \otimes A_2 : x_1 \otimes x_2 \mapsto A_1 x_1 \otimes A_2 x_2.$$

One can verify that in appropriate bases, the matrix of the operator $A_1 \oplus A_2$ has the form

$$\left\| \begin{matrix} a^{(1)} & 0 \\ 0 & a^{(2)} \end{matrix} \right\|,$$

where $\|a^{(1)}\|$ and $\|a^{(2)}\|$ are the matrices of the operators A_1 and A_2. The matrix of the operator $A_1 \otimes A_2$ when written in an appropriate basis has the property that $a_{\alpha\gamma;\beta\delta} = a^{(1)}_{\alpha\beta} \cdot a^{(2)}_{\gamma\delta}$. Thus it may be written as a block matrix:

$$\begin{pmatrix} a^{(1)}_{11} a^{(2)} & \cdots & a^{(1)}_{1n} a^{(2)} \\ \cdots & \cdots & \cdots \\ a^{(1)}_{n1} a^{(2)} & \cdots & a^{(1)}_{nn} a^{(2)} \end{pmatrix} \quad \text{or} \quad \begin{pmatrix} a^{(2)}_{11} a^{(1)} & \cdots & a^{(2)}_{1m} a^{(1)} \\ \cdots & \cdots & \cdots \\ a^{(2)}_{m1} a^{(1)} & \cdots & a^{(2)}_{mm} a^{(1)} \end{pmatrix}.$$

Problem 4. For the finite-dimensional case, prove the following equalities:

$$\operatorname{tr}(A_1 \oplus A_2) = \operatorname{tr} A_1 + \operatorname{tr} A_2, \quad \operatorname{tr}(A_1 \otimes A_2) = \operatorname{tr} A_1 \cdot \operatorname{tr} A_2.$$

3.5. Algebras

Suppose that a set A has the structure of a ring and also of a linear space over a field K. Suppose also that multiplication by elements of K depends linearly upon summands. Then A is called an *algebra over K* or more briefly a *K-algebra*.

The family of all K-algebras forms a category, the morphisms of which are algebra homomorphisms (that is, linear ring homomorphisms). Sums and products of arbitrary families of objects exist. Also, the tensor product of a finite number of objects exists in the category of K-algebras.

Suppose that $A_1, \ldots, A_n$ are K-algebras. Their *tensor product* $A_1 \otimes \cdots \otimes A_n$ is the linear space (product over K) with multiplication defined by

$$(x_1 \otimes \cdots \otimes x_n) \cdot (y_1 \otimes \cdots \otimes y_n) = x_1 y_1 \otimes \cdots \otimes x_n y_n.$$

Problem 1. If each of the algebras $A_1, \ldots, A_n$ is isomorphic to the algebra $K[t]$, then their tensor product is isomorphic to the algebra $K[t_1, \ldots, t_n]$ of polynomials in n variables with coefficients in K.

§ 3. Rings and Modules

Problem 2. Prove the following relations:

$$\mathbf{C} \underset{\mathbf{R}}{\otimes} \mathbf{C} = \mathbf{C} \oplus \mathbf{C}, \quad \mathbf{C} \underset{\mathbf{R}}{\otimes} \mathbf{H} = \mathrm{Mat}_2(\mathbf{C}), \quad \mathbf{H} \underset{\mathbf{R}}{\otimes} \mathbf{H} = \mathrm{Mat}_4(\mathbf{R}).$$

Problem 3. If A is an arbitrary K-algebra, then

$$A \underset{K}{\otimes} \mathrm{Mat}_n(K) = \mathrm{Mat}_n(A).$$

Corollary. $\mathrm{Mat}_n(K) \underset{K}{\otimes} \mathrm{Mat}_m(K) = \mathrm{Mat}_{mn}(K).$

We consider the category of K-algebras with n specified generators, the morphisms being the algebra homomorphisms that carry the generators of one algebra into the generators of another. This category contains a universal repelling object, the so-called *free algebra* with n generators $\mathscr{F}(n,K)$.

Problem 4. Let $t_1, \ldots, t_n$ be generators of $\mathscr{F}(n,K)$. Prove that a basis for the linear space $\mathscr{F}(n,K)$ is formed by all possible "words" $t_{i_1}, \ldots, t_{i_k}$, including "void word", which we denote by 1.

It is clear that $\mathscr{F}(n,K)$ is a graded algebra, if we define $\mathscr{F}^l(n,K)$ as the subspace generated by all "words" of length l.

The algebra $\mathscr{F}(n,K)$ admits another useful interpretation. Let us write the subspace $\mathscr{F}^1(n,K)$ as V. Then $\mathscr{F}^l(n,K)$ can be identified in a natural way with the tensor product $V \otimes V \otimes \cdots \otimes V$ (l factors). Hence the algebra $\mathscr{F}(n,K)$ can also be called the *tensor algebra* over V; under this interpretation we denote it by $T(V) = \bigoplus_{l=0}^{\infty} T^l(V)$. Elements of $T^l(V)$ are called *contravariant tensors* of rank l over V.

Let V^* be the dual space to V. Elements of the space $T(V) \otimes T(V^*)$ are called *mixed tensors*. More precisely, we define elements of the space $T^k(V) \otimes T^l(V^*)$ as *mixed tensors of rank* (k,l) (or k-contravariant and l-covariant tensors). Problem 5 of para. 3.3 shows that this is equivalent to the classical definition of a tensor of rank (k,l) as a multilinear function of k vectors in V and l vectors in V^*.

We say that a tensor in $T^k(V)$ is *symmetric (skew symmetric)* if the multilinear function defined by it on V^* does not change (changes its sign) upon permutation of any two of its arguments.

The set of all symmetric (skew symmetric) tensors is denoted by the symbol

$$S(V) = \bigoplus_{k=0}^{\infty} S^k(V) \quad \left(\wedge(V) = \bigoplus_{k=0}^{\infty} \wedge^k(V) \right).$$

Problem 5. Let I (J) be the two-sided ideal in $T(V)$ that is generated by all elements of the form $xy - yx$, as x and y run through V (by elements of the form x^2 as x runs through V). Prove that

$$T(V) = S(V) \oplus I = \wedge(V) \oplus J$$

(the sum indicated is the direct sum of linear spaces).

The result stated in the foregoing problem enables us to identify $S(V)$ with $T(V)/I$ and $\wedge(V)$ with $T(V)/J$, and at the same time to give $S(V)$ and $\wedge(V)$ structures as algebras over K. The algebra $S(V)$ is called the *symmetric algebra* over V. It is commutative and is isomorphic to the algebra of polynomials in n variables ($n = \dim V$). The algebra $\wedge(V)$ is called the *exterior algebra* or the *Grassmann algebra* over V.

The operation of multiplication in $\wedge(V)$ is usually denoted by $x \wedge y$ and is called the *exterior product*.

Problem 6. Prove that $\dim \wedge^k(V) = C_n^k$, where $n = \dim V$ and $C_n^k = \dfrac{n!}{k!(n-k)!}$ is the binomial coefficient.

Hint. Suppose that $x_1, \ldots, x_n$ is a basis in V. Then elements of the form $x_{i_1} \wedge \ldots \wedge x_{i_k}$, $i_1 < \ldots < i_k$ form a basis in $\wedge^k(V)$.

Suppose that Q is a *quadratic form* over V (that is, $Q(x) = B(x,x)$, where B is a certain bilinear form). Let J_Q be the two-sided ideal in $T(V)$ generated by elements of the form $x^2 - Q(x) \cdot 1$, as x runs through V. A *Clifford algebra* is the factor algebra $C(V,Q) = T(V)/J_Q$.

The structure of Clifford algebras plays a rôle in the theory of representations and also in algebraic topology. The following problems give a description of these algebras in the important case where $K = \mathbf{R}$.

In an appropriately chosen system of coordinates, the form Q has the form

$$Q(x) = x_1^2 + \cdots + x_p^2 - x_{p+1}^2 - \cdots - x_{p+q}^2, \quad p + q \leqslant n = \dim V.$$

We denote the corresponding algebra $C(V,Q)$ by $C(p,q,r)$, where $r = n - p - q$.

Problem 7. Prove that the radical of the algebra $C(p,q,r)$ is generated (as an ideal) by the basis elements $x_{p+q+1}, \ldots, x_n$ and that the factor algebra of $C(p,q,r)$ by its radical is isomorphic to $C(p,q,0)$.

Problem 8. Establish the following isomorphisms:

$$C(1,1,0) \otimes C(p,q,r) = C(p+1, q+1, r),$$
$$C(2,0,0) \otimes C(p,q,r) = C(q+2, p, r),$$
$$C(0,2,0) \otimes C(p,q,r) = C(q, p+2, r).$$

Hint. Let x_1, x_2 and $y_1, \ldots, y_n$ be canonical bases in V_1 and V_2. In the algebra $C(V_1, Q_1) \otimes C(V_2, Q_2)$, consider the set of elements

$$x_1 \otimes 1, \quad x_2 \otimes 1, \quad x_1 x_2 \otimes y_1, \ldots, x_1 x_2 \otimes y_n.$$

Problem 9. Establish the following relations:

$$C(2,0,0) = C(1,1,0) = \mathrm{Mat}_2(\mathbf{R}),$$
$$C(0,2,0) = \mathbf{H}, \quad C(1,0,0) = \mathbf{R} \oplus \mathbf{R}, \quad C(0,1,0) = \mathbf{C}.$$

For nondegenerate forms Q, we thus see that Clifford algebras are matrix algebras over **R**, **C**, or **H** or sums of such algebras. This fact is an illustration of the following general result.

Theorem 1. *Every semisimple finite-dimensional algebra is a direct sum of simple algebras* (that is, of algebras not admitting any two-sided ideals except for the zero ideal and the whole algebra). *Every simple finite-dimensional K-algebra is isomorphic to* $\mathrm{Mat}_n(D)$, *where D is a finite-dimensional skew field over K.*

We also have the following

Theorem 2. *Every finite-dimensional module over a semisimple algebra A is a direct sum of simple submodules. There exist only a finite number of equivalence classes of simple A-modules and for a simple algebra only one such equivalence class.*

We obtain an interesting infinite-dimensional algebra in the following way. Instead of Q, take a skew-symmetric bilinear form B and consider the ideal generated by all elements of the form

$$x \otimes y - y \otimes x - B(x,y) \cdot 1, \quad x, y \in V.$$

Then divide $T(V)$ by this ideal. It is known that we can select bases $p_1, \ldots, p_n, q_1, \ldots, q_n, r_1, \ldots, r_k$ such that B has the form

$$B(p_i, q_j) = \delta_{ij},$$
$$B(p_i, p_j) = B(q_i, q_j) = B(p_i, r_j) = B(q_i, r_j) = B(r_i, r_j) = 0.$$

The algebra obtained in this fashion is denoted by $R_{n,k}(K)$.

Problem 10. Establish the relation

$$R_{n_1, k_1}(K) \underset{K}{\otimes} R_{n_2, k_2}(K) = R_{n_1 + n_2, k_1 + k_2}(K).$$

§ 4. Elements of Functional Analysis

4.1. Linear Topological Spaces

Let L be a linear space over a topological field K. Suppose that:
  1) L is a topological group under addition;
  2) the mapping $K \times L \to L : (\lambda, x) \to \lambda x$ is continuous in both variables.
Then we say that L is a *topological linear space* (TLS).

Topological linear spaces over the field K form a category, the morphisms of which are linear continuous mappings. Sums, products, and projective and inductive limits are all defined in this category.

We recall that for a finite number of spaces, sums and products coincide in the following sense.

Problem 1. Let $\{L_k\}$, $k=1,\ldots,n$, be a finite family of TLS's, and (L, i_k) their sum and (M, p_k) their product. Prove that there exists an isomorphism $\tau: L \to M$, such that

$$p_k \circ \tau \circ i_l = \begin{cases} 0 & \text{if } k \neq l, \\ 1 & \text{if } k = l. \end{cases}$$

Hint. Show that L and M can both be realized as the set of all sequences $(x_1, \ldots, x_n)$, $x_i \in L_i$, with the natural structure as a TLS.

Classical functional analysis concerns itself with a study of the category of TLS's over the fields $\mathbf{R}$ and $\mathbf{C}$.

Almost all of the TLS's met with in applications are *locally convex spaces* (LCS). This means that there exists a basis of neighborhoods of zero that consist of convex sets. Locally convex spaces possess a whole series of important properties, the most important of which is the existence of a sufficiently rich set of continuous linear functionals. If L is a LCS and x is a nonzero vector in L, then there exists a continuous linear functional f on L such that $f(x) \neq 0$.

Thus every LCS can be embedded in a coordinate space K^α, where α is a sufficiently large cardinal number.

It is convenient to describe the topology of a LCS by means of a family of prenorms. A real-valued function p on a TLS is called a *prenorm* (the term *seminorm* is also used) if it satisfies the following conditions:

1) $\qquad p(x) \geq 0$,

2) $\qquad p(x+y) \leq p(x) + p(y)$,

3) $\qquad p(\lambda x) = |\lambda| p(x)$.

Suppose that p satisfies also the following property:

4) $\qquad p(x) = 0$ only for $x = 0$.

Then p is called a *norm*. We frequently use the notation $\|x\|$ to denote the norm of a vector x.

Problem 2. A net $\{x_\alpha\}_{\alpha \in A}$ in a LCS converges to x if and only if the numerical net $\{p(x_\alpha - x)\}_{\alpha \in A}$ converges to 0 for all continuous prenorms p.

Hint. Let U be a convex neighborhood of zero in L. The function

$$p_U(x) = \inf_{x \in \lambda U} \{|\lambda|\}$$

is called the *Minkowski functional* of this neighborhood. Prove that p_U is a continuous prenorm on L.

A family $\{p_\beta\}_{\beta \in B}$ of prenorms is called *determining* if it suffices to verify the condition of problem 2 only for prenorms from this family. A LCS is called

§ 4. Elements of Functional Analysis

normable (countably normable) if there exists a finite (countable) determining family of prenorms. (A finite family of prenorms may be replaced by a single prenorm, namely, its sum.)

An arbitrary family of prenorms $\{p_\beta\}_{\beta \in B}$ on a linear space L is a determining family for a certain topology, under which L is a LCS.

As a rule, we shall study complete spaces that satisfy Hausdorff's separation axiom. Let L be an arbitrary LCS. There corresponds to L a complete LCS $\hat{L}$ satisfying Hausdorff's separation axiom, which is obtained from L by factoring out the subspace L_0, which is defined as the intersection of all neighborhoods of zero, and then completing the space L/L_0.

A complete normed LCS with Hausdorff separation is called a *Banach space*.

A Banach space is called a *Hilbert space* if its norm satisfies the "parallelogram law":

$$p(x+y)^2 + p(x-y)^2 = 2p(x)^2 + 2p(y)^2. \tag{1}$$

Problem 3. Suppose that (1) holds. Define the expression

$$(x,y) = \tfrac{1}{4}[p(x+y)^2 - p(x-y)^2 + ip(x-iy)^2 - ip(x+iy)^2]. \tag{2}$$

Prove that (x,y) is linear in x, conjugate linear in y, and continuous in both variables simultaneously. Prove also that $(x,x) = p(x)^2$.

Hint. Consider the cases $\dim L = 2$ and $\dim L = 3$.

Example 1. Let $C[-1,1]$ be the family of all continuous functions on the closed interval $[-1,1]$. We set

$$p_\alpha(f) = \left[\int_{-1}^{1} |f(x)|^\alpha dx\right]^{1/\alpha}, \quad 0 < \alpha < \infty,$$

$$p_\infty(f) = \max |f(x)|.$$

Problem 4. Prove that p_α is a norm for $1 \leq \alpha \leq \infty$.

Hint. Use the inequality $(1+y)^\alpha \geq 1 + y^\alpha$ for $y \geq 0$, $\alpha \geq 1$.

Let us give $C[-1,1]$ a topology τ_α, taking as a basis of neighborhoods of zero the sets $p_\alpha(f) < \varepsilon$, $\varepsilon > 0$.

Problem 5. Prove that the space $C[-1,1]$ is complete in the topology τ_∞ but incomplete in all of the topologies τ_α, $0 < \alpha < \infty$.

Hint. Prove that the sequence $f_n(x) = \mathrm{arctg}(nx)$ is fundamental but has no limit in $C[-1,1]$.

We denote the completion of $C[-1,1]$ in the topology τ_α by the symbol $L^\alpha[-1,1]$.

Problem 6. Prove that $L^2[-1,1]$ is a Hilbert space.

Problem 7. Let α be a real number such that $0 < \alpha < 1$. Prove that the space $L^\alpha[-1,1]$ admits no nonzero continuous linear functionals.

Hint. Every function $f \in C[-1,1]$ can be represented as a sum of $2N-1$ functions $f_1, \ldots, f_{2N-1}$, which satisfy the conditions $|f_i| \leq |f|$, $f_i = 0$ outside of the interval

$$\left[-1 + \frac{i-1}{N}, \ -1 + \frac{i+1}{N}\right].$$

Use the fact that a continuous linear functional F on L^α has to satisfy the inequality

$$|F(f)| \leq C \left(\int_{-1}^{1} |f|^\alpha dx \right)^{1/\alpha}$$

and hence

$$|F(f)| \leq C \sum_{i=1}^{2N-1} \left(\int_{-1}^{1} |f_i|^\alpha dx \right)^{1/\alpha} \leq C \cdot M \cdot (2N-1)(2/N)^{1/\alpha}.$$

The most important examples of TLS's are constructed with the aid of the notion of a measure. We give here some introductory information from the theory of measure and integration, referring the reader for details to the textbooks [38] and [49], which we have already cited above. More facts can be found in the books [27] and [8].

Let X be a certain set and B a system of subsets of X closed under the operations of countable unions and intersections and complementation. Such a system is called a σ-*algebra*. We say that a *measure* μ is defined on (X, B) (or simply on X if there is no doubt as to what B is under consideration), if for every subset $E \in B$, there is defined a number $\mu(E)$ in such a way that

$$\mu(E_1 \cup E_2) = \mu(E_1) + \mu(E_2) \quad \text{if} \quad E_1 \cap E_2 = \emptyset.$$

A subset $E \in B$ is called *measurable* and $\mu(E)$ is called its *measure*. A measure μ is called *countably additive* if

$$\mu\left(\bigcup_{n=1}^{\infty} E_n\right) = \sum_{k=1}^{\infty} \mu(E_k) \quad \text{if} \quad E_i \cap E_j = \emptyset \quad \text{for} \quad i \neq j.$$

(We suppose that the series on the right side converges absolutely.)

We frequently suppose also that the measure μ assumes only nonnegative real values. Such measures are called *positive*. Given a countably additive measure, we define its *variation* as the positive measure $|\mu|$ defined by

$$|\mu|(E) = \sup \sum_{k=1}^{\infty} |\mu(E_k)|,$$

when the supremum is taken over all decompositions of E into a countable union of pairwise disjoint measurable subsets E_k.

§ 4. Elements of Functional Analysis

Every measure μ has the form $\mu = \mu_1 - \mu_2 + i\mu_3 - i\mu_4$, where $\mu_1, \mu_2, \mu_3, \mu_4$ are positive measures. These measures can be chosen in such a way that

$$\sqrt{2}\,|\mu| \geq \mu_1 + \mu_2 + \mu_3 + \mu_4 \geq |\mu|.$$

A measure μ is called *complete* if the relations

$$E_1 \in B, \quad E_2 \in B, \quad E_1 \subset E \subset E_2 \quad \text{and} \quad |\mu|(E_2 \setminus E_1) = 0$$

imply that $E \in B$. Every countably additive measure can be extended to a complete measure.

If a certain relation holds for all points $x \in X \setminus E$ and $|\mu|(E) = 0$, then one says that this relation holds *almost everywhere* on X with respect to the measure μ.

Suppose that the set X is the union of a family of pairwise disjoint subsets X_α, $\alpha \in A$. Suppose that we have a finite measure μ_α on each X_α. Then we can define on X what is called a *locally finite measure* $\mu = \sum_\alpha \mu_\alpha$. A subset E is defined as μ-measurable if its intersection with each X_α is μ_α-measurable. The measure of E is defined as the sum of the series $\sum_\alpha \mu_\alpha(E \cap X_\alpha)$, provided that the series $\sum_\alpha |\mu_\alpha|(E \cap X_\alpha)$ converges. In the contrary case we write $\mu(E) = \infty$. If the family $\{X_\alpha\}$ is countable, the measure μ is called *σ-finite*.

Sometimes locally finite measures are called simply measures, and measures in our sense are called finite measures.

A complex function f on X is called *measurable* (one also says μ-measurable or B-measurable) if for every open set $U \subset \mathbf{C}$, the set $f^{-1}(U) \subset X$ is measurable. A function f is called *simple* if it is measurable and assumes only a countable set of values $\lambda_1, \ldots, \lambda_n, \ldots$.

Let μ be a positive measure (not necessarily finite). If the series

$$\sum_k \lambda_k \mu(f^{-1}\{\lambda_k\})$$

converges absolutely, then its sum is called the *integral of f* with respect to the measure μ and is denoted by $\int_X f(x)\,d\mu(x)$ or simply $\int_X f\,d\mu$.

A nonnegative measurable function f is said to be μ-integrable if the set of all integrals of the form $\int_X g(x)\,d\mu(x)$ is bounded above, where g runs through the set of all simple functions that do not exceed f. The supremum of the set of all these integrals is called the integral of f with respect to the measure μ. A complex measurable function f is called *integrable* with respect to the (not necessarily positive) measure μ if the function $|f|$ is integrable with respect to the measure $|\mu|$. The integral $\int_X f(x)\,d\mu(x)$ is uniquely determined in this case by requiring that the integral be linear with respect to f and μ, and by agreeing that the integral for positive f and μ agrees with the integral defined above.

The set of all μ-integrable functions forms a linear space L under the prenorm

$$\|f\| = \int_X |f(x)|\,d|\mu|(x).$$

It is evident that the intersection L_0 of all neighborhoods of zero consists of the functions that are equal to zero almost everywhere. It is known that the factor space L/L_0 is complete with respect to the corresponding norm. It is denoted by $L^1(X,\mu)$.

We define the space $L^p(X,\mu)$ analogously. It consists of classes of functions for which the integral of $|f|^p$ with respect to μ is finite. For $1 \leq p \leq \infty$ this space is a Banach space with the norm

$$\|f\| = \left| \int_X |f(x)|^p d|\mu|(x) \right|^{1/p}.$$

The space $L^\infty(X,\mu)$ is defined as the Banach space obtained by factoring the space L of all bounded μ-measurable functions on X with the norm $\|f\| = \sup_{x \in X} |f(x)|$ by the closed subspace of all functions that vanish almost everywhere.

We say that a measure μ_1 is *absolutely continuous* with respect to a measure μ_2 (in symbols $\mu_1 \prec \mu_2$) if every μ_2-measurable set is μ_1-measurable and if the $|\mu_2|(E) = 0$ implies that $|\mu_1|(E) = 0$. In this case there exists a μ_2-measurable function ρ such that the equality

$$\int_X f(x) d\mu_1(x) = \int_X f(x) \rho(x) d\mu_2(x) \tag{3}$$

holds for all μ_1-integrable functions f. The function $\rho(x)$ is defined uniquely almost everywhere with respect to the measure μ_2. It is called the *derivative* of the measure μ_1 with respect to the measure μ_2 and is denoted by $d\mu_1/d\mu_2$. If μ_1 is finite, then the function $d\mu_1/d\mu_2$ is integrable with respect to the measure μ_2.

If each of the measures μ_1 and μ_2 is absolutely continuous with respect to the other, then these measures are said to be *equivalent* (in symbols $\mu_1 \sim \mu_2$).

We note that every σ-finite measure is equivalent to a finite measure. For equivalent measures, the spaces $L^p(X,\mu_1)$ and $L^p(X,\mu_2)$ are naturally isomorphic. Multiplication by $|d\mu_1/d\mu_2|^{1/p}$ carries the second space onto the first. Measures μ_1 and μ_2 are called *disjoint* if the relations $\mu_1 \succ \mu$ and $\mu_2 \succ \mu$ imply that $\mu = 0$.

Let X be a locally compact space and let $K(X)$ be the space of continuous functions on X with compact supports. We impose a topology on $K(X)$ by the following description of convergence. A net $\{f_\alpha\}$ converges to f if and only if all f_α and also f vanish outside of a certain compact set $K \subset X$ and f_α converges uniformly to f on X. It is known that every continuous linear functional on $K(X)$ has the form

$$F_\mu(f) = \int_X f(x) d\mu(x), \tag{4}$$

where μ is a certain measure defined on all Borel subsets of X and enjoying the following property.

For every measurable set E of finite measure and every $\varepsilon > 0$, there exist a compact set K and an open set U such that $K \subset E \subset U$ and $|\mu|(U \setminus K) < \varepsilon$. Such measures are called *regular*.

§ 4. Elements of Functional Analysis

Conversely, every regular measure on X defines by formula (4) a continuous linear functional on $K(X)$.

We define the *support* of a regular measure μ as the complement of the union of all open sets of measure zero. This is obviously a closed set. We denote it by the symbol $\operatorname{supp}\mu$.

Let L be a TLS. The *topological conjugate space* L' of L is the space of all continuous linear functionals on L. Plainly L' is a subspace of the algebraic conjugate space $L^* = \operatorname{Hom}_K(L, K)$. In the finite-dimensional case, L' and L^* coincide (if L satisfies Hausdorff's separation axiom). A set $E \subset L$ is called *bounded* if for every neighborhood U of zero, there exists a number $\lambda \in K$ such that $E \subset \lambda U$. If L is a LCS, then this condition is equivalent to *weak boundedness*, which is to say that $f(E)$ is a bounded set for every $f \in L'$.

There are various natural ways in which the space L' can be made into a locally convex topological linear space.

The *weak topology* corresponds to pointwise convergence: a net $\{f_\alpha\}$ converges to f provided that $f_\alpha(x)$ converges to $f(x)$ for all $x \in L$.

The *strong topology* corresponds to uniform convergence on bounded sets.

Suppose that L is a Banach space with norm p. Define a norm p' in L' by the relation

$$p'(f) = \sup_{p(x) \leq 1} |f(x)|.$$

Then L' is also a Banach space.

Let A be a continuous linear operator from L_1 into L_2. We define the adjoint operator $A': L'_2 \to L'_1$, by the formula

$$[A'f](x) = f(Ax), \quad x \in L_1, \quad f \in L'_2.$$

The operator A' is continuous if both of the spaces L'_1 and L'_2 are given either the weak or the strong topology.

The space $\mathfrak{B}(L_1, L_2)$ of all continuous linear operators from L_1 into L_2 is most frequently topologized with one of the following three topologies:

a direction A_α converges to A

1) *weakly* if for all $x \in L_1$ and all $f \in L'_2$, the numerical direction $\{f(A_\alpha x)\}$ converges to $f(Ax)$;

2) *strongly* if for all $x \in L_1$, the vector direction $A_\alpha(x)$ converges to Ax in L_2;

3) *uniformly* if for every bounded set $E \subset L$, the direction $\{f(A_\alpha x)\}$ converges to Ax uniformly for all $x \in E$.

We point out an inconsistency between these (generally accepted) definitions and the definitions given above (which are also generally accepted) of topologies in L'. Every functional $f \in L'$ can be regarded as an operator from L to K. Under this interpretation, the uniform operator topology corresponds to the strong topology in L', and the weak and strong operator topologies coincide in this case and correspond to the weak topology in L'.

Example 2. The space strongly conjugate to $L^p(X, \mu)$, for $1 \leq p < \infty$, is naturally isomorphic to the space $L^q(X, \mu)$, where $1/p + 1/q = 1$. In fact, every continuous functional on $L^p(X,\mu)$ has the form

$$F_g(f) = \int_X f(x)g(x)d\mu(x),$$

where $g \in L^q(X,\mu)$. It is also the case that the norm of the functional F_g is equal to the norm of the function g in $L^q(X,\mu)$.

We now present without proof several properties of LCS's that we shall use in the sequel.

1. *Every weakly bounded and weakly closed set is weakly compact.*

2 (the Hahn-Banach theorem). *Let K be a convex closed set containing a neighborhood of zero. For every point $x \notin K$, there exists a linear functional f such that $f(y) \leq 1$ for all $y \in K$ and $f(x) > 1$.*

3 (the Kreĭn-Mil'man theorem). *Every compact convex set is the closure of the convex hull of its extreme points.* (A point of a convex set K is called *extreme* if it is not the midpoint of any interval lying entirely in K.)

4 (Choquet's theorem). *Let K be a compact convex set. For every point $x \in K$, there exists a measure μ on the set E of extreme points such that*

$$f(x) = \int_E f(y)d\mu(y)$$

for every continuous linear functional f.

5 (the Schauder-Tihonov theorem). *Every continuous mapping of a compact convex set into itself admits a fixed point.*

6 (the Banach-Steinhaus theorem). *Let L_1 and L_2 be Banach spaces and let $S \subset \mathfrak{B}(L_1, L_2)$ be a family of operators. Suppose that the family S is bounded at every vector $x \in L_1$. Then it is uniformly bounded, which is to say that $\|A\| \leq C$ for all $A \in S$.*

Proofs of these facts can be found in the books [16], [14], and [43].

We shall also suppose that the reader is familiar with the elementary theory of Hilbert spaces: the theorem on orthogonal complements; Bessel's inequality and Parseval's equality; the existence of an orthonormal basis; the isomorphism of two spaces of the same dimension in the sense of cardinal number of a basis. All of this information can be found in the textbooks [38] and [49].

Let L_1 and L_2 be normed spaces with norms p_1 and p_2.

The tensor product $L_1 \otimes L_2$ has no *a priori* norm. Nevertheless, one can point out a natural class of norms (called *cross norms*) possessing the following properties:

1) $p(x_1 \otimes x_2) = p_1(x_1) \cdot p_2(x_2), \quad x_1 \in L_1, \quad x_2 \in L_2,$

2) $p'(f_1 \otimes f_2) = p'_1(f_1) \cdot p'_2(f_2), \quad f_1 \in L'_1, \quad f_2 \in L'_2.$

(It is obvious that every element in $L'_1 \otimes L'_2$ yields a linear functional on $L_1 \otimes L_2$.)

It turns out that there is a greatest cross norm, denoted by $p_1 \hat{\otimes} p_2$ and defined by:

$$(p_1 \hat{\otimes} p_2)(z) = \inf \sum_i p_1(x_i) \cdot p_2(y_i),$$

§ 4. Elements of Functional Analysis

where the infimum is taken over all representations of z in the form $\sum_i x_i \otimes y_i$. There is also a least cross norm $p_1 \otimes p_2$, defined by

$$(p_1 \otimes p_2)(z) = \sup |f(z)|,$$

where the supremum is taken over all f of the form $f_1 \otimes f_2$, $f_i \in L'_i$ and $p'_i(f_i) \leq 1$, $i = 1, 2$.

The completions of $L_1 \otimes L_2$ in these norms are denoted by $L_1 \hat{\otimes} L_2$ and $L_1 \check{\otimes} L_2$ respectively.

We consider the special case $L_1 = L$, $L_2 = L'$. The space $L_1 \otimes L_2$ can be identified in this case with the space of linear operators of finite rank in L: for $x \in L$ and $f \in L'$, we assign to the element $x \otimes f$ the operator

$$y \mapsto f(y) x.$$

Problem 8. Prove that the norm $p \otimes p'$ is equal to the usual operator norm.

In this case, the space $L \check{\otimes} L'$ is the norm closure of the space of operators of finite rank. All such operators are *completely continuous* (or *compact* in another terminology). That is, they carry bounded sets into sets with compact closures, which sets are also called *relatively compact*. If L is a Hilbert space, then the converse is also true: *every completely continuous operator in L is the limit in the norm of operators of finite rank*.

The identity mapping of $L \otimes L'$ onto itself can be extended to a continuous mapping $\alpha: L \hat{\otimes} L' \to L \check{\otimes} L'$. As has been recently made clear (see the works of Enflo, or Séminaire Bourbaki, Exp. 433, or Uspehi Mat. Nauk vol. 28, No. 6), this mapping may have a nonnull kernel. For the classical spaces L^p, l^p, and C, the mapping α is a monomorphism. Operators lying in the image of α have received the name *nuclear*. From the condition $\ker \alpha = 0$ it is easy to infer that the mapping $A \to \operatorname{tr} A$ can be extended to the entire class of nuclear operators. In Hilbert space, this class consists of the operators A for which the series $\sum_\alpha (A x_\alpha, x_\alpha)$ converges for an arbitrary orthonormal basis $\{x_\alpha\}$. The sum of this series is independent of the choice of the basis and coincides with the trace of A.

We can extend the definition of the products $L_1 \hat{\otimes} L_2$ and $L_1 \check{\otimes} L_2$ to arbitrary LCS's.

Suppose that the topology in the LCS L_1 is given by a family of seminorms p_α, $\alpha \in A$, and that the topology in the LCS L_2 is given by a family of seminorms p_β, $\beta \in B$. Then we can define two locally convex topologies in $L_1 \otimes L_2$. One is defined by the family of seminorms $p_\alpha \hat{\otimes} q_\beta$, $\alpha \in A$, $\beta \in B$. The second is defined by the family of seminorms $p_\alpha \otimes q_\beta$, $\alpha \in A$, $\beta \in B$. The completions of $L_1 \otimes L_2$ in these two topologies are denoted by $L_1 \hat{\otimes} L_2$ and $L_1 \check{\otimes} L_2$, and are called the *projective* and *weak* tensor products, respectively, of the spaces L_1 and L_2.

The important class of *nuclear* spaces M is characterized by the following property: for every LCS L, the tensor products $L \hat{\otimes} M$ and $L \check{\otimes} M$ coincide.

Nuclear spaces enjoy a whole series of remarkable properties, which relate them closely to finite-dimensional spaces. Thus, every bounded closed set in a nuclear space is compact; weak and strong convergence are the same in a nuclear space; every continuous operator into or out of a nuclear space is defined by a kernel. The class of nuclear spaces is closed under the formation of subspaces, of projective limits, and of countable inductive limits.

We most frequently encounter countably normed nuclear spaces. Let V be a LCS with a countable collection of prenorms $\{p_k\}$. Without loss of generality, we may suppose that $p_k \leq p_{k+1}$. (Otherwise we could replace $\{p_k\}$ by the equivalent family of prenorms $\tilde{p}_k = \sup_{i \leq k} p_i$.) It turns out that nuclearity of V is equivalent to the following property: for every index k, there exists an index $j(k)$ such that the identity mapping from $(V, p_{j(k)})$ to (V, p_k) is a nuclear operator.

As an example, consider the space $S(\mathbf{R})$ of all infinitely differentiable, rapidly decreasing functions on the line, with the collection of norms

$$p_k(f) = \sup_x \sum_{l,m=0}^{k} |x^l f^{(m)}(x)|.$$

For more about these results, see the books [23] and [45].

We note the following for the case in which L_1 and L_2 are Hilbert spaces. Among the cross norms on $L_1 \otimes L_2$ there is a natural norm defined by the scalar product. The completion of $L_1 \otimes L_2$ in this norm is called the *Hilbert tensor product* of the spaces L_1 and L_2.

In the particular case where $L_1 = L$, $L_2 = L'$, the Hilbert product of L and L' is identified with the space of *Hilbert-Schmidt operators*. One can show that an operator A is a Hilbert-Schmidt operator if and only if the operator AA^* is a nuclear operator. The norm of the operator A in the space of Hilbert-Schmidt operators can be given by the formula

$$\|A\|^2 = \operatorname{tr}(AA^*).$$

4.2. Banach Algebras

The theory of Banach algebras (originally called normed rings) was created in the early 1940's by I.M. Gel'fand. It has proved to be one of the most powerful instruments in applications to the theory of functions, the theory of operators, the theory of group representations, and other fields of mathematics.

We give here the facts about Banach algebras that we shall need in the sequel. One can find a systematic exposition in the books [9], [22], [42], and [15].

An algebra $\mathfrak{A}$ over the field $\mathbf{C}$ is called a *Banach algebra* if $\mathfrak{A}$ is simultaneously a topological ring and a Banach space.

We shall suppose that the norm in a Banach algebra satisfies the condition

$$\|xy\| \leq \|x\| \cdot \|y\|. \tag{1}$$

Problem 1. Every norm in which multiplication is continuous is proportional to a norm satisfying (1).

Hint. Using the principle of uniform boundedness (see 4.1), prove the inequality $\|xy\| \leq C\|x\| \cdot \|y\|$.

§ 4. Elements of Functional Analysis

Furthermore, we shall suppose that our algebras contain a unit element 1, and that $\|1\|=1$. The general case can be reduced to this one by the operation of "adding a unit" (see 3.1).

We remark that this operation in the category of algebras over $\mathbf{C}$ differs from the analogous operation in the category of rings. In the present case, the new algebra $\mathfrak{A}_1$ consists of elements of the form $\lambda \cdot 1 + x$, where $x \in \mathfrak{A}$, $\lambda \in \mathbf{C}$, and 1 is the adjoined unit. As norm in $\mathfrak{A}_1$, we can take the expression $\|\lambda \cdot 1 + x\| = |\lambda| + \|x\|$.

The *spectrum* of an element $x \in \mathfrak{A}$ is the set $\mathrm{Sp}\, x$ of those complex numbers λ for which the element $x - \lambda \cdot 1$ fails to have an inverse in $\mathfrak{A}$.

Problem 2. Prove that $\mathrm{Sp}\, x$ is contained in the disk with center 0 and radius $\|x\|$.

Hint. If $|\lambda| > \|x\|$, then the inverse $(x - \lambda \cdot 1)^{-1}$ is the sum of the series
$$-\sum_{n=0}^{\infty} x^n \lambda^{-n-1}.$$

The function $R_x(\lambda) = (x - \lambda \cdot 1)^{-1}$, which is defined in the complement of $\mathrm{Sp}\, x$ and has its values in the algebra $\mathfrak{A}$, is called the *resolvent* of the element x.

Problem 3. Prove that $\mathrm{Sp}\, x$ is a closed set and that $R_x(\lambda)$ is an analytic function of λ in the complement of $\mathrm{Sp}\, x$.

Hint. Prove that the sum of the series $\sum_{n=0}^{\infty} (\lambda - \lambda_0)^n R_x(\lambda_0)^{n+1}$, which converges in the disk $|\lambda - \lambda_0| < \|R_x(\lambda_0)\|^{-1}$, is the resolvent of the element x.

Problem 4. Prove that $\mathrm{Sp}\, x$ is a nonvoid compact subset of $\mathbf{C}$.

Hint. If $\mathrm{Sp}\, x = \emptyset$, then $R_x(\lambda)$ is defined and bounded in the entire complex plane. Apply Liouville's theorem to the complex-valued function $f(R_x(\lambda))$, $f \in \mathfrak{A}'$.

Corollary (the Gel'fand-Mazur theorem). *Every Banach field over $\mathbf{C}$ coincides with $\mathbf{C}$.*

In fact, if this field contains an element $x \notin \mathbf{C}$, then $\mathrm{Sp}\, x = \emptyset$.

The quantity $\rho(x) = \sup_{\lambda \in \mathrm{Sp}\, x} |\lambda|$ is called the *spectral radius* of the element x.

Problem 5. Prove that $\rho(x) = \lim_{n \to \infty} \|x^n\|^{1/n}$.

Hint. Consider the expansion of $R_x(\lambda)$ at the point $\lambda = \infty$.

Remark. It may seem surprising that the quantity $\rho(x)$, which is defined solely by the algebraic structure of $\mathfrak{A}$, can also be expressed in terms of the norm, in the choice of which there is a fair amount of arbitrariness. However, by Banach's theorem (see 4.1) two norms that yield the same topology on A are connected by the relation

$$0 < c_1 \leq \frac{\|x\|_1}{\|x\|_2} \leq c_2 < \infty.$$

It follows that

$$\lim_{n \to \infty} \left(\frac{\|x^n\|_1}{\|x^n\|_2} \right)^{1/n} = 1.$$

Problem 6. Let $\phi\colon \mathfrak{A}\to\mathfrak{B}$ be a homomorphism of Banach algebras such that $\phi(1)=1$. Prove that

$$\rho(\phi(x))\leqslant \rho(x).$$

Hint. Inverse elements go into inverse elements.

Let $P(z)=a_0+a_1 z+\ldots+a_n z^n$ be a polynomial with complex coefficients, and let x be an arbitrary element of $\mathfrak{A}$. Then one can define the element $P(x)=a_0+a_1 x+\ldots+a_n x^n\in A$.

It turns out that the correspondence $P\mapsto P(x)$ can be extended to a wider class of functions.

Let $\mathcal{O}(\mathrm{Sp}\,x)$ be the algebra of all complex-valued functions, each defined in some neighborhood of the set $\mathrm{Sp}\,x\subset\mathbf{C}$, and analytic where defined. Give this set of functions the topology of uniform convergence on compact sets.

Then there exists a continuous homomorphism $f\mapsto f(x)$ of the algebra $\mathcal{O}(\mathrm{Sp}\,x)$ into $\mathfrak{A}$, which extends the homomorphism of the algebra of polynomials described above. The element $f(x)$ can be given by the formula

$$f(x)=\frac{1}{2\pi i}\int_C f(\lambda)R_x(\lambda)d\lambda,$$

where C is an arbitrary closed contour lying in the domain of definition of f and containing in its interior the set $\mathrm{Sp}\,x$.

The process just described can be generalized, although the details are technically more complicated. Let $\{x_i\}$ be a collection of n mutually commuting elements of $\mathfrak{A}$, and let f be an analytic function of n complex variables defined in a neighborhood of a certain set $\mathrm{Sp}(x_1,\ldots,x_n)\subset\mathbf{C}^n$, which is called the joint spectrum of the set $\{x_i\}$. Then one can define the analytic function $f(x_1,\ldots,x_n)$. A precise definition of the joint spectrum will be given below.

The most important, as well as the most studied, class of Banach algebras are commutative Banach algebras.

With every such algebra $\mathfrak{A}$, we associate the set $\mathfrak{M}(\mathfrak{A})$ of all maximal ideals in $\mathfrak{A}$ distinct from $\mathfrak{A}$.

This set $\mathfrak{M}(\mathfrak{A})$ can also be defined as the set of all non-null *characters* of $\mathfrak{A}$, that is, continuous homomorphisms of $\mathfrak{A}$ onto $\mathbf{C}$. In fact, if $\chi\colon \mathfrak{A}\to\mathbf{C}$ is a character, then $\chi^{-1}(0)$ is a maximal ideal in $\mathfrak{A}$. Conversely, if J is a maximal ideal in $\mathfrak{A}$, then $\mathfrak{A}/J$ is a Banach field over $\mathbf{C}$ and so, by the Gel'fand-Mazur theorem, we have $\mathfrak{A}/J=\mathbf{C}$. Hence the canonical projection $\mathfrak{A}\to\mathfrak{A}/J$ yields a character of $\mathfrak{A}$.

Interpreting the points $\chi\in\mathfrak{M}(\mathfrak{A})$ as characters of $\mathfrak{A}$ permits us to assign to every element $x\in\mathfrak{A}$ the function $\hat{x}$ on $\mathfrak{M}(\mathfrak{A})$ defined by:

$$\hat{x}(\chi)=\chi(x). \qquad (2)$$

The function $\hat{x}$ is called the *Gel'fand transform* of the element x.

The set $\mathfrak{M}(\mathfrak{A})$ is given the weakest topology under which all of the functions $\hat{x}$, $x\in\mathfrak{A}$, are continuous.

§ 4. Elements of Functional Analysis

Problem 7. Prove that $\mathfrak{M}(\mathfrak{A})$ is compact.

Hint. The set $\mathfrak{M}(\mathfrak{A})$ is bounded and weakly closed in the conjugate space $\mathfrak{A}'$ of $\mathfrak{A}$.

Problem 8. Prove that the set of values assumed by the function $\hat{x}$ coincides with the spectrum of x.

Hint. By Zorn's lemma, every noninvertible element is contained in some maximal ideal.

Corollary 1. *For every $x \in \mathfrak{A}$, the equality*

$$\rho(x) = \sup_{\chi \in \mathfrak{M}(\mathfrak{A})} |\hat{x}(\chi)|$$

holds.

Corollary 2. *The Gel'fand transform produces a continuous homomorphism of $\mathfrak{A}$ into the algebra $C(\mathfrak{M}(\mathfrak{A}))$ of all continuous functions on $\mathfrak{M}(\mathfrak{A})$.*

The result of problem 8 suggests the following definition.

The *joint spectrum* of elements $x_1, \ldots, x_n$ of the algebra $\mathfrak{A}$ is defined as the image of $\mathfrak{M}(\mathfrak{A})$ in $\mathbf{C}^n$ under the mapping

$$\chi \mapsto (\chi(x_1), \ldots, \chi(x_n)).$$

Example 1. Let $\mathfrak{A} = C(X)$ be the algebra of continuous functions on the compact Hausdorff space X.

Problem 9. Every closed ideal $J \subset C(X)$ consists of all functions that vanish on a certain closed set $Y \subset X$.

Thus the maximal ideals of $\mathfrak{A}$ corresponds to the points of X. The space $\mathfrak{M}(\mathfrak{A})$ coincides with X, and the Gel'fand transformation is the identity mapping.

Example 2. Let $\mathfrak{A} = l^1$ be the space of summable sequences $x = (x_0, x_1, \ldots)$ with the norm $\|x\| = \sum_{i=0}^{\infty} |x_i|$, with ordinary addition, and with multiplication defined by the formula

$$(xy)_n = \sum_{k=0}^{n} x_k y_{n-k}.$$

Problem 10. The space $\mathfrak{M}(\mathfrak{A})$ is homeomorphic to the disk $|z| \leq 1$, $z \in \mathbf{C}$. Prove that the Gel'fand transform has the form

$$\hat{x}(z) = \sum_{i=0}^{\infty} x_i z^i.$$

Example 3. Let $\mathfrak{A}$ be the algebra of matrices of the form $\begin{pmatrix} \lambda & \mu \\ 0 & \lambda \end{pmatrix}$ with the usual operations and norm. One checks immediately that $\mathfrak{A}$ admits one and only one character: $\chi : \begin{pmatrix} \lambda & \mu \\ 0 & \lambda \end{pmatrix} \mapsto \lambda$. The space $\mathfrak{M}(\mathfrak{A})$ consists of a single point, and the Gel'fand transform is given by the character χ.

Problem 11. Prove that the radical of the algebra $\mathfrak{A}$ (see 3.1) consists of exactly those elements x for which $\rho(x)=0$, that is, $\mathrm{Sp}\,x=0$ and $\hat{x}=0$.

Corollary. *If A is a semisimple algebra, then it is isomorphic*[1] *to a subalgebra of $C(\mathfrak{M}(\mathfrak{A}))$.*

In fact, the Gel'fand transformation in this case is an embedding. We shall see below that in the important special case of C^*-algebras, this embedding is an isomorphism.

We say that a Banach algebra $\mathfrak{A}$ admits an *involution* if to every $x \in \mathfrak{A}$, there corresponds an element $x^* \in \mathfrak{A}$ for which the following hold:

1) $(\lambda x + \mu y)^* = \bar{\lambda} x^* + \bar{\mu} y^*$ (the bar denotes complex conjugation),

2) $(xy)^* = y^* x^*$,

3) $(x^*)^* = x$,

4)   the mapping $x \to x^*$ is continuous.

An element $x \in \mathfrak{A}$ is called *Hermitian* if $x^* = x$, *normal* if $x^* x = x x^*$, and *unitary* if $x^* x = x x^* = 1$.

Banach algebras with involution constitute a category, the morphisms of which are continuous algebra homomorphisms that commute with involution.

As an example of a Banach algebra with involution, consider the algebra $\mathfrak{B}(H)$ of all continuous operators in a Hilbert space H, with the usual operator norm and with involution defined as the mapping from an operator to its adjoint.

A morphism ϕ of the algebra $\mathfrak{A}$ into the algebra $\mathfrak{B}(H)$ is ordinarily called an *operator representation* of the algebra $\mathfrak{A}$. When we need to emphasize that we are dealing with a morphism in the category of algebras with involution, we use the term * *representation* or *symmetric representation*.

A representation ϕ is called *faithful* if $\ker \phi = 0$, that is, if ϕ is a monomorphism. Representations ϕ_i in spaces H_i, $i=1,2$ are called *equivalent* if there exists an isomorphism $\tau : H_1 \to H_2$ such that for every $x \in \mathfrak{A}$, the following diagram commutes:

$$\begin{array}{ccc} H_1 & \xrightarrow{\phi_1(x)} & H_1 \\ {\scriptstyle \tau}\downarrow & & \downarrow{\scriptstyle \tau} \\ H_2 & \xrightarrow{\phi_2(x)} & H_2 \end{array}.$$

4.3. C^*-Algebras

The problem of the existence of faithful representations for algebras with involution was solved by I. M. Gel'fand and M. A. Naĭmark. Their solution singles out in a natural way a special class of Banach algebras, the class known as C^*-algebras.

---

[1] Isomorphic as an algebra but not in general as a Banach algebra.

§ 4. Elements of Functional Analysis

An algebra with involution $\mathfrak{A}$ is called a *C*-algebra*[1] if the following identity holds:

$$\|x^*x\| = \|x\|^2 \quad \text{for all} \quad x \in \mathfrak{A}.$$

It is evident that the algebra $\mathfrak{B}(H)$ is a C*-algebra.

Theorem 1 (Gel'fand-Naĭmark). 1) *Every C*-algebra admits a faithful operator representation.* 2) *For every Banach algebra $\mathfrak{A}$ with involution there exists a C*-algebra $C^*(\mathfrak{A})$ and a morphism $\phi: \mathfrak{A} \to C^*(\mathfrak{A})$ such that every operator representation ϕ_1 of the algebra $\mathfrak{A}$ has the form $\phi_1 = \phi_2 \circ \phi$, where ϕ_2 is a representation of $C^*(\mathfrak{A})$.*

Before proving this theorem, we list a number of important properties of C*-algebras.

Problem 1. If x is an Hermitian element in a C*-algebra $\mathfrak{A}$, then $\rho(x) = \|x\|$.
Hint. Use the fact that $\|x^{2^n}\| = \|x\|^{2^n}$ and the result of problem 5 in 4.2.

Problem 2. If $\phi: \mathfrak{A}_1 \to \mathfrak{A}_2$ is a morphism of C*-algebras, then $\|\phi(x)\| \leq \|x\|$.
Hint. Use the result of problem 6 of 4.2.

Corollary. *In every C*-algebra, the norm is uniquely defined.*

Problem 3. If a is Hermitian and u is unitary in a C*-algebra $\mathfrak{A}$, then $\operatorname{Sp} a \subset \mathbf{R}$ and $\operatorname{Sp} u \subset \mathbf{T}$ (unit circle in $\mathbf{C}$).
Hint. Since every normal element (and in particular every Hermitian and every unitary element) and all functions of this element are contained in a commutative C*-subalgebra of $\mathfrak{A}$, we may suppose that $\mathfrak{A}$ itself is commutative. For a unitary element, the assertion of the problem follows from problem 8 of 4.2.
For the case of an Hermitian element a, use the fact that the element
$$e^{ita} = \sum_{k=0}^{\infty} \frac{(ita)^k}{k!} \quad \text{is unitary for all } t \in \mathbf{R}.$$

Corollary. *The Gel'fand transformation for a commutative C*-algebra $\mathfrak{A}$ is an isometric embedding of $\mathfrak{A}$ into $C(\mathfrak{M}(\mathfrak{A}))$.*

To see that the transformation is an isometry, use the result of problem 1. To see that the transformation commutes with involution, note that Hermitian elements go into real functions.
As a matter of fact, more is true.

Theorem 2 (Gel'fand). *The Gel'fand transformation for a commutative C*-algebra $\mathfrak{A}$ is an isomorphism of $\mathfrak{A}$ onto $C(\mathfrak{M}(\mathfrak{A}))$.*

Problem 4. Let $\mathfrak{A}$ be a closed C*-subalgebra in the algebra $C(X)$ of continuous functions on the compact Hausdorff space X. Let a be an Hermitian element in $\mathfrak{A}$ and f a continuous function on $\operatorname{Sp} a \subset \mathbf{R}$. Then we have $f(a) \in \mathfrak{A}$.

---

[1] The following terminology is also used: *completely regular algebra*, *star algebra* (algèbre stellaire).

Hint. Every continuous function on the compact set $\operatorname{Spa} \subset \mathbf{R}$ can be uniformly approximated by a piecewise linear function. A piecewise linear function is obtained by linear operations and superpositions from the function $x \mapsto |x|$. Finally, the function $x \mapsto |x|$ on the interval $[-N, N]$ can be represented by the uniformly convergent series

$$|x| = N\sqrt{1 - \left(1 - \frac{x^2}{N^2}\right)} = N - N \cdot \sum_{k=1}^{\infty} \frac{1 \cdot 3 \cdot 5 \ldots (2k-3)}{2 \cdot 4 \cdot 6 \ldots 2k} \left(1 - \frac{x^2}{N^2}\right)^k.$$

From problem 4, it follows that if the algebra $\mathfrak{A}$ separates the points x and y,[1] then there exists a function $a \in \mathfrak{A}$ equal to 1 in some neighborhood of x, equal to 0 in some neigborhood of y, and contained between 0 and 1 at all other points. If the algebra $\mathfrak{A}$ distinguishes arbitrary pairs of points, then, for every point x and every neighborhood U of x, there exists a function $a \in \mathfrak{A}$ equal to 1 at x and equal to 0 outside of U.

From this, we easily infer the following important fact.

Theorem 3 (the Stone-Weierstrass theorem). *If $\mathfrak{A}$ is a closed C^*-subalgebra with unit in the algebra $C(X)$, and if the elements of $\mathfrak{A}$ separate points of the compact Hausdorff space X, then $\mathfrak{A} = C(X)$.*

Theorem 2 follows from theorem 3 and the definitions of spectrum and Gel'fand transform.

For our further considerations, the following is also important.

Corollary. *If a is an Hermitian element in a C^*-algebra $\mathfrak{A}$ and $\operatorname{Spa} \subset \mathbf{R}^+$ (the positive half-line), then there exists a unique element $b \in \mathfrak{A}$ with the following properties: $\operatorname{Sp} b \subset \mathbf{R}^+$ and $b^2 = a$.*

In other words, a positive element has a unique positive square root.

We return to theorem 1. Both assertions of this theorem are proved by explicit constructions, in which the concept of a positive functional plays a basic rôle.

A functional $f \in \mathfrak{A}'$ is called *positive* if the inequality $f(x^* x) \geq 0$ holds for all $x \in \mathfrak{A}$.

Problem 5. Prove that every positive functional satisfies the inequalities
a) $|f(x)|^2 \leq f(x^* x) \cdot f(1)$,
b) $|f(x)| \leq \|x\| \cdot f(1)$.
Hint. To establish a), apply the Cauchy-Bunjakovskiĭ inequality

$$|(x, y)|^2 \leq \|x\|^2 \|y\|^2$$

to the scalar product

$$(x, y) = f(y^* x). \tag{1}$$

---

[1] An algebra $\mathfrak{A}$ is said to *separate points x and y* if there is a function $f \in \mathfrak{A}$ such that $f(x) \neq f(y)$.

§ 4. Elements of Functional Analysis

Inequality b) follows from a) and the fact that $\text{Sp}(\|x^*x\| \cdot 1 - x^*x) \subset \mathbf{R}^+$ and consequently $\|x^*x\| f(1) - f(x^*x) \geq 0$.

If ϕ is a representation of the algebra $\mathfrak{A}$ in the space H and ξ is a vector in H, then $f_\xi(x) = (\phi(x)\xi, \xi)$ is a positive functional, and every positive functional can be obtained in this way. In fact, we define in $\mathfrak{A}$ a scalar product by formula (1). Let H_f be the corresponding Hilbert space. The mapping $y \mapsto xy$ of the algebra A can be extended to an operator $\phi(x)$ in H_f. The homomorphism

$$\phi: \mathfrak{A} \to \mathfrak{B}(H_f)$$

commutes with involution, since

$$(\phi(x)y, z) = f(z^*xy) = (y, \phi(x^*)z).$$

We have obtained a representation of the algebra $\mathfrak{A}$ in the space H_f. As the element ξ, we take the image of $1 \in \mathfrak{A}$ in H_f. Then we obtain $f_\xi(x) = (\phi(x)\xi, \xi) = f(x)$.

The existence of a faithful representation of the C^*-algebra $\mathfrak{A}$ is a consequence of the following fact.

Lemma. *For every $x \neq 0$ in the C^*-algebra $\mathfrak{A}$, there exists a positive functional f such that $f(x) \neq 0$.*

The direct sum of representations, constructed as above for all positive functionals, will have null kernel in this case.

To prove the lemma, we denote by P the set of all Hermitian elements in $\mathfrak{A}$ the spectra of which lie in $\mathbf{R}^+$.

Problem 6. Prove the following: 1) P is a closed convex cone; 2) 1 is an interior point of P; 3) $P \cap -P = \{0\}$.
Hint. Use the corollary to Gel'fand's theorem.

We now consider the convex set K consisting of all points of the form $1 - y$, $y \in P$. If $\|x\| > 1$, then $x^*x \notin K$. By the Hahn-Banach theorem, there exists a functional $f \in \mathfrak{A}'$ with the following properties:

$$f(y) \leq 1 \quad \text{for all} \quad y \in K \quad \text{and} \quad f(x^*x) > 1.$$

The first property means that $f(1 - y) \leq 1$ for all $y \in P$, and the second that f is a positive functional. The lemma is proved.

To prove the second assertion of the Gel'fand-Naĭmark theorem, we define for $x \in \mathfrak{A}$:

$$\|x\|_1 = \sup \sqrt{f(x^*x)},$$

the least upper bound being taken over all positive functionals f such that $f(1) = 1$.

Problem 7. Prove that the algebra $C^*(\mathfrak{A})$ is obtained from $\mathfrak{A}$ by factoring out the ideal of all x for which $\|x\|_1 = 0$, and then completing in this norm.
Hint. Let $\phi: \mathfrak{A} \to \mathfrak{B}$ be a morphism of the algebra $\mathfrak{A}$ into an arbitrary C^*-algebra $\mathfrak{B}$. Using the methods of solution of problem 2, prove that $\|\phi(x)\| \leq \|x\|_1$.

Example 1. If $\mathfrak{A}$ is a C^*-algebra, then $C^*(\mathfrak{A}) = \mathfrak{A}$.

Example 2. Let $\mathfrak{A}$ be a commutative algebra with involution. A maximal ideal $x \in \mathfrak{M}(\mathfrak{A})$ is called *symmetric* if $\hat{x}^*(\chi) = \overline{x(\chi)}$ for all $x \in \mathfrak{A}$. The set of symmetric maximal ideals forms a compact subset $X_0 \subset X = \mathfrak{M}(\mathfrak{A})$.

Problem 8. Prove that $C^*(\mathfrak{A})$ is isomorphic to $C(X_0)$ and that the morphism $\phi: \mathfrak{A} \to C^*(\mathfrak{A})$ is obtained by composition of the Gel'fand transformation and restriction to X_0.

Hint. Every symmetric ideal defines a character χ, which is a *-representation of $\mathfrak{A}$ into $\mathbf{C}$. Hence χ has the form $\chi_0 \circ \phi$, where χ_0 is a certain character of $C^*(\mathfrak{A})$.

4.4. Commutative Operator Algebras

We now apply the information obtained above about Banach algebras to a study of algebras of operators in a Hilbert space.

Let $\mathfrak{A}$ be a commutative algebra of operators in a Hilbert space H. We shall suppose that $\mathfrak{A}$ is *symmetric*, that is, A contains the adjoint A^* of every operator $A \in \mathfrak{A}$. If $\mathfrak{A}$ is closed in the norm topology, then $\mathfrak{A}$ is a commutative C^*-algebra. We shall suppose that $\mathfrak{A}$ contains the unit operator 1. Then, in view of the results of 4.3, $\mathfrak{A}$ is isomorphic to the algebra $C(X)$ of continuous functions on the compact Hausdorff space $X = \mathfrak{M}(\mathfrak{A})$. We call two such algebras $\mathfrak{A}_1$ and $\mathfrak{A}_2$ in spaces H_1 and H_2 *spatially isomorphic* if there exists an isometric isomorphism $\tau: H_1 \to H_2$ such that the mapping $a \mapsto \tau a \tau^{-1}$ is an isomorphism of $\mathfrak{A}_1$ onto $\mathfrak{A}_2$. For each compact Hausdorff space X, we shall give a classification up to spatial isomorphism of all operator algebras which as C^*-algebras are isomorphic to $C(X)$.

It is as a matter of fact more convenient to consider a more general problem, that of describing all representations of the given C^*-algebra $\mathfrak{A} = C(X)$ up to equivalence. It is clear that the classification of faithful representations obtained in this way will solve the original problem of spatial classification of algebras that are isomorphic to $C(X)$.

Before solving the general problem, we consider the special case in which X consists of a finite number of points $x_1, \ldots, x_k$, and the space H is finite-dimensional.

The algebra $C(X)$ is isomorphic in this case to the direct sum of k copies of the field $\mathbf{C}$.

Let e_i be the function on X that is equal to 1 at x_i and to 0 at all other points. Since $e_i^* = e_i^2 = e_i$, $e_i e_j = 0$ for $i \neq j$ and $\sum_{i=1}^{k} e_i = 1$, the elements e_i under every representation must go into operators E_i such that $E_i^* = E_i^2 = E_i$, $E_i E_j = 0$ for $i \neq j$ and $\sum_{i=1}^{k} E_i = 1$.

Problem 1. Prove that the operators E_i are operators of orthogonal projection onto pairwise orthogonal subspaces $H_i \subset H$ such that $\bigoplus_{i=1}^{k} H_i = H$.

It is obvious that the numbers $n_i = \dim H_i$ are invariants of the representation and that for an arbitrary sequence $(n_1, \ldots, n_k)$ of nonnegative integers there exists

§ 4. Elements of Functional Analysis

a unique (up to equivalence) representation of the algebra with the given values of these invariants.

Thus, in the finite-dimensional case, in order to classify representations of the algebra $\mathfrak{A}$, we must give a multiplicity function $n(x)$, with nonnegative integer values, on the set X. Faithful representations correspond to strictly positive multiplicity functions.

In order to describe the situation in the infinite-dimensional case, we need some concepts from the theory of measure (see 4.1). Let μ be a positive regular measure on X. The operator of multiplication by a continuous function a in $C(X)$ can be extended to a continuous operator in $L^2(X,\mu)$, which we shall denote by $\phi_\mu(a)$. Obviously the correspondence $a \mapsto \phi_\mu(a)$ is a *-representation of the algebra $C(X)$ in the space $L^2(X,\mu)$.

If two measures μ and ν are equivalent, then the operator of multiplication by $(d\mu/d\nu)^{1/2}$ establishes an equivalence between the representations ϕ_μ and ϕ_ν.

Conversely, suppose that the representations ϕ_μ and ϕ_ν are equivalent and that the operator $\tau: L^2(X,\mu) \to L^2(X,\nu)$ establishes their equivalence. We set $f_0 \equiv 1$ and $\alpha = \tau(f_0)$. For an arbitrary continuous function f, we have

$$\tau(f) = \tau\phi_\mu(f)f_0 = \phi_\nu(f)\tau f_0 = \phi_\nu(f)\alpha = f\alpha.$$

Since τ is an isometry, the equality $d\mu = |\alpha|^2 d\nu$ obtains. Thus the measure μ is absolutely continuous with respect to the measure ν and we have $d\mu/d\nu = |\alpha|^2$. Analogously, consideration of the operator $\tau^{-1}: L^2(X,\nu) \to L^2(X,\mu)$ leads us to the conclusion that ν is absolutely continuous with respect to μ. Thus the measures μ and ν are equivalent.

By analogous reasoning, we can establish the following useful fact.

Problem 2. Every continuous operator in $L^2(X,\mu)$ that commutes with the operators in $\phi_\mu(\mathfrak{A})$ is the operator of multiplication by a certain function in $L^\infty(X,\mu)$.

In particular, it follows from this that all such operators are commuting in pairs.

A representation ϕ of the algebra $\mathfrak{A}$ in a space H is called *cyclic* if there is a vector ξ in H such that the space of all vectors of the form $\phi(a)\xi$, $a \in \mathfrak{A}$, is dense in H. The vector ξ in this case is called a *cyclic vector* or a *source*.

It is clear that all of the representations ϕ_μ constructed above in the spaces $L^2(X,\mu)$ are cyclic with source $f_0(x) \equiv 1$.

Problem 3. Show that in the finite-dimensional case, the cyclic representations are exactly those for which all of the invariants n_i are either 0 or 1.

Thus the representations ϕ_μ constructed above obviously do not exhaust all representations of the algebra $\mathfrak{A}$. We shall show nevertheless that every cyclic representation ϕ is equivalent to one of the representations ϕ_μ. In fact, let H be the space in which ϕ acts, and let ξ be a source in H. We define a continuous linear functional on $\mathfrak{A} = C(X)$, setting

$$F(a) = (\phi(a)\xi, \xi).$$

As was remarked above, F has the form

$$F(a) = \int_X a(x) d\mu(x),$$

where μ is a certain countably additive measure on X. The functional F assumes nonnegative values on nonnegative functions, for, if $a \geqslant 0$, then $a = a_1^2$, where a_1 is a real function, and

$$(\phi(a)\xi, \xi) = (\phi(a_1^2)\xi, \xi) = (\phi(a_1)\xi, \phi(a_1)\xi) \geqslant 0.$$

Hence the measure μ is positive. Finally, the mapping of $C(X)$ into H by the formula $a \mapsto \phi(a)\xi$ can be extended to a unitary operator $\tau: L^2(X, \mu) \to H$, which establishes the equivalence of ϕ_μ and ϕ.

Problem 4. Let ϕ be a representation of a C^*-algebra $\mathfrak{A}$ in a Hilbert space H. There exists a decomposition of H into a direct sum of subspaces H_α, $\alpha \in \mathfrak{A}$, invariant under $\phi(\mathfrak{A})$, such that the restriction of the operators of the representation to each H_α is a cyclic representation.

Hint. Apply Zorn's lemma to the family of subspaces of H that admit the required decomposition into a sum of cyclic subspaces.

We remark that the decomposition of a representation ϕ of the algebra $\mathfrak{A}$ into the sum of cyclic representations $\phi_\alpha = \phi_{\mu_\alpha}$, and also the corresponding collection of measures μ_α on $X = \mathfrak{M}(\mathfrak{A})$, is not unique.

In the case where the number of summands is no more than countable (for example, if the space H is separable), we will show that there exists a decomposition of ϕ into the sum of cyclic representations $\phi_i = \phi_{\mu_i}$, $i = 1, 2, \ldots$ for which the measure μ_{k+1} is absolutely continuous with respect to the measure μ_k for $k = 1, 2, \ldots$.

The measures μ_k in this case are defined by the original representation ϕ up to equivalence. The measure μ_1 is called the *basis measure* of the representation ϕ.

Thus we obtain a characteristic of the representation ϕ as a "decreasing" collection of classes of measures on the space X. This object can also be described in another way. Let ρ_k be the derivative of the measure μ_k with respect to μ_1. We define a *multiplicity function* n on X by defining the value $n(x)$ as the least upper bound of the k's for which $\rho_k(x) > 0$. It is evident that the multiplicity function is measurable, is defined uniquely almost everywhere with respect to the measure μ_1, and enables us to reconstruct all of the measures μ_k up to equivalence.

Let μ be any countably additive measure on X and let n be any μ-measurable function assuming the values $1, 2, \ldots, \infty$. We set $H_1 = L^2(X, \mu)$ and denote by H_k the subspace of H_1 consisting of all functions that vanish outside of the set $X_k = \{x : n(x) \geqslant k\}$. In the direct sum $H = \bigoplus_{k=1}^{k=\infty} H_k$, we define the representation $\phi_{\mu,n}$ of the algebra $\mathfrak{A} = C(X)$, defining $\phi_{\mu,n}(a)$ as the operator of multiplication by the function a.

Problem 5. Prove that the measure μ is a basis measure for the representation $\phi_{\mu,n}$ and that the function n is its multiplicity function. The representation $\phi_{\mu,n}$ is faithful if and only if supp $\mu = X$.

§ 4. Elements of Functional Analysis

Theorem 1. *Let $\mathfrak{A}$ be a C*-algebra and ϕ a representation of $\mathfrak{A}$ which is the sum of a finite or countably infinite number of cyclic representations. Then ϕ is equivalent to a representation of the form $\phi_{\mu,n}$. Two representations $\phi_{\mu,m}$ and $\phi_{\nu,n}$ are equivalent if and only if the measures μ and ν are equivalent and the functions m and n coincide almost everywhere.*

We shall break up the proof of this theorem into a series of problems. (A different method of proof may be found in the book of J. Dixmier [15]).

Let ϕ be a representation of a C*-algebra $\mathfrak{A}$ in a Hilbert space H. For every vector $\xi \in H$, we construct the subspace $H_\xi \subset H$, defined as the closure of the set of vectors of the form $\phi(a)\xi$, $a \in \mathfrak{A}$. The restriction of the operators in $\phi(\mathfrak{A})$ to the space H_ξ produces a cyclic representation. We denote this representation by ϕ_ξ and the corresponding measure on $X = \mathfrak{M}(\mathfrak{A})$ by μ_ξ. We introduce a partial order in H, setting $\xi > \eta$ to mean that $\mu_\xi > \mu_\eta$.

Problem 6. If $\eta \in H_\xi$, then $\eta < \xi$.
Hint. Use the natural embedding of H_η in H_ξ.

Problem 7. Let $H = \bigoplus_{k=1}^{\infty} H_{\xi_k}$ be a decomposition of H into a countable sum of cyclic subspaces. Prove that the vector $\xi = \sum_{k=1}^{\infty} \frac{1}{2^k \|\xi_k\|} \xi_k$ is a maximal element in H.
Hint. Let $\eta \in H$ and define η_k as the projection of η onto H_{ξ_k}. Then we have
$$\mu_\eta = \sum_{k=1}^{\infty} \mu_{\eta_k}.$$

Problem 8. The set of maximal vectors is everywhere dense in H.
Hint. Let $\xi \in H$ be a maximal vector. Verify that the set of maximal vectors is dense in H_ξ and that every vector of the form $\xi + \eta$, where $\eta \in H_\xi^\perp$, is maximal.

Problem 9. Suppose that there exists a countable decomposition $H = \bigoplus_{k=1}^{\infty} H_{\eta_k}$. Show that there exists a decomposition $H = \bigoplus_{k=1}^{\infty} H_{\xi_k}$ such that $\xi_k > \xi_{k+1}$ for all k.
Hint. Let $\{\zeta_i\}$ be a sequence in which each of the vectors η_i occurs an infinite number of times. Define ξ_k inductively in such a way that:

1) ξ_{k+1} is a maximal vector in $\left(\bigoplus_{i=1}^{k} H_{\xi_k}\right)^\perp$;

2) $\rho\left(\zeta_{k+1}, \bigoplus_{i=1}^{k+1} H_{\xi_i}\right) \leq \frac{1}{k+1}$.

The first assertion of theorem 1 follows from the solution of problem 9. The second can be obtained from the following assertion.

Problem 10. Every operator from $L^2(X,\mu)$ to $L^2(X,\nu)$ that commutes with multiplication by continuous functions is an operator of multiplication by a certain measurable function α that satisfies the condition $|\alpha|^2 \nu < \mu$.

In particular, if the measures μ and ν are disjoint, then the operator in question is zero.

Hint. Consider the argument that precedes problem 2.

A different method of proof of uniqueness is to give a definition of the basis measure μ and the multiplicity function n (or the corresponding decreasing collection of measures μ_i) in invariant terms. This turns out to be possible.

Problem 11. Let $\xi_1, \ldots, \xi_{k-1}$ be a finite collection of vectors from the space H of the representation ϕ, and let η be a maximal vector in $\left(\bigoplus_{k=1}^{k-1} H_{\xi_k} \right)^\perp$.

Prove that $\mu_\eta \succ \mu_k$ and that there exist a collection $\tilde{\xi}_1, \ldots, \tilde{\xi}_k$ and a maximal vector $\eta \in \left(\bigoplus_{i=1}^{k-1} H_{\tilde{\xi}_i} \right)^\perp$, such that $\mu_\eta \sim \mu_k$.

Hint. Use the part of theorem 1 already proved, reducing the problem to the case $\phi = \phi_{\mu,n}$.

We point out the analogy between the characterization just obtained for the measures μ_k and the minimax characterization of the eigenvalues of Hermitian matrices.

Problem 12 (Courant's principle). Let A be an Hermitian operator in a finite-dimensional Hilbert space H, and $\lambda_1 \geq \lambda_2 \geq \ldots \geq \lambda_n$ its eigenvalues. Then we have

$$\lambda_k = \min_{\xi_1, \ldots, \xi_{k-1}} \max_{\eta \in (\xi_1, \ldots, \xi_{k-1})^\perp} \frac{(A\eta, \eta)}{(\eta, \eta)}.$$

(An analogous assertion holds also in the infinite-dimensional case, if A is a compact Hermitian operator.)

We shall show how to derive the classical spectral theorem from theorem 1.

Let X be a compact Hausdorff space and H a Hilbert space. Suppose that to every Borel set $E \subset X$ there is assigned a projection operator $P(E)$ in the space H in such a way that

1) $P\left(\bigcup_{i=1}^\infty E_i \right) = \sum_{i=1}^\infty P(E_i)$, if $E_i \cap E_j = \emptyset$ for $i \neq j$,

2) $P(\emptyset) = 0$, $P(X) = 1$.

We then say that we have a *projection measure* P on X. For every continuous function f on X, we can define the operator

$$\int_X f(x) \, dP(x),$$

where the integral is defined as the limit (in the norm) of the Riemann integral sums $\sum f(x_k) P(\Delta_k)$.

Theorem 2. *Let A be a normal (in particular, an Hermitian or unitary) operator in a Hilbert space H. There exists a unique projection measure P on the spectrum of the operator A such that:*

1) *the operators $P(E)$ commute with all operators that commute with A and with A^*;*

2) *the equality*

$$A = \int x \, dP(x)$$

obtains.

For the proof, we consider the C^*-algebra $\mathfrak{A}$ generated by the operator A. It is evident that $\mathfrak{M}(\mathfrak{A})$ can be identified with the spectrum of A, since every character of the algebra $\mathfrak{A}$ is defined by its value at A. By theorem 1, we may suppose that H is the space of one of the representations $\phi_{\mu,n}$ of the algebra $\mathfrak{A}$. In this case, we can take $P(E)$ to be the operator of multiplication by the characteristic function of the set E.

4.5. Continuous Sums of Hilbert Spaces and von Neumann Algebras

The operation of direct sum of Hilbert spaces admits a further generalization. Let there be given a family $\{H_x\}_{x \in X}$ of Hilbert spaces, and on the set X let there be defined a certain measure μ. It is natural to try to define a new space $H = \int_X H_x \, d\mu(x)$, the elements of which are to be functions f on X, assuming values in H_x for all $x \in X$, and the scalar product being defined by the formula

$$(f_1, f_2) = \int_X (f_1(x), f_2(x))_{H_x} \, d\mu(x). \tag{1}$$

The difficulty consists in the fact that the integrand in (1) may be nonmeasurable. In the case where all of the spaces H_x are separable, one can introduce an appropriate definition of *measurable* vector-functions $f(x)$, which will guarantee the measurability of the numerical functions $x \mapsto (f_1(x), f_2(x))$.

For example, suppose that all of the spaces H_x coincide with a single separable space H. Then we can define a vector-function $f(x)$ as measurable if for a certain basis $\{e_k\}$ (and hence for all bases) in H, the numerical functions $x \mapsto (f(x), e_k)$ are measurable.

A more general case is that in which all of the spaces H_x are subspaces of a single separable space H, and such that the orthogonal projection P_x onto H_x is a measurable operator function, that is, for every basis $\{e_k\}$ in H, the numerical functions $x \mapsto (P_x e_k, e_j)$ are measurable.

Finally, we can consider as *a priori* measurable all vector-functions in a certain countable family Γ having the following properties:
1) the functions $x \mapsto (f_1(x), f_2(x))$ are measurable for all $f_1, f_2 \in \Gamma$,
2) for almost all $x \in X$, the vectors $f(x)$, $f \in \Gamma$, generate H_x.

Then a vector-function $f(x)$ is defined as measurable if its scalar products with all vector-functions in Γ are measurable.

Although the last method appears more general, it is actually equivalent to the preceding one. (In fact, it is equivalent to a special case of the preceding method, where H_x is the subspace spanned by the first $n(x)$ basis vectors in H. Here $n(x)$ is a measurable function on X, assuming the values $1, 2, \ldots, \infty$).

Once an appropriate concept of measurable vector-function has been adopted, we define the *continuous sum* $H = \int_X H_x d\mu(x)$ as the set of equivalence classes of measurable vector-functions f with summable square of norms.

An operator A in the space $H = \int_X H_x d\mu(x)$ is called *decomposable* if there exists a family $\{A(x)\}$ of operators in the spaces H_x such that

$$(Af)(x) = A(x)f(x) \quad \text{almost everywhere on } X. \tag{2}$$

A family of operators $\{A(x)\}$ is called measurable if for every measurable vector-function $f(x)$, the vector-function $x \mapsto A(x)f(x)$ is measurable. If the numerical function $a(x) = \|A(x)\|$ belongs to $L^\infty(X, \mu)$, then the family $\{A(x)\}$ defines by formula (2) a decomposable operator A with norm $\|A\|_H = \|a\|_{L^\infty(X,\mu)}$.

A special case of decomposable operators are *diagonal* operators, for which all $A(x)$ are scalar: $A(x) = a(x) \cdot 1$.

It is evident that decomposable operators form a subalgebra $\mathscr{R}$ in $\mathfrak{B}(H)$, and that diagonal operators are a commutative subalgebra $\mathscr{D}$ of $\mathscr{R}$.

If X is a compact Hausdorff space and μ is a regular measure on X, then we denote by $\mathscr{D}_0$ the subalgebra of $\mathscr{D}$ consisting of *continuously diagonal* operators (for which $a(x)$ is a continuous function). We shall suppose that $\operatorname{supp} \mu = X$.

Let H be an arbitrary separable Hilbert space, $\mathscr{R}$ a subalgebra of $\mathfrak{B}(H)$, $\mathscr{D}$ a commutative subalgebra of $\mathscr{R}$ and $\mathscr{D}_0$ a subalgebra of $\mathscr{D}$.

Theorem 1. *The subalgebras $\mathscr{R}$, $\mathscr{D}$, and $\mathscr{D}_0$ can play the rôles of decomposable, diagonal, and continuously diagonal operators for a certain realization of H in the form of a continuous sum of Hilbert spaces if and only if the following conditions are satisfied:*

1) *$\mathscr{R}$ is closed in the strong (or, what is the same, in the weak) operator topology;*
2) *$\mathscr{D}$ coincides with the center of $\mathscr{R}$;*
3) *$\mathscr{D}_0$ is norm closed and dense in $\mathscr{D}$ with respect to the strong (or, what is the same, the weak) topology.*

For the proof it is convenient to use the concept of a von Neumann algebra.

A symmetric algebra $\mathfrak{A}$ of operators in a Hilbert space H is called a *von Neumann algebra* if one of the following conditions is satisfied.

1) $\mathfrak{A}$ contains 1 and is weakly closed.
2) $\mathfrak{A}$ contains 1 and is strongly closed.
3) $\mathfrak{A}$ coincides with its bicommutant.

We recall that the *commutant* of an algebra $\mathfrak{A}$ is defined as the set $\mathfrak{A}^!$ of all operators that commute with all of the operators in $\mathfrak{A}$. The *bicommutant* is the commutant of the commutant.[1] It is evident that condition 3) implies 1) and that condition 1) implies 2). Let us show that 2) implies 3).

It suffices to show that if $\mathfrak{A}$ contains the identity operator, then $\mathfrak{A}^{!!} = (\mathfrak{A}^!)^!$ coincides with the strong closure of $\mathfrak{A}$. Suppose that $A \in \mathfrak{A}^{!!}$ and let $\xi_1, \ldots, \xi_n$ be any finite collection of vectors in H. We must show that for an arbitrary $\varepsilon > 0$, there

---

[1] The commutant $\mathfrak{A}^!$ is ordinarily written as $\mathfrak{A}'$. We have chosen the symbol $\mathfrak{A}^!$ in order to avoid confusion with our symbol for the topological dual space to $\mathfrak{A}$.

§ 4. Elements of Functional Analysis

exists an operator $A_0 \in \mathfrak{A}$ such that $\|A_0 \xi_i - A \xi_i\| < \varepsilon$ for $i = 1, 2, \ldots, n$. We consider first the case $n = 1$.

Let E be the closed subspace in H generated by vectors of the form $A_0 \xi$, $A_0 \in \mathfrak{A}$. The operator P of orthogonal projection onto E lies in $\mathfrak{A}'$ (since E is invariant under $\mathfrak{A}$). Hence P commutes with the operator $A \in \mathfrak{A}''$. This means that $A\xi = AP_\xi = PA_\xi \in E$, as we required. The general case is reduced to the one just examined by the following device. We consider the space $\tilde{H} = H \otimes \mathbf{C}^n = H \oplus \cdots \oplus H$ (n terms), and the algebra $\tilde{\mathfrak{A}} = \mathfrak{A} \otimes 1_{\mathbf{C}^n}$.

Problem 1. Let $\mathfrak{A}$ be a symmetric algebra of operators in the Hilbert space H_1. If H_2 is any other Hilbert space, then we have the following relations for algebras of operators in the Hilbert tensor product $H_1 \otimes H_2$:

$$(\mathfrak{A} \otimes 1_{H_2})' \supset \mathfrak{A}' \otimes \mathfrak{B}(H_2), \quad (\mathfrak{A} \otimes \mathfrak{B}(H_2))' = \mathfrak{A}' \otimes 1_{H_2}.$$

Hint. Use the matrix description of the tensor product of operators (see 3.4).

In our present case, we thus have $(\tilde{\mathfrak{A}})'' = \mathfrak{A}'' \otimes 1_{\mathbf{C}^n}$. Applying the part of the assertion already proved to the vector $(\xi_1, \ldots, \xi_n) \in \tilde{H}$, we complete the proof in the general case.

Problem 2. If $\mathfrak{A}$ is a von Neumann algebra, then its center $Z(\mathfrak{A})$ coincides with the center $Z(\mathfrak{A}')$ of the algebra $\mathfrak{A}'$ and with the intersection $\mathfrak{A} \cap \mathfrak{A}'$.

Thus the conditions of theorem 1 can be formulated as follows: $\mathscr{D}_0$ is an arbitrary commutative C*-algebra of operators, $\mathscr{R} = \mathscr{D}'_0$, $\mathscr{D} = \mathscr{D}'''_0$.

We shall prove the necessity of these conditions. The closedness of $\mathscr{D}_0$ in norm is evident, since the norm of the operator of multiplication by the continuous function $a(x)$ in $L^2(X, \mu)$ is equal to the maximum of the modulus of a on $\operatorname{supp} \mu = X$.

Suppose that $A \in \mathscr{D}'_0$. We shall show that A is a decomposable operator. Since A commutes with multiplication by all continuous functions, it also commutes with weak limits of these operators, in particular, with multiplication by bounded measurable functions. Suppose that

$$H = \int_X H_x \, d\mu(x)$$

and let X_k be the subset of X on which $\dim H_x = k$.

The operator A commutes with multiplication by the characteristic functions of the sets X_k. This allows us to reduce the problem to the case in which $X = X_k$.

In this case, all of the spaces H_x can be identified with a fixed space L, and the space H with the Hilbert tensor product $L \otimes L^2(X, \mu)$. Let $e_1, \ldots, e_k$ be a basis in L. We must prove that there exist measurable functions $a_{ij}(x)$ such that the operator A acts according to the formula

$$A : e_i \otimes f(x) \mapsto \sum_j e_j \otimes a_{ij}(x) f(x). \tag{1}$$

We define $a_{ij}(x)$ from the relation

$$A(e_i \otimes 1) = \sum_j e_j \otimes a_{ij}(x). \tag{2}$$

(We note that every vector in $L \otimes L^2(X,\mu)$ has the form of the right side of the equality (2).)

Then the equality (1) follows from (2) and the fact that A commutes with multiplication by $f(x)$. It follows that $\mathscr{R} \supset \mathscr{D}_0'$ and consequently $\mathscr{R} = \mathscr{D}_0'$.

Suppose now that $A \in \mathscr{D}_0'' = \mathscr{R}'$. Since $\mathscr{D}_0 \subset \mathscr{D}_0'$, we see that $\mathscr{D}_0'' \subset \mathscr{D}_0' = \mathscr{R}$. Hence A is decomposable and so is given by a collection of functions $\{a_{ij}(x)\}$. The operator $A \in \mathscr{R}'$ must commute with the decomposable operator B_{kl}, given by the collection

$$b_{ij}(x) = \begin{cases} 1 & \text{for } i=k, \ j=l \\ 0 & \text{otherwise.} \end{cases}$$

From this it follows that $a_{ij}(x)=0$ almost everywhere for $i \neq j$ and that $a_{ii}(x) = a_{jj}(x)$ almost everywhere for arbitrary i and j. This means that A is a diagonal operator, that is, $\mathscr{D}_0'' = \mathscr{D}$.

To prove the second part of the theorem, it suffices to apply theorem 1 from 4.4 to the algebra $\mathscr{D}_0$ and to note that the space of the representation $\phi_{\mu,n}$ is the continuous sum $\int_X H_x d\mu(x)$, where $X = \mathfrak{M}(\mathscr{D}_0)$ and $\dim H_x = n_x$.

The theorem is proved.

A von Neumann algebra $\mathfrak{A}$ is called an *algebra of type I* if it is isomorphic (as a C^*-algebra) to the algebra of all decomposable operators in a certain continuous sum of Hilbert spaces $\int_X H_x d\mu(x)$. We denote this algebra by the symbol $\int_X \mathfrak{B}(H_x) d\mu(x)$ and call it a *continuous product of the algebras* $\mathfrak{B}(H_x)$.

An algebra $\mathscr{A}$ of type I is called *homogeneous of degree n* if $\dim H_x \equiv n$. In other words, a homogeneous algebra of type I and degree n is isomorphic to the algebra of all measurable essentially bounded functions on a certain space X with a measure μ, assuming values in the algebra $\mathfrak{B}(H)$ of all bounded operators in an n-dimensional Hilbert space H.

Problem 3. If X consists of a finite or countable set of points of positive measure, then the algebra $\mathfrak{A} = \int_X \mathfrak{B}(H_x) d\mu(x)$ is isomorphic to the product (in the category of C^*-algebras) of the family of algebras $\mathfrak{B}(H_x)$, $x \in X$.

Hint. Let p_x denote the correspondence sending each operator function into its value at the point $x \in X$. Verify that the collection $\{p_x\}$ is a family of canonical projections of $\mathfrak{A}$ onto $\mathfrak{B}(H_x)$.

Problem 4. Prove that every algebra of type I is isomorphic to a product of homogeneous algebras.

Hint. Let $X = \bigcup_k X_k$ be the decomposition of X into sets of constancy of the function $n(x) = \dim H_x$. Prove that $\mathfrak{A}$ is a product of the algebras $\mathfrak{A}_k = \int_{X_k} \mathfrak{B}(H_x) d\mu(x)$.

In the sequel, we shall require a classification of algebras of type I up to spatial isomorphism (see 4.4).

Let $\{H_x\}$ and $\{L_x\}$ be two measurable families of separable Hilbert spaces on a space X with measure μ. In the space $V = \int_X H_x \otimes L_x d\mu(x)$, we consider the algebra

§ 4. Elements of Functional Analysis

of all decomposable operators of the form $A = \{A_x \otimes 1\}$, $A_x \in \mathfrak{B}(H_x)$. We denote this algebra by the symbol $\int_X \mathfrak{B}(H_x) \otimes 1 \, d\mu(x)$. It is clear that it is isomorphic to the continuous product $\int_X \mathfrak{B}(H_x) \, d\mu(x)$.

Theorem 2. *Every von Neumann algebra, isomorphic as a C*-algebra to a continuous product $\int \mathfrak{B}(H_x) d\mu(x)$, is spatially isomorphic to the algebra*

$$\int_X \mathfrak{B}(H_x) \otimes 1 \, d\mu(x)$$

in the space $\int_X H_x \otimes L_x \, d\mu(x)$.

Proof. First of all we note that it suffices to consider the case of a homogeneous algebra of degree n. In fact, let $\mathfrak{A} = \prod_k \mathfrak{A}_k$ be a decomposition of $\mathfrak{A}$ into a product of homogeneous components and let E_k be the element of $\mathfrak{A}$ defined by the conditions

$$p_j(E_k) = \begin{cases} 1 & \text{if } j = k, \\ 0 & \text{if } j \neq k. \end{cases}$$

Then $\{E_k\}$ is a family of orthogonal idempotents (see 8.3), yielding a decomposition of V into the Hilbert sum of subspaces $V_k = E_k V$, so that the restriction of $\mathfrak{A}$ to V_k is isomorphic to $\mathfrak{A}_k$.

Thus we may suppose that $\mathfrak{A}$ is homogeneous of degree n and isomorphic to an algebra of operator functions on X with values in $\mathfrak{B}(H)$, $\dim H = n$.

Applying theorem 1 to the subalgebra $C(X) \subset \mathfrak{A}$, which enters in the rôle of $\mathcal{D}_0$, we may suppose that the space V in which $\mathfrak{A}$ acts has the form $V = \int_X V_x \, d\nu(x)$.

Problem 5. Prove that the measure ν is equivalent to the measure μ that figures in the condition of theorem 2.

Hint. The algebras $L^\infty(X, \mu)$ and $L^\infty(X, \nu)$ are both isomorphic to $\mathcal{D}_0''$.

In the sequel we shall suppose that $\nu = \mu$. We choose a basis $\{\xi_i\}$ in H and denote by E_{ik} the element of $\mathfrak{A}$ corresponding to the constant function on X, the value of which is the operator e_{ik} that carries ξ_i into ξ_k and carries all other vectors of the basis into zero.

The relations

$$E_{ij} E_{kl} = \delta_{jk} E_{il}, \quad \sum_i E_{ii} = 1$$

imply that for almost all $x \in X$, there exists a decomposition $V_x = \bigoplus_k V_x^k$ such that the operators $(E_{ii})_x$ are projections onto the subspaces V_x^i, and $(E_{jk})_x$ gives an isomorphism of V_x^k onto V_x^j. Let L_x be a Hilbert space isomorphic to each of the spaces V_x^i, $i = 1, 2, \ldots, n$. Then V_x can be identified with $H_x \otimes L_x$, and the operator $(E_{ik})_x$ with the operator $e_{ik} \otimes 1$.

From this we infer without difficulty the assertion of the theorem.

Problem 6. Prove that

$$\left(\int_X \mathfrak{B}(H_x) \otimes 1 \, d\mu(x)\right)' = \int_X 1 \otimes \mathfrak{B}(L_x) \, d\mu(x).$$

Corollary. *If $\mathfrak{A}$ is an algebra of type I, then $\mathfrak{A}'$ is also of type I.*

A von Neumann algebra is called a *factor* if its center consists of scalar operators. The foregoing results can be formulated in the following way. Every factor of type I is isomorphic as a C^*-algebra to the algebra $\mathfrak{B}(H)$ of all bounded operators in a certain Hilbert space H and is spatially isomorphic to the algebra $\mathfrak{B}(H) \otimes 1$ in a certain space of the form $H \otimes L$. Every von Neumann algebra of type I is spatially isomorphic to a continuous product of factors of type I.

More information about von Neumann algebras can be found in the books [15] and [42].

§ 5. Analysis on Manifolds

5.1. Manifolds

Sets with structure locally like Euclidean spaces are called manifolds. This property enables us to introduce local systems of coordinates on manifolds and to employ the apparatus of mathematical analysis. A precise definition of manifold follows.

A topological space M is called a *manifold* if for every point $x \in M$, there exists a neighborhood U homeomorphic to an open set in $\mathbf{R}^n$. The number n is called the *dimension* of the manifold M at the point x.

The validity of the definition of dimension follows from a well-known theorem of topology:

A nonvoid open set in $\mathbf{R}^n$ cannot be homeomorphic to an open set in $\mathbf{R}^m$ for $m \neq n$.

Problem 1. Prove that a connected manifold has the same dimension at all of its points.

A *local system of coordinates*, or briefly a *chart*, on a manifold M is an open set $U \subset M$ together with a fixed homeomorphism α of this set onto an open set in $\mathbf{R}^n$. The coordinates of the point $\alpha(x) \in \mathbf{R}^n$ are called *local coordinates* of the point $x \in U$.

A collection of charts which together cover the entire manifold M is called an *atlas*.

Example 1. Geographic latitude and longitude on the surface of a sphere. Here M is the unit sphere in $\mathbf{R}^3$, defined by the equation $x^2 + y^2 + z^2 = 1$, and U the region obtained from the sphere by removing the meridian $y = 0$, $x \leq 0$. The homeomorphism α connects the points $(x, y, z) \in M$ and $(\phi, \psi) \in \mathbf{R}^2$ by the formulas

$$x = \cos\phi \cos\psi, \quad y = \sin\phi \cos\psi, \quad z = \sin\psi.$$

§ 5. Analysis on Manifolds

Thus the region U goes into the open rectangle

$$|\phi|<\pi, \quad |\psi|<\pi/2 \quad \text{in} \quad \mathbf{R}^2.$$

Example 2. Stereographical projection of a sphere onto a plane. Here U is obtained by removing the point $(0,0,1)$ from M. The homeomorphism α establishes a correspondence between $(x,y,z)\in M$ and $(u,v)\in \mathbf{R}^2$ by the formulas

$$\frac{2x}{1-z}, \quad v=\frac{2y}{1-z}.$$

Example 3. As M we take the surface of the cube in $\mathbf{R}^3$ given by the inequalities $|x|\leqslant 1, |y|\leqslant 1, |z|\leqslant 1$. On M we define an atlas of eight charts, taking as U_i the star of the i-th vertex, that is, the union of this vertex, the three open edges issuing from it, and the three open faces meeting at this vertex. As the homeomorphism α_i we take the projection of U_i onto the plane perpendicular to the main diagonal passing through the i-th vertex. Obviously $\alpha_i(U_i)$ will be the interior of a regular hexagon in $\mathbf{R}^2$.

In the sequel, we shall suppose that all manifolds under consideration satisfy Hausdorff's separation axiom and possess a countable basis of open sets.

We obtain an example of a manifold without Hausdorff separation as follows. Consider two replicas of a straight line and glue them together along a certain half-line. This example can be generalized by taking several "ordinary" manifolds and gluing them together along certain open sets.

The *line of Aleksandrov* is a manifold lacking a countable basis. Let T be the set of all countable ordinal numbers (that is, equivalence classes of well-ordered countable sets). In the product $T\times[0,1)=A$, one can define a lexicographic order: $(t_1,x_1)<(t_2,x_2)$ if $t_1<t_2$ or $t_1=t_2$ and $x_1<x_2$. As a basis of a topology in A, take the intervals

$$I(b,c)=\{a\in A; b<a<c\} \quad \text{and} \quad I(b)=\{a\in A; a>b\}.$$

Consider a mapping ϕ of an open set $U\subset \mathbf{R}^n$ into the space $\mathbf{R}^m$. We say that ϕ belongs to the *class* C^k, where $k=0,1,2,\ldots,\infty,\omega$, if the coordinates of the point $\phi(x)$, as functions of the coordinates of the point x, have k continuous derivatives (for positive integers k), are continuous for $k=0$, are infinitely differentiable for $k=\infty$, or are analytic for $k=\omega$.

Let (U,α) and (V,β) be two charts of the manifold M. We say that these charts are *k-smoothly connected* if the mappings $\alpha\circ\beta^{-1}$ and $\beta\circ\alpha^{-1}$ belong to the class C^k.

We shall suppose that the conditions of the definition are fulfilled if $U\cap V=\emptyset$, that is, two nonintersecting charts are k-smoothly connected for arbitrary k.

A manifold M is called *a smooth manifold of class C^k* if there is given an atlas A on M consisting of k-smoothly connected charts. A chart on M is called *admissible* if it is k-smoothly connected with all charts in A.

Problem 2. Prove that two charts on the sphere in examples 1 and 2 are ω-smoothly connected with each other.

Problem 3. Prove that charts on the surface of the cube in example 3 are k-smoothly connected with each other, where

$$k = \begin{cases} 0 & \text{if the } i\text{-th and } j\text{-th vertices are ends of a diagonal of a face,} \\ \omega & \text{in other cases.} \end{cases}$$

If M and N are manifolds of class C^k, then it makes sense to speak of l-smoothness of mappings of M into N for $l \leqslant k$. Specifically, we denote by $C^l(M, N)$ the set of all mappings $\phi: M \to N$ such that for arbitrary admissible charts (U, α) on M and (V, β) on N, the mapping $\beta \circ \phi \circ \alpha^{-1}$ belongs to the class C^l.

In particular, $C^l(M, \mathbf{R})$ or simply $C^l(M)$ is the set of all l-smooth real functions on M, and $C^l(\mathbf{R}, M)$ is the set of all l-smooth curves on M.

A one-to-one mapping $\phi: M \to N$ is called a *diffeomorphism* if ϕ and ϕ^{-1} are smooth mappings. In this case, we say that M and N are *diffeomorphic*.

It is known that on every manifold of dimension less than 7, one can introduce in one and only one way the structure of a smooth manifold. (We recall that all manifolds under consideration have a countable basis. It is known that a C^ω-structure can be introduced in nonunique fashion on the Aleksandrov line.) On the seven-dimensional sphere S^7, smoothness can be introduced in a nonunique way, and among eight-dimensional manifolds there are some that admit no smoothness at all.

We shall give several examples of smooth manifolds. It is simple enough to show that the following are C^ω-manifolds: the numerical spaces $\mathbf{R}^n$ and $\mathbf{C}^n$; the sphere $S^n \subset \mathbf{R}^{n+1}$, given by the equation $\sum_{i=1}^{n+1} x_i^2 = 1$; and the torus $\mathbf{T}^n \subset \mathbf{C}^n$ given by the equations $|z_i| = 1$, $i = 1, \ldots, n$. If M and N are smooth manifolds, then the product $M \times N$ is also a smooth manifold, if as an atlas on $M \times N$ we take the set of charts of the form $(U \times V, \alpha \times \beta)$, where (U, α) and (V, β) are admissible charts on M and N. The manifold $M \times N$ is the product of M and N in the category of smooth manifolds (morphisms are smooth mappings).

Example 4. Real projective space $\mathbf{RP}^n$, which is defined as the set of all lines going through the origin of coordinates in $\mathbf{R}^{n+1}$. A generic basis element for the family of open sets in $\mathbf{RP}^n$ is the set of lines having nonvoid intersection with a certain open set in $\mathbf{R}^{n+1}$. We shall show that $\mathbf{RP}^n$ can be made into a C^ω-manifold. A line in $\mathbf{RP}^n$ is defined by a direction vector $(a_1, \ldots, a_{n+1})$, where the numbers a_i are not all equal to zero and are defined up to a common multiple. Let U_i denote the set of lines for which $a_i \neq 0$, and define a homeomorphism $\alpha_i: U_i \to \mathbf{R}^n$, which carries the line with direction vector $(a_1, \ldots, a_{n+1})$ into the point

$$\left(\frac{a_1}{a_i}, \ldots, \frac{a_{i-1}}{a_i}, \frac{a_{i+1}}{a_i}, \ldots, \frac{a_{n+1}}{a_i} \right) \in \mathbf{R}^n.$$

Problem 4. Show that the collection $\{(U_i, \alpha_i)\}$ is an atlas consisting of ω-smoothly connected charts on $\mathbf{RP}^n$.

More general is the following

§ 5. Analysis on Manifolds

Example 5. Let $\mathbf{G}_{n,k}^{\mathbf{R}}$ be the set of k-dimensional subspaces in n-dimensional real space. An element of $\mathbf{G}_{n,k}^{\mathbf{R}}$ is generated by a system of k linearly independent vectors

$$\xi_i = (\xi_i^1, \ldots, \xi_i^n), \quad i = 1, 2, \ldots, k,$$

that is, it is defined by a matrix $\Xi = \|\xi_i^j\|$ of order $k \times n$ and rank k. Obviously matrices Ξ and Ξ' define one and the same element of $\mathbf{G}_{n,k}^{\mathbf{R}}$ if $\Xi' = C\Xi$, where C is a nonsingular matrix of order $k \times k$.

Let $J = \{j_1, \ldots, j_k\}$ be a subset of $\{1, 2, \ldots, n\}$. We denote by U_J the subset of $\mathbf{G}_{n,k}^{\mathbf{R}}$ the elements of which are defined by matrices Ξ with the condition $\det \|\xi_j^m\|_{j \in J} \neq 0$. Replacing Ξ by $\Xi' = C\Xi$, we can obtain the conditions

$$\tilde{\xi}_{j_r}^m = \begin{cases} 0 & \text{for } m \neq r, \\ 1 & \text{for } m = r. \end{cases}$$

The remaining matrix elements $\tilde{\xi}_j^i$, $j \notin J$, give a mapping of U_J into $\mathbf{R}^{k(n-k)}$ which we denote by α_J.

Problem 5. Prove that the collection $\{(U_J, \alpha_J)\}$ is an atlas of ω-smoothly connected charts on $\mathbf{G}_{n,k}^{\mathbf{R}}$.

Problem 6. Prove that $\mathbf{G}_{n,k}^{\mathbf{R}}$ and $\mathbf{G}_{n,n-k}^{\mathbf{R}}$ are diffeomorphic.

The manifolds $\mathbf{G}_{n,k}^{\mathbf{R}}$ are called *real Grassmann manifolds*. One can define in just the same way *complex Grassmann manifolds* $\mathbf{G}_{n,k}^{\mathbf{C}}$.

We obtain an important class of manifolds in the following way. Let $f_1, \ldots, f_m$ be a collection of k-smooth functions, defined in a certain region $U \subset \mathbf{R}^n$. We denote by M the set of all solutions of the system

$$f_1(x) = \cdots = f_m(x) = 0.$$

We need the following fact from analysis, which follows from the implicit function theorem.

Theorem 1. *Suppose that the rank of the matrix $F(x) = \|\partial f_i(x)/\partial x_j\|$ assumes a constant value r in some neighborhood of the set M. Then one can define uniquely on M the structure of a k-smooth $(n-r)$-dimensional manifold with the following properties:*
1) *restrictions to M of functions in $C^k(U)$ belong to $C^k(M)$;*
2) *as an admissible chart in a neighborhood of a point $x \in M$, we can take the projection of this neighborhood onto an arbitrary $(n-r)$-dimensional plane transversal to the rows of the matrix $F(x)$ (that is, containing no nonnull linear combination of rows of $F(x)$).*

Example 6. Let $\text{Mat}_n(K)$ be the set of all matrices of order n with elements in K. In the examples that we shall need, K will be the field of real numbers $\mathbf{R}$, or the field of complex numbers $\mathbf{C}$, or the skew field of quaternions $\mathbf{H}$. If $A \in \text{Mat}_n(K)$, then A' is the transposed matrix with elements $a'_{ij} = a_{ji}$, and A^* is the adjoint matrix with elements $a_{ij}^* = \bar{a}_{ji}$ (the bar denotes the complex or quaternion conjugate).

We denote by $GL(n, K)$ the open subset of $\text{Mat}_n(K)$ defined by the condition $\det A \neq 0$.[1]

Let $A \in \text{Mat}_n(K)$. We consider in $GL(n, K)$ the subsets defined by the following equations:

$$X^{-1} A X = A, \tag{1}$$

$$X' A X = A, \tag{2}$$

$$X^* A X = A. \tag{3}$$

Each of these matrix equations is equivalent to a system of n^2 ordinary equations in n^2 unknowns, the matrix elements x_{ij} of the matrix X.

Problem 7. Prove that the following mappings of $GL(n, K)$ into $\text{Mat}_n K$ have constant rank:

$$X \mapsto X^{-1} A X - A,$$
$$X \mapsto X' A X - A,$$
$$X \mapsto X^* A X - A.$$

Hint. Prove that if $C \in GL(n, K)$, then the mappings $X \to CX$ and $X \to XC$ are diffeomorphisms of $\text{Mat}_n(K)$. Use also the fact that the rank of a mapping does not change under composition of this mapping with a diffeomorphism.

The statements of problem 7 and theorem 1 show that the subsets defined by equations (1), (2), (3) are smooth manifolds. We note some special cases of this construction.

Let 1_n be the unit matrix of order n.

Equation (2) for $A = 1_n$ yields the set $O(n, K)$ of *orthogonal* matrices ($K = \mathbf{R}$ or $\mathbf{C}$). For $A = 1_n$ and $K = \mathbf{C}$, equation (3) yields the set $U(n)$ of *unitary* matrices. Finally, let A be the skew symmetric matrix of order $2n$ $\begin{pmatrix} 0 & 1_n \\ -1_n & 0 \end{pmatrix}$ and let $K = \mathbf{R}$ or $\mathbf{C}$. The equation (2) defines the set $Sp(2n, K)$ of *symplectic* matrices.

Problem 8. Suppose that $K = \mathbf{R}$ and that $A = \begin{pmatrix} 1_n & 1_n \\ -1_n & 1_n \end{pmatrix} \in \text{Mat}_{2n}(\mathbf{R})$. Prove that equation (2) defines the set $O(2n, \mathbf{R}) \cap Sp(2n, \mathbf{R})$, which is homeomorphic to $U(n)$.

Hint. Consider the mapping of $\text{Mat}_n(\mathbf{C})$ into $\text{Mat}_{2n}(\mathbf{R})$ defined by the formula

$$X + iY \mapsto \begin{pmatrix} X & Y \\ -Y & X \end{pmatrix}, \quad X, Y \in \text{Mat}_n(\mathbf{R}).$$

---

[1] In the case $K = \mathbf{H}$, we can define $\det A$ by embedding $\mathbf{H}$ in $\text{Mat}_2(\mathbf{C})$ or in $\text{Mat}_4(\mathbf{R})$ (see 3.2).

§ 5. Analysis on Manifolds

If $A = \begin{pmatrix} 1_p & 0 \\ 0 & -1_q \end{pmatrix}$, and $K = \mathbf{R}$, then equation (2) defines the set $O(p,q)$ of *pseudo-orthogonal* matrices of type (p,q). For $K = \mathbf{C}$ and this same matrix A, equation (3) gives the set $U(p,q)$ of *pseudo-unitary* matrices of type (p,q). For $K = \mathbf{H}$ and this matrix A, (3) yields the set $Sp(p,q)$ of *pseudo-symplectic* matrices of type (p,q).

We say that two charts (U,α) und (V,β) on a manifold M are *positively connected* if the mapping $\alpha\beta^{-1}$ has a positive Jacobian throughout its entire domain of definition. An atlas is called *positive* if it consists of positively connected charts. Two positive atlases are *equivalent* if their union is a positive atlas.

Problem 9. Prove that for an arbitrary connected manifold M, the number of equivalence classes of positive atlases is equal either to 0 or 2.

In the first case, the manifold M is called *nonorientable* and in the second, *orientable*. The choice of one of the two classes of positive atlases is called an *orientation of M*.

Problem 10. For what n and k is the manifold $\mathbf{G}_{n,k}^{\mathbf{R}}$ orientable?

Answer. For even n (if $0 < k < n$).

There is another way of defining the structure of a k-smooth manifold on a topological space M: for every open subset $U \subset X$, one can specify the ring $C^k(U)$ of k-smooth functions on U. It is evident that the correspondence $U \mapsto C^k(U)$ produces a contravariant functor from the category of open sets (morphisms are embeddings) to the category of rings. Such functors are called *presheaves* (more exactly, *presheaves of rings*). One defines in like fashion *presheaves of groups, of algebras, of sets*, and so on.

Let F be a presheaf. Suppose that for every open set $U \subset X$ and every covering of U by open subsets U_α, $\alpha \in A$, the following sequence is exact:

$$0 \longrightarrow F(U) \overset{i}{\longrightarrow} \prod_{\alpha \in A} F(U_\alpha) \overset{j}{\longrightarrow} \prod_{\alpha,\beta \in A} F(U_\alpha \cap U_\beta); \tag{4}$$

here the morphism i corresponds to embedding of U_α in U, and the morphism j is the difference of morphisms corresponding to the embeddings of $U_\alpha \cap U_\beta$ in U_α and U_β. Then the presheaf F is called a *sheaf*.

The intuitive meaning of exactness of this sequence is seen from the example of the presheaf $U \mapsto C^k(U)$ (which is in fact a sheaf). Namely, exactness in the first member means that every smooth function on U is completely defined by its restrictions to U_α, $\alpha \in A$. Exactness in the second member means that if in each U_α there is given a smooth function f_α in such a way that the functions f_α and f_β coincide on $U_\alpha \cap U_\beta$, then all of the f_α are obtained by restrictions of a single smooth function f on U.

A topological space X with a sheaf of rings $\mathcal{O}$ defined on it is called a *ring space*, and the sheaf $\mathcal{O}$ is called a *structure sheaf*.

We can now formulate the definition of a smooth manifold in the following way.

A manifold of class C^n is a ring space $(X, \mathcal{O})$ which is locally isomorphic to an open ball in $\mathbf{R}^n$, equipped with the sheaf of k-smooth functions.

In this definition, we may replace $\mathbf{R}^n$ by $\mathbf{C}^n$ and the sheaf of k-smooth functions by the sheaf of holomorphic functions (i.e., complex analytic functions). Doing this, we obtain the definition of a *complex manifold*.

We note that the classical concept of an *algebraic manifold* over K is subsumed under the general scheme of a ring space. A local model of an algebraic manifold is a subset of K^n, given by a system of algebraic equations, and equipped with the sheaf of everywhere defined rational functions and the Zariski topology. The same is true of its contemporary variant, the concept of a scheme under the field K. A local model is the set of simple ideals of a certain K-algebra.

We now consider the ring space $(X, \mathcal{O})$ locally isomorphic to a neighborhood of zero in the space $L = \mathbf{R}^k \oplus \mathbf{C}^l$, and equipped with the sheaf of functions infinitely differentiable in the real coordinates in $\mathbf{R}^k$ and holomorphic in the complex coordinates in $\mathbf{C}^l$.

Suppose that there exists a smooth k-dimensional real manifold Y and a mapping $p: X \to Y$ such that every "fiber" $p^{-1}(y)$, $y \in Y$, is an l-dimensional complex manifold and such that the structure sheaf $\mathcal{O}$ on X consists of smooth functions that are holomorphic along each fiber. Then we say that the space $(X, \mathcal{O})$ is a mixed manifold of type (k, l).

Problem 11. Prove that the space Y and the projection p are defined uniquely by $(X, \mathcal{O})$.

Hint. Let $\bar{\mathcal{O}}$ be the sheaf of functions that are the complex conjugates of functions in $\mathcal{O}$, and let A be the algebra $\mathcal{O}(X) \cap \bar{\mathcal{O}}(X)$. Then $Y = \mathfrak{M}(A)$ and p is the natural mapping of X into Y that associates to the point $x \in X$ the ideal $p(x) \subset A$ of all functions vanishing at x.

Problem 12. Let $(X, \mathcal{O})$ be the ring space arising from $L = \mathbf{R}^k \oplus \mathbf{C}^l$ by factoring with respect to a discrete group Γ of parallel translations in L. Prove that $(X, \mathcal{O})$ will be a mixed manifold if and only if the projection Γ_0 of the group Γ in the group of parallel translations of $\mathbf{R}^k$ is discrete.

Hint. Let $A = \mathcal{O}(X) \cap \bar{\mathcal{O}}(X)$. Verify that A is isomorphic to the algebra of smooth functions on $\mathbf{R}^k$ that are invariant under Γ_0.

5.2. Vector Fields

One of the basic concepts in the theory of smooth manifolds in the notion of tangent vector. The intuitive sense of this concept is the velocity of a point moving on the manifold. A precise definition can be given as follows. A *smooth curve* on a manifold M is any smooth mapping of $\mathbf{R}$ into M. We will denote the parameter in $\mathbf{R}$ by the letter t and the point of the manifold by the letter x. A smooth curve $x(t)$ is given in a neighborhood of a point $x_0 = x_0(t)$ by a collection of $n = \dim M$ smooth functions $x^i(t)$ (local coordinates of the point $x(t)$ in a certain chart containing the point x_0). We shall say that a curve $x(t)$ starts at a point x_0 if $x(0) = x_0$. We say that two smooth curves $x(t)$ and $y(t)$ starting at the point x_0 are equivalent if $|x^i(t) - y^i(t)| = o(t)$ as $t \to 0$, $i = 1, 2, \ldots, n$.

It is not hard to show that this definition of equivalence does not depend upon the choice of a chart that contains the point x_0.

§ 5. Analysis on Manifolds

A *tangent vector* to the manifold at the point x_0 is an equivalence class of smooth curves starting at the point x_0.

The expression "X is a tangent vector to the curve $x(t)$ at $t=0$", or the expression $X = x'(0)$, means that the curve $x(t)$ belongs to the class X.

We choose an arbitrary chart (U, α) for which the point x_0 is the origin of coordinates, and for simplicity we suppose that $\alpha(U) = \mathbf{R}^n$. The set of smooth curves in U is a linear space under the operations of coordinatewise addition and multiplication by a number:

$$(x+y)^i(t) = x^i(t) + y^i(t), \quad (\lambda x)^i(t) = \lambda \cdot x^i(t).$$

Problem 1. Verify that these operations preserve the relation of equivalence defined above and hence produce a linear space structure (over $\mathbf{R}$) in the set $T_x M$ of all tangent vectors to the manifold M at the point x.

Problem 2. Prove that the dimension of the space $T_x M$ is equal to the dimension of the manifold M at the point x.

The definition we have adopted of tangent vector agrees with one's intuitive picture of velocity, but is awkward in various other connections. For example, the definition of the sum of two tangent vectors and of the product of a vector by a number require the introduction of local coordinates.

We shall now present a different definition, which is in no way dependent upon local coordinates. To do this, we note that every smooth curve $x(t)$ starting at the point x_0 defines a continuous functional F on the space $C^\infty(M)$ by the formula

$$F(f) = \frac{d}{dt} f(x(t)) \bigg|_{t=0}. \tag{1}$$

Problem 3. Show that two functionals of the above form coincide if and only if the corresponding curves are equivalent.

Thus, to every tangent vector X there corresponds a functional F, which is ordinarily called the *derivative along the vector* X.

Problem 4. Prove that a functional F on $C^\infty(M)$ can be represented in the form (1) for a certain curve $x(t)$ starting at x_0 if and only if the equality

$$F(f_1 f_2) = F(f_1) f_2(x_0) + f_1(x_0) F(f_2) \tag{2}$$

holds for arbitrary functions f_1 and f_2 in $C^\infty(M)$.

These results show that our definition of tangent vector is equivalent to the following:

a *tangent vector* to the manifold M at the point x_0 is a linear functional on $C^\infty(M)$ that satisfies condition (2).

Under this definition, the structure of a linear space in the set of tangent vectors is obtained in an obvious way, since linear functionals can be added and multiplied by numbers.

For practical computations with tangent vectors, it is useful to introduce coordinates in the space $T_x M$.

Let (U, α) be an arbitrary chart on M. For an arbitrary point $x \in U$ we can define n tangent vectors $\partial/\partial x^1, \ldots, \partial/\partial x^n$, defining the corresponding functionals as partial derivatives in the local coordinates at the point x.

Problem 5. Prove that for every point $x \in U$, the vectors $\partial/\partial x^i$, $i = 1, \ldots, n$, form a basis in the space $T_x M$.

The coordinates of a vector $\xi \in T_x M$ thus depend upon the choice of a chart. If U and V are two charts covering the point x, then we obtain two bases $\{\partial/\partial x^i\}$, $\{\partial/\partial y^i\}$, in the space $T_x M$, where $\{x^i\}$, $\{y^i\}$ are local coordinates in these charts.

The matrix of transition from one basis to another obviously has the form $a_j^i = \frac{\partial x^i}{\partial y^j}$, since $\frac{\partial f}{\partial y^j} = \sum_i \frac{\partial f}{\partial x^i} \frac{\partial x^i}{\partial y^j}$. This law of transformation is sometimes used in the definition of a tangent vector.

According to this definition, a *tangent vector* on the manifold M at the point x is the association with each chart U containing x of a collection of numbers $\{\xi^i(U)\}$ in such a way that if $\{x^i\}$ are local coordinates in U and $\{y^i\}$ are local coordinates in V, then the identity

$$\xi^i(U) = \sum_j \frac{\partial x^i}{\partial y^j} \xi^j(V). \tag{3}$$

holds.

Now let M and N be two smooth manifolds and let $f: M \to N$ be a smooth mapping. Under the mapping f, the set of smooth curves starting at a point $x \in M$ goes into a set of smooth curves starting at the point $y = f(x) \in N$. It can be verified with no difficulty that equivalent curves go into equivalent curves. Thus we obtain a mapping $f_*(x): T_x M \to T_y N$.

Problem 6. Prove that f_* is a linear mapping and compute its matrix in the bases $\{\partial/\partial x^i\}$ in $T_x M$ and $\{\partial/\partial y^j\}$ in $T_y N$.

Problem 7. If M, N, L are three smooth manifolds and $f: M \to N$, $g: N \to L$ are two smooth mappings, then we have

$$(g \circ f)_* = g_* \circ f_*.$$

This is what is called the *chain rule* for computing the partial derivatives of composite functions. Writing each mapping by means of the matrices in natural bases, we come to the usual form of the chain rule:

$$\frac{\partial z^i}{\partial x^k} = \sum_j \frac{\partial z^i}{\partial y^j} \frac{\partial y^j}{\partial x^k}.$$

The mapping $f_*(x): T_x M \to T_y N$ is called the *derivative* of the mapping $f: M \to N$ at the point x. We consider the special case in which $N = \mathbf{R}$. Since $\mathbf{R}$ has a standard coordinate, all of the spaces $T_y \mathbf{R}$ are identified in a natural way with $\mathbf{R}$. Thus the derivative mapping f_* gives a linear functional on $T_x M$. This functional is called df and is called the *differential* of the function f.

§ 5. Analysis on Manifolds

If at every point of a manifold M there is defined a vector tangent to the manifold, we say that a *vector field* is defined on M. In a local system of coordinates, a vector field can be written in the form

$$\xi = \sum_i \xi^i(x) \frac{\partial}{\partial x^i}.$$

A vector field ξ is called *smooth* if its coordinate functions $\xi^i(x)$ are smooth functions. Plainly this definition of smoothness of a vector field does not depend upon the choice of a chart.

Just as vectors arise from considering smooth curves, vector fields are connected with mappings of a manifold onto itself. Let $\phi_t: M \to M$ be a one-parameter family of smooth mappings of the manifold M. We require that ϕ_t depend smoothly on the parameter t (that is, that the mapping $\mathbf{R} \times M \to M$ defined by the formula $(t, x) \mapsto \phi_t(x)$ be smooth). We also require that for $t = 0$, we obtain the identity mapping $\phi_0(x) \equiv x$.

Then, to every point $x \in M$, there corresponds a smooth curve $x(t) = \phi_t(x)$ starting at the point x, and thus we have a tangent vector $x'(0)$.

We have obtained a vector field called the *derivative* of the family ϕ_t.

The question arises as to whether or not every vector field is the derivative of a certain family of mappings of M into itself. The answer to this question is affirmative. For compact manifolds, one can prove a stronger statement.

Theorem 1. *Every smooth vector field ξ on a compact manifold M is the derivative of a certain one-parameter group $\{\phi_t\}$ of mappings of M onto itself.*

For the proof, we consider the following differential equation on M:

$$x'(t) = \xi(x(t)), \quad x(0) = x_0. \tag{4}$$

In a local system of coordinates this differential equation is written in the form of a system of equations:

$$\frac{dx^i}{dt} = \xi^i(x^1, \ldots, x^n), \quad x^i(0) = x_0^i. \tag{4'}$$

According to a well-known theorem from the theory of ordinary differential equations, there is a solution $x(t, x_0)$ of the equation (4), depending continuously on the initial conditions, for a certain neighborhood of an arbitrary point $x_0 \in M$ and a certain interval $(-\varepsilon, \varepsilon)$ of variation of t (this interval depends upon x_0).

In view of the compactness of M, there exists an interval $(-\varepsilon_0, \varepsilon_0)$ in which the solutions $x(t, x_0)$ are defined for any location whatever of the initial point x_0.

Problem 8. Prove that the mappings $\phi_t: x_0 \to x(t, x_0)$ satisfy the relation

$$\phi_t \circ \phi_s = \phi_{t+s}$$

for sufficiently small t and s.

Hint. Verify that the curves $x_1(t)=\phi_t(\phi_s(x))$ and $x_2(t)=\phi_{t+s}(x)$ satisfy (4) for $x_0=\phi_s(x)$.

The mappings ϕ_t are defined for $|t|\geqslant \varepsilon_0$ from the group law. The theorem is proved.

For noncompact manifolds M, the assertion of theorem 1 fails.

Problem 9. Prove that there is no group of transformations corresponding to the vector field $\xi=(1+x^2)\dfrac{d}{dx}$ on the line.

Hint. Show that under the mapping $x\mapsto y=\operatorname{arctg} x$ the line goes into the interval $(-\pi/2, \pi/2)$ and the field ξ into the field $\eta=d/dy$.

Nevertheless a somewhat weaker assertion does hold.

Problem 10. Let ξ be a vector field on the manifold M. Prove that there exists a smooth function a on $M\times \mathbf{R}$ such that:
1) the equation

$$x'(t)=a(x,t)\,\xi(x), \qquad x(0)=x_0 \tag{5}$$

admits a solution for all $x_0\in M$ and all $t\in \mathbf{R}$;
2) for every point $x\in M$, there exists an $\varepsilon>0$ such that $a(x,t)\equiv 1$ for $|t|<\varepsilon$.

Hint. Use the fact that M admits a countable basis and consequently is the union of a countable set of compact subsets.

It is evident that the field ξ is a derivative for the family of mappings $\phi_t \colon x_0\mapsto x(t,x_0)$, where $x(t,x_0)$ is a solution of the equation (5).

As in the case of vectors, one can give a second definition of vector fields that does not use local coordinates. For this we note that every smooth vector field ξ on M defines a linear operator in $C^\infty(M)$, which associates with a function f and a point $x\in M$ the derivative of f along the vector $\xi(x)$. We shall denote this linear operator by the same letter that we use to denote the vector field itself.

It follows from the equality (2) that operators corresponding to vector fields have the property that

$$\xi(f_1 f_2)=\xi f_1\cdot f_2+f_1\cdot \xi f_2. \tag{6}$$

Problem 11. Prove that every operator in $C^\infty(M)$ that satisfies (4) corresponds to a smooth vector field.

Thus we obtain an equivalent definition of a vector field as an operator in $C^\infty(M)$.

In addition to the usual operations of a linear space (addition and multiplication by a number), vector fields admit still another important operation, that of commutation.

Given vector fields ξ and η, we define the *commutator* of these vector fields as the vector field $[\xi,\eta]$ defined by the formula

$$[\xi,\eta]=\xi\eta-\eta\xi. \tag{7}$$

§ 5. Analysis on Manifolds

Problem 12. Verify that the operator (7) does in fact correspond to a vector field.

Hint. Use the results of problem 11.

Problem 13. Suppose that in a certain chart, we have

$$\xi = \sum_i \xi^i(x) \frac{\partial}{\partial x^i}, \quad \eta = \sum_i \eta^i(x) \frac{\partial}{\partial x^i}.$$

Find an expression for the field $[\xi, \eta]$ in the same chart.

Answer. We have $[\xi, \eta] = \sum_i \zeta^i(x) \frac{\partial}{\partial x^i}$, where

$$\zeta^i(x) = \sum_j \left(\xi^j(x) \frac{\partial \eta^i(x)}{\partial x^j} - \eta^j(x) \frac{\partial \xi^i(x)}{\partial x^j} \right).$$

The derivative of a function f along a vector field ξ and the commutator of vector fields ξ and η are special cases of what are called *Lie operators* L_ξ, corresponding to the vector field ξ.

Consider the category of manifolds, with morphisms taken as smooth homeomorphisms, usually called diffeomorphisms. Let F be a functor from this category to the category of linear topological spaces. If ξ is a vector field on the manifold M and $\{\phi_t\}$ is a certain family of diffeomorphisms of M, for which ξ is the derivative field, then in the space $F(M)$ we obtain a family of linear mappings $F(\phi_t)$. A vector $f \in F(M)$ is said to be *differentiable along the field* ξ if the vector function $t \mapsto F(\phi_t)f$ is differentiable for $t=0$ and the derivative $\frac{d}{dt}(F(\phi_t)f)|_{t=0} = L_\xi f$ for which ξ is the derivative field does not depend upon the choice of the family $\{\phi_t\}$.

Example. Let F_0 be the contravariant functor given by the following formulas: $F_0(M) = C^k(M)$, $k > 0$, $F_0(\phi) = f \mapsto f \circ \phi$.
Then we have $L_\xi f = \xi f$.

Problem 14. Let F_1 be the contravariant functor that associates with the manifold M the space $\text{Vect } M$ of all smooth vector fields on M, and associates with the diffeomorphism ϕ the mapping $F_1(\phi): \eta \mapsto \phi_*^{-1} \circ \eta \circ \phi$ (that is, $[F_1(\phi)\eta](x) = \phi_*(x)^{-1} \eta(\phi(x))$).
Prove that $L_\xi \eta = [\xi, \eta]$.
Hint. Use the identity

$$F_0(\phi)\eta F_0(\phi^{-1}) = F_1(\phi)(\eta)$$

and the equality

$$(L_\xi(\eta))(f) = L_\xi L(\eta) f + \eta(L_\xi f), \tag{8}$$

which follows from it.

5.3. Differential Forms

A *tensor field* on a manifold M is a function which associates with every point $x \in M$ a certain tensor on the space $T_x M$ tangent to M and the point x.

Special cases of tensor fields are functions (tensor fields of rank 0) and vector fields (contravariant tensor fields of rank 1).

If U is a local system of coordinates on M, then one can choose simultaneously a basis $\partial/\partial x^1, \ldots, \partial/\partial x^n$ in all of the tangent spaces $T_x M$, $x \in U$, and in all of the spaces $T_x^* M$ a dual basis $dx^1, \ldots, dx^n$.

The value of the covector dx^i at the vector $\partial/\partial x^j$ is equal to Kronecker's symbol

$$\delta_j^i = \begin{cases} 0 & \text{for } i \neq j, \\ 1 & \text{for } i = j. \end{cases}$$

Every mixed tensor field of rank (k, l) can be decomposed using the basis fields

$$\frac{\partial}{\partial x^{i_1}} \otimes \cdots \otimes \frac{\partial}{\partial x^{i_k}} \otimes dx^{j_1} \otimes \cdots \otimes dx^{j_l}.$$

A tensor field is said to be *smooth* if the coefficients in its decomposition in basis fields are smooth functions. It is clear that this property of the field does not depend upon the choice of an (admissible) system of coordinates.

Skew-symmetric covariant tensor fields are called *differential forms*. Differential forms form a noncommutative algebra relative to exterior multiplication. Furthermore, we can define in the space of differential forms what is called the *operator of exterior differentiation*, or the *differential d*, which enjoys the following properties:

a) $d(\omega_1 + \omega_2) = d\omega_1 + d\omega_2$;
b) $d(\omega_1 \wedge \omega_2) = d\omega_1 \wedge \omega_2 + (-1)^k \omega_1 \wedge d\omega_2$, \hfill (1)

if ω_1 is a form of degree k;

c) on forms of degree 0 (that is, functions), the operator d coincides with the ordinary differential (see 5.2);

d) $d^2 = 0$.

The properties listed above define the operator d uniquely, since for every form ω, having the expression

$$\omega = \sum_{i_1, \ldots, i_k} a_{i_1 \ldots i_k}(x) \, dx^{i_1} \wedge \ldots \wedge dx^{i_k}$$

in a local system of coordinates, we must have

$$d\omega = \sum_{i_1, \ldots, i_k} d(a_{i_1 \ldots i_k}(x) \, dx^{i_1} \wedge \ldots \wedge dx^{i_k})$$
$$= \sum_{i_1, \ldots, i_k} da_{i_1 \ldots i_k} \wedge dx^{i_1} \wedge \ldots \wedge dx^{i_k}$$

(since $d(dx^i) = 0$).

§ 5. Analysis on Manifolds

Problem 1. Prove that the correspondence $\omega \mapsto d\omega$ defined by the last equality does not depend upon the choice of a system of coordinates and enjoys the properties (1).

It is clear that the operator d raises by one the the degree of a differential form. If ξ is a vector field on M, then one can define an operator $\iota(\xi)$, which decreases the degree of a differential form by one, in the following way.

If ω is a form of degree k, then $\iota(\xi)\omega$ is the form of degree $k-1$ which assumes at the vectors $\xi_1, \ldots, \xi_{k-1}$ in $T_x M$ the value $k[\omega(x)](\xi(x), \xi_1, \ldots, \xi_{k-1})$.

Problem 2. Prove that $\iota(\xi)$ is the unique operator on the space of differential forms on M that has the following properties:
  a) $\iota(\xi)(\omega_1 + \omega_2) = \iota(\xi)\omega_1 + \iota(\xi)\omega_2$;
  b) $\iota(\xi)(\omega_1 \wedge \omega_2) = \iota(\xi)\omega_1 \wedge \omega_2 + (-1)^k \omega_1 \wedge \iota(\xi)\omega_2$ (2)
if ω_1 is a form of degree k;
  c) if ω is a form of degree 1, then $\iota(\xi)\omega = \omega(\xi)$;
  d) if ω is a form of degree 0, then $\iota(\xi)\omega = 0$.

Problem 3. Prove that the operators $\iota(\xi)$ and d are connected with the Lie operator L_ξ by the identity

$$L_\xi = d \circ \iota(\xi) + \iota(\xi) \circ d. \tag{3}$$

An important property of the operator d is the fact that it commutes with smooth mappings.

Problem 4. Let $f : M \to N$ be a smooth mapping. For every differential form ω on N, we define the form $f^*\omega$ on M by the equality

$$[(f^*\omega)(x)](\xi_1, \ldots, \xi_k) = [\omega(f(x))](f_*(x)\xi_1, \ldots, f_*(x)\xi_k).$$

Prove that $d \circ f^* = f^* \circ d$.

Hint. Prove that f^* commutes with exterior multiplication and use the formulas (1), which reduce the problem to the case $k=0$ or $k=1$ and $\omega = dx$.

The assertion of problem 4 can also be deduced from the following explicit expression for the differential $d\omega$ of a form ω of degree k:

$$(k+1)d\omega(\xi_0, \ldots, \xi_k) = \sum_{i=0}^{k} (-1)^i \xi_i \omega(\xi_0, \ldots, \hat{\xi}_i, \ldots, \xi_k)$$

$$+ \sum_{0 \le i < j \le k} (-1)^{i+j} \omega([\xi_i, \xi_j], \xi_0, \ldots, \hat{\xi}_i, \ldots, \hat{\xi}_j, \ldots, \xi_k). \tag{4}$$

Here $\xi_0, \ldots, \xi_k$ are arbitrary smooth vector fields, and the symbol $\hat{\ }$ means that the corresponding argument must be omitted.

We will prove the equality (4) by induction on the degree of the form. We rewrite the left side as $[i(\xi_0)d\omega](\xi_1, \ldots, \xi_k)$ and use the identity (3). We obtain

$$(L_{\xi_0}\omega)(\xi_1, \ldots, \xi_k) - d[i(\xi_0)\omega](\xi_1, \ldots, \xi_k). \tag{5}$$

Using the identity

$$L_{\xi_0}[\omega(\xi_1,\ldots,\xi_k)] = (L_{\xi_0}\omega)(\xi_1,\ldots,\xi_k) + \sum_{i=1}^{k} \omega(\xi_1,\ldots,L_{\xi_0}\xi_i,\ldots,\xi_k)$$

and our inductive hypothesis, we easily rewrite the expression (5) in such a way that it coincides with the right side of the desired equality (4). It remains to verify the correctness of the formula (4) for $k=0$, which is trivial.

A form ω is said to be *closed* if $d\omega = 0$, and *exact* if $\omega = d\omega'$ for a certain form ω'.

Problem 5. Closed forms are a ring with respect to exterior multiplication, in which the set of exact forms is a two-sided ideal.

The corresponding factor ring is denoted by $H^*(M, \mathbf{R})$ (or $H^*(M, \mathbf{C})$ if we are considering forms with complex coefficients). It is called the *de Rham cohomology ring* of the manifold M. It admits a natural grading (by the degrees of forms):

$$H^*(M, \mathbf{R}) = \sum_{k=0}^{\dim M} H^k(M, \mathbf{R}).$$

de Rham's Theorem. *The groups $H^k(M, K)$, $K = \mathbf{R}, \mathbf{C}$, coincide with the cohomology groups of the manifold M with coefficients in K (see 1.3).*

The great rôle played by differential forms in analysis is explained by the fact that they arise naturally when we try to carry over the operation of integration to manifolds. In fact, suppose that ω is a differential form of degree k on M and that N is a certain k-dimensional submanifold in M. We suppose that there exists a smooth mapping ϕ of a certain region $D \subset \mathbf{R}^k$ into the manifold M for which $\phi(D) = N$. Then the form $\phi^*(\omega)$ can be written as $a(x^1,\ldots,x^k)dx^1 \wedge \ldots \wedge dx^k$. The classical formula for change of variables shows that the quantity

$$I = \int\ldots\int_D a(x^1,\ldots,x^k)dx^1 \ldots dx^k$$

is defined up to its sign by the manifold N and the form ω (that is, does not depend upon the choice of the parametrizing mapping ϕ).

In order to define the sign of I uniquely, we must fix an orientation on N (see 5.1).

The quantity I is called the *integral of the form ω on the oriented manifold N* and is denoted by $\int_N \omega$.

In the general case, where N is not covered by a single chart, we proceed as follows. Let N be an oriented manifold and $\{U_\alpha, \phi_\alpha\}$ a positive atlas belonging to the class singled out.

Problem 6 (The theorem on partitions of unity). *If $\{U_\alpha\}_{\alpha \in A}$ is an open covering of a manifold M of class C^k, then there exists a family of functions $\{f_\alpha\}_{\alpha \in A}$ of class C^l, $l = \min(k, \infty)$, having the following properties:*

§ 5. Analysis on Manifolds

1) $f_\alpha = 0$ *outside of* U_α;
2) $0 \leq f_\alpha \leq 1$ *on* M;
3) $\sum_{\alpha \in A} f_\alpha \equiv 1$ *on* M.

Corollary. *A form ω on M is the sum of forms ω_α such that $\omega_\alpha = 0$ outside of U_α.*

We define

$$\int_N \omega = \sum_{\alpha \in A} \int_{U_\alpha} \omega_\alpha.$$

One can show that this quantity does not depend upon the choice of the covering U_α and the partition of ω into the sum of the ω_α.

A *manifold with boundary* is a set M each point of which has a neighborhood diffeomorphic to a neighborhood of zero either in $\mathbf{R}^n$ or in $\mathbf{R}^n_+$ (the half space defined by the condition $x^n \geq 0$). The set of points of the second type is called the *boundary* of the manifold M and is denoted by ∂M.

If a manifold with boundary M is orientable, then its boundary ∂M is also orientable and for an appropriate choice of orientations, the formula

$$\int_M d\omega = \int_{\partial M} \omega$$

holds for every differential form of degree $n-1$.

A proof of this formula and other complementary facts can be found in the books [56], [51].

5.4. Bundles

The majority of linear spaces comprising the field of application of functional analysis can be considered from a unifying point of view, namely, as spaces of cross sections of vector bundles on smooth manifolds. Examples are spaces of functions, of vector fields, of differential forms, and others.

In this section we shall acquaint the reader with the basic definitions and facts of the theory of vector bundles, referring for details to the book of D. Husemoller [32].

We also introduce the definition of cohomologies with coefficients in a sheaf and the generalized theorem of Dolbeault.

The reader can find a proof of this theorem and supplementary information from the theory of sheaves in the monograph of R. Godement [24].

A *vector bundle* over a smooth manifold M is a pair $\mathscr{E} = (E, p)$, where E is a certain manifold and p is a mapping of E into M that enjoys the following properties:

1) the inverse image $p^{-1}(x)$ of each point $x \in M$ has the structure of a linear space over K, where $K = \mathbf{R}$ or $\mathbf{C}$;
2) every point $x \in M$ has a neighborhood U such that $p^{-1}(U)$ admits a diffeomorphism ϕ onto $U \times K^n$, which commutes with projection onto U and is linear in each fiber $p^{-1}(x)$.

The manifold E is called the *space of the bundle*, the space K^n a *fiber*, the manifold M the *base*, and the mapping p the *projection* of the bundle.

The set of all bundles with fiber K^n over M forms a category, the morphisms of which are the commutive diagrams

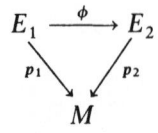

$$\begin{array}{ccc} E_1 & \xrightarrow{\phi} & E_2 \\ & \searrow p_1 \quad \swarrow p_2 & \\ & M & \end{array}$$

where ϕ is a certain diffeomorphism, linear on each fiber.

As an example of a bundle, we mention the direct product $M \times K^n$ with the natural projection onto the first factor. This bundle and all bundles equivalent to it are called *trivial*.

Every bundle can be "pasted together" from trivial bundles, in the following fashion. Let $\{U_\alpha\}_{\alpha \in A}$ be a covering of M by open sets. In the intersections $U_\alpha \cap U_\beta$, we specify smooth operator functions $g_{\alpha\beta}$ with values in $\operatorname{Aut} K^n$, which satisfy the following conditions:

$$\left. \begin{array}{l} g_{\alpha\alpha} \equiv 1 \quad \text{in} \quad U_\alpha, \quad g_{\alpha\beta} \circ g_{\beta\alpha} \equiv 1 \quad \text{in} \quad U_\alpha \cap U_\beta, \\ g_{\alpha\beta} \circ g_{\beta\gamma} \circ g_{\gamma\alpha} \equiv 1 \quad \text{in} \quad U_\alpha \cap U_\beta \cap U_\gamma. \end{array} \right\} \tag{1}$$

These functions are called *transition functions*. Let $\tilde{E}$ be the union of the manifolds $U_\alpha \times K^n$. We say that points $(x_\alpha, y_\alpha) \in U_\alpha \times K^n$ and $(x_\beta, y_\beta) \in U_\beta \times K^n$ are equivalent if $x_\alpha = x_\beta$ and $y_\alpha = g_{\alpha\beta}(x_\beta) y_\beta$. (The fact that this is actually an equivalence relation is implied by, and also implies, the conditions (1) on the functions $g_{\alpha\beta}$.) Let E denote the factor space by this equivalence relation, and let p denote the projection of E onto M [this is the factor mapping of the natural projection of $\tilde{E}$ onto M, carrying (x_α, y_α) into x_α].

Problem 1. Prove that $\mathscr{E} = (E, p)$ is a vector bundle over M and that every bundle over M is equivalent to a bundle of this type.

Hint. Consider a covering of M by open sets U_α for which, by condition 2), there exists a diffeomorphism $\phi_\alpha : p^{-1}(U_\alpha) \to U_\alpha \times K^n$, and set $g_{\alpha\beta} = \phi_\beta \circ \phi_\alpha^{-1}$.

As an example of the practical realization of this construction (for $M = S^1$, $K^n = \mathbf{R}^1$), consider the construction of a "Möbius strip" from two strips of paper (Fig. 1).

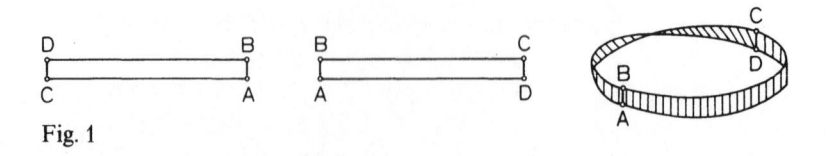

Fig. 1

Problem 2. Suppose that $f_\alpha : U_\alpha \to \operatorname{Aut} K^n$ are arbitrary smooth operator functions. Prove that the bundles $\mathscr{E}$ and $\mathscr{E}'$ defined by the transition functions $g_{\alpha\beta}$ and $g'_{\alpha\beta} = f_\alpha^{-1} \circ g_{\alpha\beta} \circ f_\beta$ are equivalent.

§ 5. Analysis on Manifolds

Hint. The desired equivalence $\phi: E_1 \to E_2$ is the factor mapping of the mapping $\tilde{\phi}: \tilde{E}_1 \to \tilde{E}_2$, defined by the formula $\tilde{\phi}(x_\alpha, y_\alpha) = (x_\alpha, f_\alpha(x_\alpha) y_\alpha)$.

A *cross section* of a bundle E over an open set $U \subset M$ is a smooth mapping $s: U \to E$ with the property that $p \circ s = 1$. (Sometimes the image of U under the mapping s is also called a cross section.)

The set of all cross sections of $\mathscr{E}$ over U is denoted by $\Gamma(\mathscr{E}, U)$. If $\mathscr{E}$ is a trivial bundle, then $\Gamma(\mathscr{E}, U)$ is identified in a natural way with the smooth vector functions on U with values in K^n.

In the general case, where the bundle $\mathscr{E}$ is defined by a covering $\{U_\alpha\}$ and transition functions $g_{\alpha\beta}$, a cross section s is given by the collection of smooth vector functions

$$s_\alpha: U \cap U_\alpha \to K^n, \tag{2}$$

which are called *components of s* and which satisfy the conditions

$$s_\alpha(x) = g_{\alpha\beta}(x) s_\beta(x), \quad x \in U \cap U_\alpha \cap U_\beta. \tag{3}$$

We note that $\Gamma(\mathscr{E}, U)$ is a module over the ring $\mathcal{O}(U)$ of smooth (numerical) functions on U: the cross section $f \cdot s$ has components $(f \cdot s)_\alpha = f \cdot s_\alpha$.

Problem 3. Prove that a bundle $\mathscr{E}$ with fiber K^n is trivial if and only if there exist n cross sections $s_1, \ldots, s_n$ in $\Gamma(\mathscr{E}, M)$ which are linearly independent at every point $x \in M$.

Hint. The necessity of the condition is obvious. Its sufficiency follows from mapping $M \times K^n$ into E by the formula

$$(x, (\lambda_1, \ldots, \lambda_n)) \mapsto (\lambda_1 s_1 + \ldots + \lambda_n s_n)(x).$$

Let TM be the set of all tangent vectors to the manifold M (that is, $TM = \bigcup_{x \in M} T_x M$; see 5.2).

Mapping the vector $\xi \in T_x M$ into the point $x \in M$, we obtain a projection $p: TM \to M$.

Problem 4. Prove that $\tau(M) = (TM, p)$ is a vector bundle over M with fiber $\mathbf{R}^n$, $n = \dim M$, and that $\Gamma(\tau(M), M)$ is identifiable in a natural way with the space of smooth vector fields on M.

Hint. Let $\{(U_\alpha, \phi_\alpha)\}$ be an atlas on M. Consider the bundle defined by the covering $\{U_\alpha\}$ and the transition functions $g_{\alpha\beta} = (\phi_\alpha \circ \phi_\beta^{-1})_*$.

The bundle $\tau(M)$ is called the *tangent bundle* of the manifold M. Analogous definitions can be given of the *cotangent bundle* $\tau^*(M)$ and *tensor bundles* $\tau^{k,l}(M)$, the cross sections of which are fields of mixed tensors of rank (k, l) on M.

The operations of *direct sum* and *tensor product* are defined for vector bundles over M. These are based on direct sum and tensor product of the corresponding transition functions. For example, the bundle $\tau^{k,l}(M)$ is equivalent to the tensor product of k copies of $\tau(M)$ and l copies of $\tau^*(M)$.

These operations can be carried over to equivalence classes of bundles.

Let $\mathcal{K}(M)$ be the set of equivalence classes of complex vector bundles over M. We consider the category of all mappings of $\mathcal{K}(M)$ into commutative rings that carry direct sum and tensor product of classes of bundles into sum and product of elements of the ring. In this category there exists a universal repelling object $K(M)$. The correspondence $M \mapsto K(M)$ turns out to be a contravariant functor from the category of manifolds to the category of commutative rings. This is what is called the *K*-functor (or *KO*-functor, if one is considering real bundles), which plays a great rôle in topological applications.

Let M be a complex manifold. Among complex vector bundles over M one can single out the class of *holomorphic bundles*, in which the space of the bundle is a complex manifold, and the diffeomorphisms $\phi: p^{-1}(U) \to U \times \mathbf{C}^n$ are holomorphic mappings. It is clear that a bundle $\mathscr{E}$ is holomorphic if and only if one can construct it with the help of holomorphic transition functions.

For holomorphic bundles, it is natural to consider *holomorphic cross sections* over open sets $U \subset M$ (that is, holomorphic mappings $s: U \to E$, having the property that $p \circ s = 1$). Such cross sections are given by a collection of holomorphic mappings $s_\alpha: U_\alpha \to \mathbf{C}^n$ that satisfy conditions (3). If $\mathscr{E}$ is a holomorphic bundle, then the symbol $\Gamma(\mathscr{E}, U)$ will denote the set of holomorphic cross sections of $\mathscr{E}$ over U. The correspondence $U \mapsto \Gamma(\mathscr{E}, U)$, just as in the real case, defines a sheaf of modules over the structure sheaf $\mathscr{O}$ of the manifold M.

As an example, we give a description of all one-dimensional holomorphic bundles over the Riemann sphere $\mathbf{CP}^1$.

Problem 5. Every one-dimensional holomorphic bundle $\mathscr{E}$ over $\mathbf{C}$ is trivial.

Hint. Triviality of $\mathscr{E}$ is equivalent to the existence of a cross section $s \in \Gamma(\mathscr{E}, \mathbf{C})$ everywhere different from zero. A proof of this last (far from trivial) assertion can be found in [26].

The Riemann sphere $\mathbf{CP}^1$ can be covered by two neighborhoods U_1, U_2 each isomorphic to $\mathbf{C}$, the canonical parameters of which are connected by the relation $z_1 \cdot z_2 = 1$. In view of problem 5, every bundle $\mathscr{E}$ over $\mathbf{CP}^1$ is given by a transition function $g(z)$, holomorphic and different from zero in $U_1 \cap U_2 = \mathbf{C} \setminus \{0\}$.

Problem 6. Every function $g(z)$, holomorphic and different from zero in $\mathbf{C} \setminus \{0\}$, has the form

$$g_1(z) \cdot z^k \cdot g_2(z^{-1}),$$

where g_1 and g_2 are functions on $\mathbf{C}$ that are holomorphic and everywhere different from zero.

In view of problems 5 and 6, every holomorphic bundle $\mathscr{E}$ over $\mathbf{CP}^1$ is equivalent to one of the bundles $\mathscr{E}_k$ given by the transition function $g(z) = z^k$, $k \in \mathbf{Z}$.

Problem 7. Prove that $\tau(\mathbf{CP}^1) \approx \mathscr{E}_2$, $\tau^*(\mathbf{CP}^1) \approx \mathscr{E}_{-2}$.

Hint. Vector fields on $\mathbf{C}$ have the form $a(z)\dfrac{d}{dz}$, and covector fields the form $-b(z)dz$, where a and b are entire functions.

It is evident that the tensor product of the bundles $\mathscr{E}_k$ and $\mathscr{E}_l$ is equivalent to the bundle $\mathscr{E}_{k+l}$. Hence the equivalence classes of one-dimensional holomorphic bundles over $\mathbf{CP}^1$ under tensor products form a cyclic group.

§ 5. Analysis on Manifolds

The fact that this group is isomorphic to $\mathbf{Z}$ (that is, all $\mathscr{E}_k$'s are inequivalent in pairs) follows from the next assertion.

Problem 8. Prove that

$$\dim_{\mathbf{C}} \Gamma(\mathscr{E}_k, \mathbf{CP}^1) = \begin{cases} k+1 & \text{for } k \geq 0, \\ 0 & \text{for } k < 0. \end{cases}$$

Now let M be a mixed manifold. Among the complex vector bundles over M, one can single out the class of *partially holomorphic bundles* and their *partially holomorphic cross sections*. This can be done in such a way that the correspondence $U \mapsto \Gamma(\mathscr{E}, U)$ is a sheaf of modules over the structure sheaf of the mixed manifold M.

In terms of transition functions and components of a cross section, this means that

$$g_{\alpha\beta} \in \mathcal{O}(U_\alpha \cap U_\beta) \underset{\mathbf{C}}{\otimes} \operatorname{Aut} \mathbf{C}^n, \quad s_\alpha \in \mathcal{O}(U_\alpha) \underset{\mathbf{C}}{\otimes} \mathbf{C}^n.$$

In appropriate coordinates $x_1, \ldots, x_k, z_1, \ldots, z_l$, the functions $g_{\alpha\beta}$ and s_α will be smooth functions of the real variables x_i and holomorphic functions of the complex variables z_j.

The definition just given leads to the usual definition of bundle in the case $l = 0$ and to the definition of a holomorphic bundle in the case $k = 0$.

Let F be a sheaf of abelian groups on the manifold M and $\mathfrak{U} = \{U_\alpha\}_{\alpha \in A}$ a covering of M by open sets. We define an n-dimensional cochain of the covering $\mathfrak{U}$ with coefficients in F as a mapping c that carries the collection of indices $(\alpha_0, \alpha_1, \ldots, \alpha_n)$ into the element $c(\alpha_0, \ldots, \alpha_n) \in F(U_{\alpha_0} \cap \ldots \cap U_{\alpha_n})$ and is skew-symmetric (that is, changes sign under permutation of any two arguments). The set of n-dimensional cochains is denoted by $C^n(\mathfrak{U}, F)$.

The coboundary operator $d: C^n(\mathfrak{U}, F) \to C^{n+1}(\mathfrak{U}, F)$ is defined by the formula

$$dc(\alpha_0, \ldots, \alpha_{n+1}) = \sum_{i=0}^{n+1} (-1)^i F(\phi_i) c(\alpha_0, \ldots, \hat{\alpha}_i, \ldots, \alpha_{n+1}).$$

Here ϕ_i is the embedding of $U_{\alpha_0} \cap \ldots \cap U_{\alpha_{n+1}}$ into $U_{\alpha_0} \cap \ldots U_{\alpha_i} \ldots \cap U_{\alpha_{n+1}}$.

If F is a *constant sheaf* (that is, $F(U) = \Pi$ for $U \neq \emptyset$ and $F(\phi) \equiv 1$ for every embedding $U \subset V$), then the definition just given coincides with the definition of cochains and coboundary operator in 1.3. The definitions given there of coboundaries, cocycles, and cohomology can be carried over with no change to the case of general sheaves.

The theorem of Leray is also true for arbitrary sheaves and is one of the basic instruments for computing cohomologies.

There is another method of computation, based on the consideration of the groups $H^k(M, F)$ as functors from the category of sheaves on M to the category of groups. We will illustrate this method by a special example—the sheaf of partially holomorphic cross sections of a bundle over a mixed manifold.

Let $\mathscr{E}$ be a partially holomorphic bundle over a mixed manifold of type (k, l). We recall that locally the manifold M is constructed as a direct product of a neighborhood of zero in $\mathbf{R}^k$ and a neighborhood of zero in $\mathbf{C}^l$. We denote by $h(M)$

the complex l-dimensional bundle over M, the fiber of which is the tangent space to the "complex part" of the manifold M. We then set

$$\lambda^r(\mathscr{E}) = \mathscr{E} \otimes \wedge^r \bar{h}^*(M),$$

where the bar denotes complex conjugation, * denotes transition to the dual space, and $\wedge^r$ is the r-th exterior power. Cross sections of the bundle $\lambda^r(\mathscr{E})$ are called *differential forms of type $(0,r)$ on M with values in the bundle $\mathscr{E}$*. In local coordinates they have the form

$$s = \sum_{i_1,\ldots,i_r} s_{i_1\ldots i_r} d\bar{z}_{i_1} \wedge \ldots \wedge d\bar{z}_{i_r},$$

where $s_{i_1\ldots i_r}$ are smooth functions of the variables $x_1,\ldots,x_k$, $u_1, v_1,\ldots,u_l$, v_l ($z_j = u_j + iv_j$).

We can define the coboundary operator

$$\bar{\partial}_r : \Gamma(\lambda^r(\mathscr{E}), M) \to \Gamma(\lambda^{r+1}(\mathscr{E}), M),$$

which in local coordinates acts according to the formula

$$\bar{\partial}_r s = \sum_{i_1,\ldots,i_r} \bar{\partial} s_{i_1\ldots i_r} \wedge d\bar{z}_{i_1} \wedge \ldots \wedge d\bar{z}_{i_r}.$$

Here we define $\bar{\partial} s = \sum_{j=1}^{l} \dfrac{\partial s}{\partial \bar{z}_j} d\bar{z}_j$, and $\dfrac{\partial}{\partial \bar{z}_j} = \dfrac{\partial}{\partial u_j} + i \dfrac{\partial}{\partial v_j}$.

The generalized theorem of Dolbeault. *We have the isomorphism*

$$H^r(M, \Gamma_{\mathscr{E}}) = \ker \bar{\partial}_r / \mathrm{im}\, \bar{\partial}_{r-1},$$

where $\Gamma_{\mathscr{E}}$ is the sheaf of partially holomorphic cross sections of the bundle $\mathscr{E}$.

Problem 9. Prove that the space $H^0(M, \Gamma)$ coincides with the space of partially holomorphic cross sections of the bundle $\mathscr{E}$.

Hint. The condition $\partial f / \partial \bar{z} = 0$ is equivalent to the Cauchy-Riemann equations, which guarantee that f is holomorphic in z.

Problem 10. Compute $H^r(\mathbf{CP}^1, \Gamma_{\mathscr{E}_k})$ (see problems 6–8).

Hint. Since $\dim_{\mathbf{C}} \mathbf{CP}^1 = 1$ and $H^0(\mathbf{CP}^1, \Gamma_{\mathscr{E}_k}) = \Gamma(\mathscr{E}_k, \mathbf{CP}^1)$ (see problem 9), it suffices to consider the case $r=1$. Setting $z = \rho e^{i\phi}$ and decomposing in a Fourier series in ϕ, we reduce the problem to the following.

Let $b(\rho)$ be a smooth function on $(0, \infty)$ satisfying the two boundary conditions
1) $b(\rho) = O(\rho^{|m|})$ for $\rho \to 0$,
2) $b(\rho) = O(\rho^{k-2-|k-2-m|})$ for $\rho \to \infty$.

Find the set of all such functions $b(\rho)$ for which there exists a solution $a(\rho)$ of the equation

$$a' - \frac{m-1}{\rho} a = b$$

which satisfies the following boundary conditions:
3) $a(\rho) = O(\rho^{|m-1|})$ for $\rho \to 0$,
4) $a(\rho) = O(\rho^{k-|k-m+1|})$ for $\rho \to \infty$.

Answer.

$$\dim_{\mathbf{C}} H^1(\mathbf{CP}^1, \Gamma_{\mathscr{E}_k}) = \begin{cases} 0 & \text{for } k \geq 0, \\ -(k+1) & \text{for } k < 0. \end{cases}$$

§ 6. Lie Groups and Lie Algebras

6.1. Lie Groups

A set G is called a Lie group if it is a topological group and a smooth manifold for which the mapping $G \times G \xrightarrow{\phi} G$, given by $\phi(x, y) = xy^{-1}$ is smooth.

We shall not define more precisely the notion of smoothness entering into this definition. This is because the theorem of Gleason-Montgomery-Zippin (giving an affirmative solution to Hilbert's fifth problem) shows that on every Lie group of class C^0 one can impose the structure of a manifold of class C^ω which is consistent with the group structure.

Remark. Smoothness of the mapping ϕ in the definition is equivalent to smoothness of the two following mappings:

$$G \times G \xrightarrow{\psi} G: \psi(x, y) = xy$$

and

$$G \xrightarrow{\varepsilon} G: \varepsilon(x) = x^{-1}.$$

Examples of Lie groups

1. The real line **R** with the operation of addition.

2. The circumference **T** with the operation of multiplication (we suppose that **T** is the unit circle in the complex plane **C**).

3. The set $GL(n, \mathbf{R})$ of all nonsingular matrices of order n (regarded as a region in $\mathbf{R}^{n^2}$) with the operation of matrix multiplication. In the special case $n = 1$, we obtain $\mathbf{R}^* = \mathbf{R} \setminus \{0\}$ with the operation of multiplication.

4. If G_1 and G_2 are Lie groups, then their product $G_1 \times G_2$ is also a Lie group. In particular, $\mathbf{R}^n$ and $\mathbf{T}^n$ are Lie groups.

5. The group of affine transformation of the line, equipped with the natural topology, is a Lie group.

In fact, every affine transformation g has the form $g(x) = ax + b$, where $a \in \mathbf{R}^*$ and $b \in \mathbf{R}$.

The coordinates (a,b) allow us to identify the manifold of the group with the region in $\mathbf{R}^2$ obtained by removing the line $a=0$. In these coordinates the mapping ϕ appearing in the definition of a Lie group is given by the formula

$$\phi(a_1,b_1;a_2,b_2)=(a_1 a_2^{-1}; b_1 - b_2 a_1 a_2^{-1}).$$

6. The group $\mathbf{H}^*$ of nonzero quaternions (see 3.2) with the operation of multiplication.

7. The manifolds in matrix spaces defined in 5.1 are Lie groups with respect to the operation of matrix multiplication.

In fact, it is easy to verify that equations (1), (2), (3) from 5.1 define groups, that is, if matrices x and y satisfy these equations, then the matrices x^{-1} and xy also satisfy them. We shall verify that the mapping $(x,y) \mapsto z = xy^{-1}$ is smooth.

By the theorem of analysis cited in 5.1, we can take as local coordinates in neighborhoods of the points x, y, and $z = xy^{-1}$ a certain collection of matrix elements (different collections for each of the neighborhoods). The remaining matrix elements will be smooth functions of those we have picked out. The coordinates of the point z are evidently smoothly expressible in terms of the matrix elements of the points x and y^{-1}, which in turn are smoothly expressible in terms of the coordinates of x and y. (We use the fact that in the set of non-singular matrices, the matrix elements of y^{-1} depend smoothly upon the matrix elements of y.)

Problem 1. Prove that one can introduce the structure of a Lie group on the three-dimensional sphere S^3.

Hint. Use the realization of S^3 as the set of quaternions of norm 1.

Problem 2. Prove that there is no Lie group structure on the two-dimensional sphere S^2.

Hint. Use the following topological lemma. *Every continuous mapping of the two-dimensional sphere into itself which is homotopic to the identity mapping has a fixed point.*

Among all of the examples we have constructed, there are two one-dimensional Lie groups: $\mathbf{R}$ and $\mathbf{T}$.

It turns out that there are no other one-dimensional Lie groups.

Problem 3. Every one-dimensional Lie group G is isomorphic to $\mathbf{R}$ or $\mathbf{T}$.

Hint. Suppose first that the group G is diffeomorphic to $\mathbf{R}$ and that $x \in G$ is an element different from the identity. Prove that for every rational number $r = m/n$, there exists a unique element $y \in G$ with the property that $y^n = x^m$. (Use the fact that the product of two elements of G depends monotonely upon each factor.) Denote this element by x^r. Prove that the mapping $\mathbf{Q} \to G : r \mapsto x^r$ can be extended to the desired isomorphism of $\mathbf{R} \to G$.

In the case where G is diffeomorphic to $\mathbf{T}$, prove that for every integer $m>0$, there exists an embedding $\mathbf{Z}_m \to G$ and that these embeddings can be chosen so that the mapping $\bigcup_m \mathbf{Z}_m = \mathbf{Q}/\mathbf{Z} \to G$ obtained from them can be extended to the desired isomorphism $\mathbf{T} = \mathbf{R}/\mathbf{Z} \to G$.

§ 6. Lie Groups and Lie Algebras

This result has very wide applications, since every Lie group contains one-dimensional subgroups. The assertion of problem 3 enables us to choose a *canonical parameter* on these subgroups, which is connected with the law of multiplication in the group in a very natural way.

Problem 4. Find the general form of a homomorphism of G_1 into G_2, where each of the groups G_i may be either **R** or **T**.
  Answer: a) $\mathbf{R} \mapsto \mathbf{R}$: $x \to ax$, $a \in \mathbf{R}$;
  b) $\mathbf{R} \mapsto \mathbf{T}$: $x \to e^{i\lambda x}$, $\lambda \in \mathbf{R}$;
  c) $\mathbf{T} \mapsto \mathbf{R}$: $z \to 0$;
  d) $\mathbf{T} \mapsto \mathbf{T}$: $z \to z^n$, $n \in \mathbf{Z}$.

Let G be a Lie group. The connected component of the identity in G is defined as the set G^o of all elements of G that can be connected with the identity by a continuous curve.

Problem 5. Prove that G^o is a connected, open, and closed subgroup of G and that G^o is uniquely defined by these properties.

If G is a connected but not simply connected Lie group, then one can construct what is called the *simply connected covering group* $\tilde{G}$ of G. By definition, $\tilde{G}$ consists of the equivalence classes of continuous mappings ϕ of the interval $[0,1]$ into G that satisfy the condition $\phi(0) = e$.

Mappings ϕ_1 and ϕ_2 are taken to be equivalent if there exists a family of mappings $\phi_t : [0,1] \to G$ such that $\phi_t(0) \equiv e$, $\phi_t(1) \equiv \phi_0(1)$, and such that the mapping

$$[0,1] \times [0,1] \to G : (t,x) \to \phi_t(x)$$

is continuous.

We denote the class of the mapping ϕ by $[\phi]$. We introduce the topology of uniform convergence in the set of mappings ϕ, and in $\tilde{G}$ the corresponding factor topology.

We define a projection $p : \tilde{G} \to G$, setting $p([\phi]) = \phi(1)$.

Problem 6. Prove that the mapping p is a local homeomorphism.

We define a law of multiplication in $\tilde{G}$, setting $[\phi_1] \cdot [\phi_2] = [\phi]$, where

$$\phi(x) = \begin{cases} \phi_1(2x), & \text{if } x \leq \tfrac{1}{2}, \\ \phi_1(1)\phi_2(2x-1), & \text{if } x \geq \tfrac{1}{2}. \end{cases}$$

Problem 7. Verify that this law gives the structure of a Lie group to $\tilde{G}$, and that the projection p is a homomorphism of $\tilde{G}$ onto G.
  Hint. Use the result of the foregoing problem.

Problem 8. Prove that the group $\tilde{G}$ is connected and simply connected.

We shall see *infra* that the group $\tilde{G}$ is uniquely defined by the properties listed in problems 6, 7 and 8. (See the monodromy theorem in 6.3.) We suggest that the reader formulate the definition of universal covering group as a universal object in an appropriate category.

To conclude this section, we state without proof two more important properties of Lie groups.

Theorem 1. *Every closed subgroup of a Lie group is itself a Lie group.*

Theorem 2. *If G is a Lie group and H is a closed subgroup of G, then G/H is a smooth manifold on which the group G acts by smooth transformations. (In particular, if H is a normal subgroup, then G/H is a Lie group.)*

Proofs of these theorems, and other facts about Lie groups, can be found in the books [44], [29], and [11].

The Gleason-Montgomery-Zippin theory appears in the Russian language in an article by B. M. Gluškov, *The structure of locally-bicompact groups and Hilbert's fifth problem*, in Uspehi matem. Nauk, 1957, vol. XII, part 2, pages 3–41.

6.2. Lie Algebras

The concept of a Lie algebra (formerly the name *infinitesimal Lie group* was used) arose in the study of Lie groups, but later became the object of an independent theory. We shall here give a number of facts from this theory.

Let L be a linear space over a field K. Suppose that L admits an operation, denoted by $[x, y]$, and called *commutation*, which satisfies the following conditions:
1) bilinearity;
2) antisymmetry: $[x, x] = 0$ for all $x \in L$;
3) the Jacobi identity:

$$[[x,y],z] + [[y,z],x] + [[z,x],y] = 0$$

for all $x, y, z \in L$.

Then L is called a *Lie algebra*.

Problem 1. Prove that the equality $[x, y] = -[y, x]$ for all $x, y \in L$ follows from the first two properties.

In order to define the structure of a Lie algebra on the space L, it suffices to know the commutators of each pair of basis vectors $x_1, \ldots, x_n$, that is, the coefficients c_{ij}^k in the expressions

$$[x_i, x_j] = \sum_{k=1}^{n} c_{ij}^k x_k.$$

These coefficients are called the *structure constants* of the algebra L. In accordance with problem 1, the identities $c_{ij}^k + c_{ji}^k = 0$ hold, so that it suffices to know the $\dfrac{n^2(n-1)}{2}$ structure constants c_{ij}^k for which we have $i < j$.

§ 6. Lie Groups and Lie Algebras

Problem 2. Prove that if the elements x, y, z are linearly dependent, then the Jacobi identity follows from the first two conditions.

From problem 2, we infer the following. To check that a collection of numbers $\{c_{ij}^k\}_{i<j}$ is the set of structure constants of a certain Lie algebra, if suffices to verify $n(n-1)(n-2)/6$ equalities—the Jacobi identities for triples x_i, x_j, x_k for which $i<j<k$. We note that the Jacobi identity is a vector equality, which is equivalent to n ordinary equalities, so that the manifold of all collections of structure constants is given by $n^2(n-1)(n-2)/6$ quadratic equations in $n^2(n-1)/2$ independent variables.

Example. The space $\mathbf{R}^3$ with vector multiplication as the commutator is a Lie algebra over $\mathbf{R}$. Indeed, properties 1) and 2) are evident, and it suffices to prove the Jacobi identity for the case in which x, y, z are an orthonormal basis with positive orientation. In this case we have $[x, y] = z$, $[y, z] = x$, $[z, x] = y$ and the Jacobi identity holds.

An even simpler example is any linear space with zero commutator. Such a Lie algebra is called *commutative* or *abelian*.

A great stock of examples is constructed in the following way.

Problem 3. Let A be an associative algebra over K. For $x, y \in A$, we define

$$[x, y] = xy - yx.$$

Prove that the space A with this commutation operation is a Lie algebra.

Problem 4. By a differentiation of a (not necessarily associative) algebra A over K, we mean a linear mapping $D: A \to A$, for which the identity

$$D(xy) = D(x)y + xD(y)$$

holds for all $x, y \in A$. Prove that the space of all differentiations of the algebra A is a Lie algebra over K with the operation

$$[D_1, D_2] = D_1 D_2 - D_2 D_1.$$

Corollary. *The set* Vect M *of smooth vector fields on a manifold* M *is a Lie algebra over* $\mathbf{R}$ *with respect to the operation of commutation.*

In this case, we take $C^\infty(M)$ to be the algebra A.

Lie algebras over a field K form a category. The morphisms of this category are *Lie algebra homomorphisms*, that is, linear mappings $\phi: L_1 \to L_2$ that have the property

$$\phi([x, y]) = [\phi(x), \phi(y)]$$

for all $x, y \in L$.

A homomorphism of L_1 into L_2 is also called a *representation* of L_1 in L_2. Suppose in particular that L_2 is a Lie algebra of linear operators in a certain linear space H (with respect to the operation $[A, B] = AB - BA$). Then ϕ is said to be a *linear representation* of L_1 in the space H. The word "linear" is often omitted in this expression.

We have

Ado's theorem. *Every finite-dimensional Lie algebra admits a faithful linear representation.*

This important theorem permits us to reduce the proofs of theorems about Lie algebras to the case of matrix Lie algebras.

The *center* of a Lie algebras L is the set of all elements $x \in L$ such that $[x,y]=0$ for all $y \in L$.

Problem 5. Let L be a Lie algebra. For each $x \in L$, we define the operator $\operatorname{ad} x$ in L by the formula

$$\operatorname{ad} x(y) = [x,y].$$

Prove that $\operatorname{ad} x$ is a differentiation of the Lie algebra L, that the mapping $x \mapsto \operatorname{ad} x$ is a representation of L in the space L (it is called the *regular representation*) and that the kernel of this representation is the center of L.

Example. Let $L = \mathbf{R}^3$ be the Lie algebra considered above. In this case, the regular representation has the form

$$\operatorname{ad}(ax+by+cz) = \begin{pmatrix} 0 & c & -b \\ -c & 0 & a \\ b & -a & 0 \end{pmatrix}.$$

Thus the Lie algebra L is isomorphic to the Lie algebra of all skew-symmetric matrices of the third order.

We now list the Lie algebras of small dimensions over $\mathbf{R}$.

I. $\dim L = 1$. There is only one one-dimensional Lie algebra, the line $\mathbf{R}$ with null commutator.

II. $\dim L = 2$. Let x and y be a basis in L. If $[x,y]=0$, then the commutator of any two elements is equal to 0. If however $[x,y]=z \neq 0$, then the commutator of any two vectors is proportional to z. There exists in particular a vector t such that $[t,z]=z$.

Thus there exist two nonisomorphic Lie algebras of dimension 2.

III. $\dim L = 3$. We consider the space $L_1 \subset L$ generated by all commutators. If $\dim L_1 = 0$, then all commutators are equal to 0.

If $\dim L_1 = 1$, then $[x,y] = B(x,y)z$, where z is a fixed vector and $B(x,y)$ is a skew-symmetric bilinear form in L. Two cases are possible:

1) $B(x,z)=0$ for all $x \in L$; then one can choose a basis x,y,z in L with the commutation relations $[x,y]=z$, $[x,z]=[y,z]=0$.

2) There exists a vector $x \in L$ such that $B(x,z)=1$; in this case there exists a basis x,y,z with the commutation relations $[x,y]=[y,z]=0, [x,z]=z$.

Suppose that $\dim L_1 = 2$. We note that the subspace L_1 is itself a Lie algebra, since L_1 contains all commutators. But we already know that there are exactly two two-dimensional Lie algebras, with the commutation relations $[x,y]=0$

§ 6. Lie Groups and Lie Algebras

and $[x,y] = y$, respectively. Let z be a vector which with x and y forms a basis in L. The operator $\operatorname{ad} z$ is a differentiation of the algebra L_1.

In the case $[x,y] = y$, it follows from this that the operator $\operatorname{ad} z$ has the form $x \mapsto ay$, $y \mapsto by$, which contradicts the condition $\dim L_1 = 2$.

In the case $[x,y] = 0$, there are no restrictions of any kind: $\operatorname{ad} z$ can be an arbitrary matrix A of the second order. We note however that the condition $\dim L_1 = 2$ implies that A is nonsingular.

Problem 6. Prove that matrices A and B define isomorphic Lie algebras if and only if there exist a nonsingular matrix C and a nonzero number c such that

$$cA = CBC^{-1}. \tag{1}$$

Thus, in the case $\dim L_1 = 2$ we have obtained an entire family of Lie algebras, parametrized by nonsingular matrices of the second order, up to the equivalence defined by the formula (1).

Finally we consider the case $\dim L_1 = 3$, that is, $L_1 = L$. Here we quote without proof the final answer (which follows from the theory of semisimple Lie algebras, to be taken up *infra*). There are two nonisomorphic three-dimensional Lie algebras for which $L_1 = L$. Their commutation relations have the forms

$$[x,y] = z, \quad [y,z] = x, \quad [z,x] = y$$

and

$$[x,y] = 2y, \quad [x,z] = -2z, \quad [y,z] = x.$$

The first algebra is the Lie algebra of skew-symmetric matrices of order 3, already discussed. The second algebra is easily shown to be isomorphic to the Lie algebra of matrices of the second order with zero trace:

$$x \leftrightarrow \begin{pmatrix} 1 & 0 \\ 0 & -1 \end{pmatrix}, \quad y \leftrightarrow \begin{pmatrix} 0 & 1 \\ 0 & 0 \end{pmatrix}, \quad z \leftrightarrow \begin{pmatrix} 0 & 0 \\ 1 & 0 \end{pmatrix}.$$

In principle it would be possible to classify Lie algebras of dimension 4 by the same crude arguments, but the needed computations and the final answer are too cumbersome to be given here.

Instead of this, we shall introduce some general concepts, which facilitate the classification of Lie algebras of arbitrary dimension.

A subspace M in a Lie algebra L is called a *subalgebra* (or an *ideal*) if $[M,M] \subset M$ (or $[L,M] \subset M$). (Here and in the sequel, the expression $[L,M]$ denotes the linear hull of all vectors of the form $[x,y]$ as x runs through L and y runs through M.) If M is an ideal in L, then the factor space L/M is provided in a natural way with the structure of a Lie algebra, which is called the *factor algebra* of the algebra L by the ideal M.

In every Lie algebra L we can define two sequences of subspaces:

$$L_1 = [L,L], \quad L_2 = [L,L_1], \ldots, L_{n+1} = [L,L_n], \ldots,$$
$$L^1 = [L,L], \quad L^2 = [L^1, L^1], \ldots, L^{n+1} = [L^n, L^n], \ldots$$

It is clear that the following inclusions hold:

$$L_1 \supset L_2 \supset L_3 \supset \ldots \supset L_n \supset \ldots$$
$$\parallel \quad \cup \quad \cup \qquad \quad \cup$$
$$L^1 \supset L^2 \supset L^3 \supset \ldots \supset L^n \supset \ldots$$

Problem 7. Prove that all of the subspaces L^n and L_n are ideals in L and that the factor algebras L_n/L_{n+1} and L^n/L^{n+1} are commutative.

If $\dim L < \infty$, then the sequences $\{L_n\}$ and $\{L^n\}$ are stabilized: beginning with a certain n, we have $L_n = L_{n+1} = \ldots = L_\infty$ and $L^n = L^{n+1} = \ldots = L^\infty$.

A Lie algebra is called *solvable* (or *nilpotent*) if $L^\infty = \{0\}$ (or $L_\infty = \{0\}$).

The smallest number n for which $L^n = \{0\}$ (or $L_n = \{0\}$), is called the *rank of solvability* (or *rank of nilpotency*) of the algebra L.

As an example of a solvable (or nilpotent) Lie algebra, consider the algebra $T(n, K)$ (or $T_o(n, K)$) of matrices of the n-th order with coefficients in K satisfying the condition $a_{ij} = 0$ for $i > j$ (or for $i \geq j$).

One of the most important properties of solvable Lie algebras is

Lie's theorem. *Let ρ be a finite-dimensional representation of a solvable Lie algebra L in a space V over an algebraically closed field K. Then the representation ρ is equivalent to a triangular representation, that is, a representation such that all of the matrices $\rho(x)$, $x \in L$, belong to $T(n, K)$.*

A proof follows from the following problems.

Problem 8. If $A_1, \ldots, A_n$ are pairwise commuting operators in a space V over an algebraically closed field, then there exists a vector $\xi \in V$ which is an eigenvector for all of the operators A_k.

Hint. For a single operator A_1 there is such a vector, since the equation $\det(A - \lambda \cdot 1) = 0$ has at least one root λ_1.

Prove that the space V_1 of all solutions of the equation

$$A_1 \xi = \lambda_1 \xi$$

is invariant under all of the operators $A_2, \ldots, A_n$, and use induction.

Problem 9. Under the hypotheses of Lie's theorem, prove that all of the operators $\rho(x)$, $x \in L$ have a common eigenvector.

Hint. Use the method of solution of problem 8 and induction on the rank of solvability of L.

Lie's theorem is proved from the result of problem 9 by induction on the dimension of V.

Corollary. *If L is a solvable Lie algebra, then the subalgebra $[L, L]$ is nilpotent.*

The name "solvable algebra" comes from solvable Lie groups, which are so named from their connection with differential equations that are solvable by quadratures. (In the same way, finite solvable group are connected with algebraic equations solvable in radicals.) In this connection, see the book of I. Kaplansky [35].

§ 6. Lie Groups and Lie Algebras

The name "nilpotent Lie algebra" is justified by the following circumstance.

Engel's theorem. *A Lie algebra is nilpotent if and only if the operator* $\operatorname{ad} x$ *is nilpotent for all* $x \in L$ *(that is,* $(\operatorname{ad} x)_n = 0$ *for a certain* n*).*

The classes of nilpotent and solvable Lie algebras can be defined in a different way. Consider the increasing chains of ideals defined by the following rules:

$$M_0 = M^0 = \{0\},$$

M_{n+1} is the maximal of the ideals J containing M_n and having the property that the factor algebra J/M_n is commutative;
M^{n+1} is the ideal containing M^n such that the factor algebra M^{n+1}/M^n is the center of the Lie algebra L/M^n.

The following inclusions are obvious:

$$M^1 \subset M^2 \subset \ldots \subset M^n \subset ,$$
$$\cap \quad \cap \quad \quad \cap$$
$$M_1 \subset M_2 \subset \ldots \subset M_n \subset .$$

If $\dim L < \infty$, then we have $M^n = M^{n+1} = \ldots = M^\infty$ and $M_n = M_{n+1} = \ldots = M_\infty$, beginning with a certain n.

Problem 10. Prove that the algebra L is solvable (nilpotent) if and only if $M_\infty = L$ ($M^\infty = L$).

A Lie algebra is called *simple* if it contains no nontrivial ideals and is called *semisimple* if it contains no nontrivial commutative ideals.

Problem 11. Prove that every finite-dimensional Lie algebra L contains a solvable ideal R such that the factor algebra $S = L/R$ is semisimple.

Hint. Consider a maximal solvable ideal (that is, a solvable ideal contained in no larger solvable ideal).

As a matter of fact, a stronger assertion is true.

Theorem (E. Cartan-Levy-Mal'cev). *All solvable ideals of the algebra L are contained in a single solvable ideal R (called the* radical *of the algebra L). There exists a semisimple subalgebra $S \subset L$, such that $L = S \oplus R$ (direct sum of linear spaces). Any two subalgebras S with this property are carried the one into the other by an automorphism of the algebra L that preserves R.*

Thus the study of general Lie algebras can be reduced in large measure to the study of solvable and semisimple algebras.

Although solvable Lie algebras would seem to have the simplest structure, they have not been classified up to this time.

It is known that this problem contains as a special case the problem of the simultaneous reduction to canonical form of k matrices, a problem that is unsolved even for $k = 2$.

A useful tool for the study of Lie algebras is the *bilinear Killing form*

$$B(x, y) = \operatorname{tr}(\operatorname{ad} x \cdot \operatorname{ad} y). \tag{2}$$

Problem 12. Prove that the Killing form is invariant under all automorphisms of a Lie algebra.

Problem 13. Prove that the Killing form is identically zero if and only if the Lie algebra is nilpotent.
Hint. Use Engel's theorem.

Problem 14 (Cartan's criterion). The Killing form is nonsingular if and only if the Lie algebra is semisimple.
Hint. If J is an ideal in a Lie algebra L, then the Killing form of the Lie algebra J is the restriction to J of the Killing form of the algebra L. One of the assertions of the problem follows from this, and a proof of the other reduces to the case of a simple algebra. Now use the fact that the kernel of the Killing form is an ideal in a Lie algebra.

A complete classification exists for semisimple Lie algebras.

It is known that every semisimple Lie algebra is a direct sum of simple ideals.

Over the field **C**, there exist four infinite series of simple Lie algebras and five "exceptional" simple Lie algebras. Over the field **R**, there exist 12 infinite series of simple algebras and 23 "exceptional" simple algebras.

The algebras of the infinite series are called the *classical simple algebras*. It is a remarkable fact that all of these algebras can be realized as algebras of matrices with entries in **R**, **C**, or **H**, satisfying simple algebraic conditions.

Let 1_n denote the unit matrix of order n. We set

$$J_{2n} = \begin{pmatrix} 0 & 1_n \\ -1_n & 0 \end{pmatrix}, \quad I_{p,q} = \begin{pmatrix} 1_p & 0 \\ 0 & -1_q \end{pmatrix}.$$

We now list the classical simple algebras:

1) A_n is the algebra of all complex matrices of the $(n+1)$st order having trace zero $(n = 1, 2, \ldots)$;

2) B_n is the algebra of skew-symmetric complex matrices of order $2n+1$ $(n=2,3,\ldots)$ (B_1 is isomorphic to A_1);

3) C_n is the algebra of *canonical* complex matrices of order $2n$, that is, matrices satisfying the equation $X'J_{2n} + J_{2n}X = 0$) $(n = 3, 4, \ldots)$ $(C_1 \approx A_1 \approx B_1, C_2 \approx B_2)$.

4) D_n is the algebra of skew-symmetric complex matrices of order $2n$ $(n=4,5,\ldots)$ (D_1 is commutative, and $D_2 \approx A_1 \oplus A_1, D_3 \approx A_3$).

All of these algebras are also simple algebras over **R**. The remaining real classical algebras remain simple upon complexification (that is, tensor multiplication by **C** over **R**). They are called *absolutely simple*. An absolutely simple real algebra L is called the *real form* of a simple complex algebra L_C if $L_C = L \underset{R}{\otimes} C$.

The real forms of the algebras A_n are the following:

a) the algebra of real matrices of order $n+1$ with trace 0;

b) the algebra of complex matrices of order $n+1$ that satisfy the conditions $\operatorname{tr} X = 0$ and $X^* I_{p,q} + I_{p,q} X = 0$, where $p + q = n+1, p \geq q$.

c) (for odd n) the algebra of quaternion matrices of order $(n+1)/2$ that satisfy the condition $\operatorname{tr}(X + X^*) = 0$.

§ 6. Lie Groups and Lie Algebras

The real forms of the algebra B_n are the algebras consisting of all real matrices of order $2n+1$ that satisfy the condition $X'I_{p,q}+I_{p,q}X=0$, where $p+q=2n+1$, $p>q$.

The real forms of the algebra C_n:

a) the algebra of real canonical matrices of order $2n$;

b) the algebra of quaternion matrices of order n that satisfy the condition $X^*I_{p,q}+I_{p,q}X=0$, where $p+q=n$, $p\geq q$.

The real forms of the algebra D_n:

a) the algebra of real matrices of order $2n$ that satisfy the condition $X'I_{p,q}+I_{p,q}X=0$, where $p+q=2n$, $p\geq q$.

b) (for even n) the algebra of quaternion matrices of order n that satisfy the condition $X^*J_{2n}+J_{2n}X=0$.

Many facts of the theory of semisimple Lie algebras were first established for the classical algebras. Only later was the machinery developed that is needed for the study of arbitrary semisimple Lie algebras.

The structure of semisimple Lie algebras turns out to be connected with a very beautiful geometric object—the system of roots in euclidean space $\mathbf{R}^n$. We shall give here some definitions and important facts relating to this connection.

A finite set Σ of vectors in $\mathbf{R}^n$ is called a *root system* if

1) $\dfrac{2(x,y)}{(x,x)} \in \mathbf{Z}$ for all $x, y \in \Sigma$,

2) $y - \dfrac{2(x,y)}{(x,x)} x \in \Sigma$ for all $x, y \in \Sigma$.

Here the symbol (x,y) denotes the usual scalar product in $\mathbf{R}^n$: $(x,y) = \sum_{k=1}^{n} x_k y_k$. Geometrically, these conditions mean the following.

1) The angles between vectors in Σ can be only $0, \pi/6, \pi/4, \pi/3, \pi/2, 2\pi/3, 3\pi/4, 5\pi/6, \pi$, and the ratio of the squares of the lengths of two nonparallel and nonperpendicular vectors is equal to 1, 2, or 3, depending upon the angle between them.

2) The system of vectors Σ is invariant under the mappings $\sigma(x): y \mapsto y - \dfrac{2(x,y)}{(x,x)} x$, which are reflections in hyperplanes orthogonal to the vectors $x \in \Sigma$.

The finite group W generated by the reflections $\sigma(x)$, $x \in \Sigma$, is called the *Weyl group* of the system Σ. We say that a vector $x \in \mathbf{R}^n$ is *regular* if it is orthogonal to no vector in Σ. The set of regular vectors falls into $|W|$ connected components, called *Weyl chambers*. Let us fix a certain Weyl chamber C. The set C is a convex cone bounded by pieces of hyperplanes $(x, x_i)=0$, $x_i \in \Pi$, where Π is a certain subsystem of Σ, called a *system of simple roots*. Every root $x \in \Sigma$ is an integral linear combination of simple roots x_i with coefficients of a single sign. Thus we have $\Sigma = \Sigma_+ \cup \Sigma_-$, where $\Sigma_+ (\Sigma_-)$ is the set of positive (negative) roots, having positive (negative) coefficients.

We call a system Σ *nonsingular* if Σ generates the entire space $\mathbf{R}^n$, and *reduced* if $x \in \Sigma$ implies that $2x \notin \Sigma$.

One can show that a nonsingular reduced root system Σ is completely defined by its subsystem of simple roots Π and that a set $\Pi \subset \mathbf{R}^n$ is a system of simple roots if and only if Π possesses the following properties:

1) $\dfrac{2(x,y)}{(x,x)}$ is a nonpositive integer for arbitrary $x, y \in \Pi$;

2) Π is a basis in $\mathbf{R}^n$.

It is convenient to present systems Π in the form of *Dynkin diagrams* by the following rule. The vertices of the diagram correspond to the elements of Π. Vertices a_i and a_j are connected by a single (or a double, or a triple) line if the corresponding vectors form an angle of 120° (or 135° or 150°). The double and triple lines are made into arrows that point to the smaller root.

With every semisimple complex Lie algebra L, we can associate a certain nonsingular reduced root system.

In this correspondence, the root systems corresponding to simple Lie algebras are indecomposable (that is, are not the union of two orthogonal subsystems).

Problem 15. All two-dimensional nonsingular reduced root systems are listed in the accompanying Fig. 2.

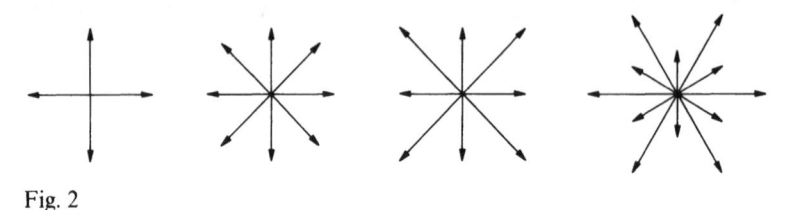

Fig. 2

Hint. Use the fact that all two-dimensional Dynkin schemes are listed in Fig. 3.

Fig. 3

These systems correspond to the classical algebras $A_1 \oplus A_1$, A_2, B_2 and the exceptional algebra G_2.

The correspondence referred to above is constructed in the following way.

Let L be a Lie algebra and $x \in L$. We consider a polynomial in λ, defined by

$$\det(\operatorname{ad} x - \lambda \cdot 1) = \sum_{l=0}^{\dim L} p_l(x) \lambda^l.$$

The coefficients $p_l(x)$ of this polynomial are evidently polynomial functions on L.

The *rank of the algebra* L (denoted by $\operatorname{rg} L$) is the smallest of the numbers l for which the coefficient $p_l(x)$ is not identically equal to zero. The elements $x \in L$ for which $p_{\operatorname{rg} L}(x) \neq 0$ are called *regular*.

Problem 16. Prove that the rank of the algebras A_n, B_n, C_n, D_n is equal to n.

Hint. Use the fact that the regular elements form an open dense subset of L, invariant under all automorphisms of L.

§ 6. Lie Groups and Lie Algebras

A *Cartan subalgebra* of the algebra L is a maximal solvable subalgebra containing a regular element. The dimension of a Cartan subalgebra is equal to the rank of L.

In semisimple Lie algebras, Cartan subalgebras are commutative and coincide with the centralizers of their regular elements.

If H is a Cartan subalgebra in a complex semisimple Lie algebra L, then all of the operators ad x, $x \in H$, can simultaneously be reduced to diagonal form. Thus, in an appropriate basis x_α, we have the relations

$$[x, x_\alpha] = \lambda_\alpha(x) x_\alpha, \quad x \in H.$$

The eigenvalues $\lambda_\alpha(x)$ are obviously linear functionals on H. Since the Killing form on L is nonsingular, one can find elements $\mu_\alpha \in L$ such that

$$\lambda_\alpha(x) = B(\mu_\alpha, x).$$

Let R be the real subspace of L generated by all of the μ_α.

One can show that R is an l-dimensional subspace of H, and the restriction of the form B to R is positive-definite. Let Σ be the set of nonzero vectors μ_α (the number of these is $\dim L - \mathrm{rg}\, L$). Then Σ is a nonsingular reduced root system in R.

The structure of the Lie algebra L is completely defined by the following relations:

$$[x_\alpha, x_\beta] = \begin{cases} N_{\alpha\beta} x_{\alpha+\beta}, & \text{if } \alpha+\beta \in \Sigma, \\ B(x_\alpha, x_{-\alpha}) \mu_\alpha, & \text{if } \alpha+\beta = 0, \\ 0 & \text{in all remaining cases.} \end{cases}$$

(The constants $N_{\alpha\beta}$ can be made integers under an appropriate normalization of the x_α and are constructed from the root system.)

A detailed exposition of the theory of Lie algebras, and in particular, a classification of semisimple Lie algebras, can be found in the books [34], [46].

6.3. The Connection between Lie Groups and Lie Algebras

There are many ways to construct a Lie algebra from a Lie group. We present here four methods.

The first method is based on a study of the function $\psi(x, y)$, which gives the law of multiplication in the Lie group G in a neighborhood of the group identity. We choose an arbitrary chart for which the group identity is the origin. Then the function ψ has the properties that $\psi(x, 0) = \psi(0, x) = x$. Consequently, the Taylor series expansion of this function has the form

$$\psi(x, y) = x + y + B(x, y) + \ldots , \tag{1}$$

where $B(x,y)$ is a bilinear vector form in the vectors x and y, and the series of dots denotes the terms of order ≥ 3.

We now define the commutator in $\mathbf{R}^n$ (where $n = \dim G$) by the formula

$$[x,y] = B(x,y) - B(y,x). \tag{2}$$

We shall show that the Jacobi identity holds. We use the equality $\psi(\psi(x,y),z) = \psi(x,\psi(y,z))$ (associativity of group multiplication). Expanding both sides of this equality in a Taylor series and equating terms, linear in each of the vectors x, y, z, we obtain

$$B(B(x,y),z) = B(x,B(y,z)). \tag{3}$$

On the other hand, substituting the expression (2) in the first term of the Jacobi identity, we obtain

$$[[x,y],z] = B(B(x,y),z) - B(z,B(x,y)) + B(z,B(y,x)) - B(B(y,x),z).$$

In view of (3), this is equal to

$$B(x,B(y,z)) - B(z,B(x,y)) + B(z,B(y,x)) - B(y,B(x,z)).$$

In this expression, we now permute x, y, z cyclically and add the quantities so obtained. Obviously all of the terms cancel, and so the Jacobi identity holds.

Problem 1. Show that the choice of a different chart leads to an isomorphic Lie algebra, and that the matrix of the operator establishing the isomorphism has the form $a^i_j = \partial x^i / \partial y^j$, where $\{x^i\}, \{y^j\}$ are the coordinates in the two charts.

Our second method introduces the structure of a Lie algebra in the tangent space $T_e G$ to the group G at the point e. In the sequel we shall denote this space by $\mathfrak{g}$. Let $x(t), y(t)$ be two smooth curves issuing from the point e. We consider the curve

$$z(t) = x(\tau) * y(\tau) * x(\tau)^{-1} * y(\tau)^{-1}, \tag{4}$$

where $\tau = \operatorname{sgn} t \cdot \sqrt{|t|}$.

Problem 2. Prove that the curve $z(t)$ belongs to the class $C^1(\mathbf{R}, G)$ and hence has a tangent vector $z'(t)$. We set

$$[x'(0), y'(0)] = z'(0),$$

where $x'(0), y'(0)$ are tangent vectors to the curves $x(t)$ and $y(t)$ respectively. Under the operation (4), $\mathfrak{g}$ becomes a Lie algebra.

Hint. Use the decomposition (1) and the formula

$$x^{-1} = -x + B(x,x) + \cdots, \tag{5}$$

§ 6. Lie Groups and Lie Algebras

which follows from it. As before, the sequence of dots denotes the terms of order $\geqslant 3$. Show that if $x(t)=ta+o(t)$, $y(t)=tb+o(t)$, then $z(t)=tB(a,b)-tB(b,a)+o(t)$. (From this it is evident that the Lie algebra constructed in the tangent space is isomorphic to the Lie algebra constructed by the first method.)

The third method also produces the structure of a Lie algebra in the space $\mathfrak{g}$. It is based on what is called the *adjoint representation* of the group G in the space $\mathfrak{g}$.

We denote by $A(g)$ the diffeomorphism of G given by the formula

$$A(g): h \mapsto ghg^{-1}, \quad h, g \in G. \tag{6}$$

Evidently $A(g)$ is an automorphism of the group G (we recall that automorphisms of this sort are called *inner automorphisms*. The point e remains fixed under all of the automorphisms $A(g)$.

We consider the derived mapping $A(g)_*(e): \mathfrak{g} \to \mathfrak{g}$, which is ordinarily denoted by $\mathrm{Ad}\, g$. In view of the chain rule (see 5.2), the mapping $g \mapsto \mathrm{Ad}\, g$ is a homomorphism of G into the group $\mathrm{Aut}\,\mathfrak{g}$ of linear invertible mappings of the space $\mathfrak{g}$ onto itself. This homomorphism is called the *adjoint representation*. Since the adjoint representation is a smooth mapping of G into $\mathrm{Aut}\,\mathfrak{g}$, we can consider the derived mapping $\mathrm{Ad}_*(e): \mathfrak{g} \to \mathrm{End}\,\mathfrak{g}$. (It is evident that $\mathrm{End}\,\mathfrak{g}$ is the tangent space to $\mathrm{Aut}\,\mathfrak{g}$, since $\mathrm{Aut}\,\mathfrak{g}$ is an open set in the linear space $\mathrm{End}\,\mathfrak{g}$.)

The mapping $\mathrm{Ad}_*(e)$ is denoted by ad and assigns to every element $X \in \mathfrak{g}$ a linear mapping $\mathrm{ad}\, X$ of the space $\mathfrak{g}$ into itself. We set

$$[X, Y] = \mathrm{ad}\, X\, Y. \tag{7}$$

Problem 3. Prove that the operation defined by formula (7) makes $\mathfrak{g}$ into a Lie algebra and that this Lie algebra coincides with the algebra constructed by the second method.

Hint. Consider the mapping $\alpha: G \times G \to G$, defined by the formula $\alpha(x, y) = xyx^{-1}$. The derived mapping $\alpha_*(e, e): \mathfrak{g} \times \mathfrak{g} \to \mathfrak{g}$ coincides with the commutator defined by formula (7). The assertion of the problem follows from the explicit form of the expansion of the function α in a Taylor series (up to terms of the second order). This explicit form is easily obtained from formula (1). (See also problem 11 in 5.2.)

Example 1. Let $G = GL(n, \mathbf{R})$ be the group of all nonsingular real matrices of order n. The space $\mathfrak{g}$ here coincides with the space $\mathrm{Mat}_n(\mathbf{R})$ of all real matrices of order n (with the matrix $A \in \mathrm{Mat}_n(\mathbf{R})$ we associate the class of equivalent curves of the form $x(t) = 1_n + tA + o(t)$). The adjoint representation has the form

$$\mathrm{Ad}\, x: A \mapsto xAx^{-1}, \quad x \in G, \quad A \in \mathfrak{g}. \tag{8}$$

To compute the operator ad, we set $x(t) = 1_n + tB$.
We then have

$$x(t) A x(t)^{-1} = (1_n + tB) A (1_n - tB + o(t)) = A + t(BA - AB) + o(t),$$

from which we see that ad $BA = BA - AB$, which is to say that

$$[B,A] = BA - AB. \tag{9}$$

The fourth method consists in constructing a Lie algebra from differential operators on the group. We recall that the space $\operatorname{Vect} G$ of smooth vector fields on the Lie group G is an (infinite-dimensional) Lie algebra (see problem 4 in 6.2 and its corollary). Let $\mathfrak{g}_L$ denote the subspace of $\operatorname{Vect} G$ consisting of the vector fields that are invariant under left translations on G.

Problem 5. Prove that $\mathfrak{g}_L$ is a subalgebra of $\operatorname{Vect} G$, isomorphic to the Lie algebras constructed above.

Hint. Prove that a vector field $X \in \operatorname{Vect} G$ is left invariant if and only if the equality $L(g)X = XL(g)$ holds, where $L(g)$ denotes the operator on $C^\infty(G)$ that acts according to the formula $[L(g)f](h) = f(g^{-1}h)$.

To prove the second assertion of the problem, verify that the mapping assigning to a vector field its value at e is an isomorphism of the subspace $\mathfrak{g}_L \subset \operatorname{Vect} G$ onto $\mathfrak{g} = T_e G$. The fact that this mapping is an isomorphism of Lie algebras can be inferred from the following assertion, which also has independent interest.

Problem 6. Prove that every left invariant vector field on G is the derived vector field (see 5.2) for a certain family of right translations on G.

Thus, with every Lie group G, we associate a well-defined Lie algebra $\mathfrak{g}$. We shall show that to every smooth homomorphism $\phi: G_1 \to G_2$ of one Lie group into another, there corresponds a homomorphism of the corresponding Lie algebras. In fact, we can take the derived mapping $\phi_*(e)$ for our mapping of $\mathfrak{g}_1$ into $\mathfrak{g}_2$.

It follows immediately from the second method of constructing a Lie algebra from a Lie group that $\phi_*(e)$ is a homomorphism of Lie algebras.

The result we have obtained can be briefly expressed as follows. We have constructed a functor from the category of Lie groups to the category of Lie algebras.

We note that in the construction of a Lie algebra from a Lie group we have used only the knowledge of the law of multiplication in an arbitrarily small neighborhood of the identity. This lies behind the introduction of a new concept.

Consider a pair (V, ψ) where V is an open subset in $\mathbf{R}^n$ that contains the origin of coordinates and ψ is a smooth mapping of $V \times V$ into $\mathbf{R}^n$ that satisfies the following conditions:

1) $\psi(x, \psi(y,z)) = \psi(\psi(x,y), z)$,
2) $\psi(0, x) = \psi(x, 0) = x$,
3) there exists a smooth mapping $\varepsilon: V \to \mathbf{R}^n$ such that $\psi(x, \varepsilon(x)) = \psi(\varepsilon(x), x) = 0$.

(All of the equalities are postulated to hold in the region where all expressions appearing in the equalities are defined.) Then we say that (V, ψ) is a *local Lie group*.

Given a Lie group, we can produce an infinite set of local Lie groups from it in the following way. Let (U, α) be a chart on G containing the identity as the origin of coordinates. We consider a neighborhood $V \subset U$ for which the inclusions $V \cdot V \subset U$ and $V^{-1} \subset U$ hold. (As usual, $V \cdot V$ denotes the set of all products xy,

§ 6. Lie Groups and Lie Algebras

with $x \in V$ and $y \in V$, while V^{-1} denotes the set of elements of the form x^{-1} for $x \in V$.) The existence of such a neighborhood follows from the continuity of the mapping $(x,y) \mapsto xy$ and $x \mapsto x^{-1}$.

Identifying V with the region $\alpha(V) \subset \mathbf{R}^n$ with the aid of the mapping α and setting $\psi(x,y) = xy$, $\varepsilon(x) = x^{-1}$, we obviously obtain a local Lie group.

Two local Lie groups (V_1, ψ_1) and (V_2, ψ_2) are called *isomorphic* if there exist neighborhoods of zero $V_1' \subset V_1$ and $V_2' \subset V_2$, and also a diffeomorphism $\alpha: V_1' \to V_2'$ such that the diagram

$$\begin{array}{ccc} V_1' \times V_1' & \xrightarrow{\psi_1} & \mathbf{R}^n \supset V_1' \\ {\scriptstyle \alpha \times \alpha} \downarrow & & \downarrow {\scriptstyle \alpha} \\ V_2' \times V_2' & \xrightarrow{\psi_2} & \mathbf{R}^n \supset V_2' \end{array}$$

is commutative (that is, we have $\alpha(\psi_1(x,y)) = \psi_2(\alpha(x), \alpha(y))$, wherever both sides are defined).

It is clear that all local groups obtained by the above construction from a single Lie group G are isomorphic with each other.

It is nevertheless possible that the local groups constructed for two Lie groups G_1 and G_2 may be isomorphic, while the groups themselves are nonisomorphic. In this case, the groups G_1 and G_2 are said to be *locally isomorphic*.

Example 2. Let $G_1 = \mathbf{R}$, $G_2 = \mathbf{T}$. As $V_1 \subset G_1$, we take the interval $(-\pi, \pi)$ and as $V_2 \subset G_2$ we take the complement of the point $-1 \in \mathbf{T}$. The mapping $\alpha(x) = e^{ix}$ establishes a diffeomorphism between V_1 and V_2 that commutes with multiplication. Thus G_1 and G_2 are locally isomorphic.

Theorem 1. *Every Lie algebra is the Lie algebra of a certain local Lie group. Local Lie groups are isomorphic if and only if the corresponding Lie algebras are isomorphic.*

We shall prove this theorem in the following section.

To describe the connection between Lie groups and local Lie groups, we need the following fact.

Monodromy theorem. *Let G be a connected and simply connected Lie group, and let H be an arbitrary Lie group. Then every local homomorphism of G into H (that is, a homomorphism of the corresponding local groups) can be extended in one and only one way to a global homomorphism of G into H.*

Also, we shall make use of the famous theorem of E. Cartan asserting that *every Lie algebra is the Lie algebra of a certain Lie group*.

Theorem 2. *To every Lie algebra $\mathfrak{g}$ there corresponds a unique connected and simply connected Lie group G, for which $\mathfrak{g}$ is its Lie algebra. All connected Lie groups having this property have the form G/D, where D is a discrete normal subgroup contained in the center of G.*

Proof. Let G_0 be any Lie group with Lie algebra $\mathfrak{g}$ (this group exists by E. Cartan's theorem). We shall suppose that G_0 is connected, since otherwise we could consider in place of G_0 the connected component of the identity, which will not affect the Lie algebra (or the corresponding local group).

Let G denote the simply connected covering group of the group G_0 (see 6.1). The projection $p: G \to G_0$ is a homomorphism of Lie groups, and $D = p^{-1}(e)$ is a discrete normal subgroup of G.

By the result of problem 1 from 2.6, D belongs to the center of the group G.

Thus every connected Lie group G_0 having $\mathfrak{g}$ for its Lie algebra can be represented in the form G/D, where G is a simply connected Lie group with the same Lie algebra. If G_1 and G_2 are two simply connected Lie groups with the same Lie algebra, then by theorem 1, they are locally isomorphic. By the monodromy theorem, a local isomorphism α can be extended to a global homomorphism $\alpha_1: G_1 \to G_2$ and the inverse mapping α^{-1} to a global homomorphism $\alpha_2: G_2 \to G_1$. The mappings $\alpha_1 \alpha_2$ and $\alpha_2 \alpha_1$ coincide with the identity mapping in a neighborhood of the identity. Since G_1 and G_2 are connected, these mappings are everywhere the identity mapping. This means that G_1 and G_2 are isomorphic. The theorem is proved.

Problem 7. Prove that the groups G/D_1 and G/D_2, where G is a simply connected Lie group, and D_1 and D_2 are discrete normal subgroups, are isomorphic if and only if there exists an automorphism of the group G carrying D_1 into D_2.

Hint. Use the monodromy theorem.

We shall make use of the above results to list all connected Lie groups of dimension ≤ 3. We listed the corresponding Lie algebras in 6.2. We indicate the corresponding simply connected groups.

I. $\dim G = 1$: $G = \mathbf{R}$.
II. $\dim G = 2$:
  a) $G = \mathbf{R}^2$ with the operation of addition;
  b) G is the group of affine mappings of the line onto itself that preserve orientation (in the coordinates a, b of example 5 of 6.1, this group is given by the condition $a > 0$).
III. $\dim G = 3$:
  a) $\dim [\mathfrak{g},\mathfrak{g}] = 0$, $G = \mathbf{R}^3$ with the operation of addition;
  b) $\dim [\mathfrak{g},\mathfrak{g}] = 1$, $[\mathfrak{g},[\mathfrak{g},\mathfrak{g}]] = \{0\}$, and G is the Lie group of triangular matrices of the form

  $$\begin{pmatrix} 1 & x & z \\ 0 & 1 & y \\ 0 & 0 & 1 \end{pmatrix}, \quad x, y, z \in \mathbf{R};$$

  c) $\dim [\mathfrak{g},\mathfrak{g}] = \dim [\mathfrak{g},[\mathfrak{g},\mathfrak{g}]] = 1$, and G is the product of $\mathbf{R}$ and the group from example II b);
  d) $\dim [\mathfrak{g},\mathfrak{g}] = 2$.

Let A be the nonsingular matrix that defines $\mathfrak{g}$ according to problem 6 from 6.2. We set $e^{tA} = \sum_{k=0}^{\infty} \frac{t^k A^k}{k!} = \begin{pmatrix} a(t) & b(t) \\ c(t) & d(t) \end{pmatrix}$. As G we can take $\mathbf{R}^3$ with the law of multiplication

$$(x,y,z)(x',y',z') = (x + a(z)x' + b(z)y', y + c(z)x' + d(z)y', z + z').$$

§ 6. Lie Groups and Lie Algebras

e) $[\mathfrak{g},\mathfrak{g}]=\mathfrak{g}$ and $\mathfrak{g}=\mathbf{R}^3$ with the vector product for commutator, and G is the group $SU(2)$ of unitary matrices of the second order with determinant 1 (isomorphic to the group of quaternions of norm 1—see example 6 from 6.1);

f) $\mathfrak{g}=[\mathfrak{g},\mathfrak{g}]$ is the algebra of real matrices with trace zero, and G is the simply connected covering group of the group $SL(2,\mathbf{R})$ of real matrices of the second order with determinant 1. (One can show that this group cannot be realized as a subgroup of $GL(n,\mathbf{C})$, regardless of the size of the positive integer n.)

It remains for us to list all discrete normal subgroups of the groups described above, up to equivalence in the sense of problem 7. Here it is useful to recall that these normal subgroups belong to the center of the group.

Problem 8. Prove that the centers of the groups listed above have the following forms:

I. $Z(G)=G$.
II. a) $Z(G)=G$,
   b) $Z(G)=\{e\}$.
III. a) $Z(G)=G$,
   b) $Z(G)=\left\{\begin{pmatrix} 1 & 0 & z \\ 0 & 1 & 0 \\ 0 & 0 & 1 \end{pmatrix}\right\}$, $z\in\mathbf{R}$,
   c) $Z(G)=\mathbf{R}\times\{e\}$,
   d) $Z(G)=\begin{cases} \{e\}, & \text{if } A\not\sim\begin{pmatrix} 0 & a \\ -a & 0 \end{pmatrix}, \quad a\in\mathbf{R}^*, \\ \left\{\left(0,0,\dfrac{2\pi n}{a}\right)\right\}, \quad n\in\mathbf{Z}, & \text{if } A\sim\begin{pmatrix} 0 & a \\ -a & 0 \end{pmatrix}, \quad a\in\mathbf{R}^*; \end{cases}$
   e) $Z(G)=\{1,-1\}$,
   f) $Z(G)=p^{-1}(\{1,-1\})$, where $p:G\to SL(2,\mathbf{R})$ is the natural projection.

We leave it to the reader to complete the classification of Lie groups of dimension ≤ 3. In particular, one finds along the way a simple proof of theorem 1 in 6.1 which classifies one-dimensional Lie groups (under the hypothesis that the group is smooth, of course).

For proofs of the theorems quoted in this section, see the books [44] and [11].

6.4. The Exponential Mapping

An important rôle in the study of Lie groups is played by *one-parameter subgroups*, that is, subgroups that are one-dimensional Lie groups. One can choose the parameter t so that the equalities

$$x(0)=e, \quad x(t)x(s)=x(t+s) \tag{1}$$

hold.

Since $x(t)$ is a smooth curve, the tangent vector

$$X = x'(0) \in \mathfrak{g} \tag{2}$$

is defined.

Theorem 1. *For every vector $X \in \mathfrak{g}$, there exists a unique smooth curve $x(t)$ satisfying the conditions (1) and (2).*

We agree on the following notation. If ξ is a tangent vector to the group G at the point g_0, then $g\xi$ (or ξg) denotes the tangent vector to G at the point gg_0 (or $g_0 g$) obtained from ξ by the mapping $L(g)_*(g_0)$ (or $R(g)_*(g_0)$), where $L(g)$ (or $R(g)$) is left (or right) translation by the element g.

Problem 1. Prove the identity

$$g\xi g^{-1} = Ad\, g\, \xi, \quad g \in G, \quad \xi \in \mathfrak{g}. \tag{3}$$

Hint. See the third method of constructing the Lie algebra of a Lie group, as given in 6.2.

Problem 2. Prove that the following identities follow from (1):

$$x(t)x'(0) = x'(t), \quad x'(0)x(t) = x'(t). \tag{4}$$

Hint. Rewrite the conditions of the problem in the form

$$L(x(t))x(s) = x(t+s) \quad \text{or} \quad R(x(t))x(s) = x(s+t).$$

Now let X be a fixed element of $\mathfrak{g}$. We consider vector fields on G given by the formulas

$$\xi_X(g) = gX, \quad \eta_X(g) = Xg.$$

Problem 3. Prove that a vector field on G is left invariant (or right invariant) if and only if it has the form $\xi_X(g)$ [or $\eta_X(g)$] for a certain $X \in \mathfrak{g}$.

Hint. Write the condition for left invariance (or right invariance) of the field $\xi(g)(\eta(g))$ in the form

$$g\xi(h) = \xi(gh), \quad \eta(h)g = \eta(hg)$$

and prove the identities

$$g_1(g_2 \xi) = (g_1 g_2)\xi, \quad (\eta g_1)g_2 = \eta(g_1 g_2).$$

We are now in a position to prove theorem 1. It follows from the equalities (4) that a curve $x(t)$ that satisfies conditions (1) and (2) is tangent at each of its points to the vector field $\xi_X(g)$. (It is also tangent to the field $\eta_X(g)$, since, as shown in problem 1, these fields coincide on the curve $x(t)$.) The existence of a local one-dimensional group with tangent vector X now follows from the theorem

§ 6. Lie Groups and Lie Algebras

of existence and uniqueness of a solution for a system of ordinary differential equations (see 5.2).

Problem 4. Show that every local one-dimensional subgroup of a Lie group G can be extended to a Lie subgroup.

Hint. Suppose that $x(t)$ is defined for $|t|<a$ and satisfies condition (1) for $|t|+|s|<a$.

For every $t \in \mathbf{R}$, we set $x(t)=x(t/N)^N$, where the number N is chosen so large that $|t/N|<a$. Prove that $x(t)$ is independent of the choice of N and that (1) holds for all t and s.

Now let **G** be a Lie group and $\mathfrak{g}$ its Lie algebra. We define the *exponential mapping* $\exp: \mathfrak{g} \to G$ as follows:

$$\exp X = x(1), \tag{5}$$

where $x(t)$ is the one-parameter group that corresponds to X by theorem 1. (This mapping is also called *canonical*.)

One of the most important properties of the exponential mapping is that it is a functor. This is expressed by the commutativity of the following diagram:

$$\begin{array}{ccc} G_1 & \xrightarrow{\phi} & G_2 \\ \exp \uparrow & & \uparrow \exp \\ \mathfrak{g}_1 & \xrightarrow{\phi_*(e)} & \mathfrak{g}_2 \end{array}$$

The proof of this fact is immediate from the definition of the exponential mapping.

Corollary 1. *Finite-dimensional representations of a local Lie group are in one-to-one correspondence with the representations of its Lie algebra.*

Corollary 2. *For a connected and simply connected Lie group, the finite-dimensional representations are in one-to-one correspondence with the representations of its Lie algebra.*

(This follows from the monodromy theorem—see 6.3.)

We consider more closely the case in which $G=GL(n,\mathbf{R})$ is the group of nonsingular matrices of the n-th order. In this case, the algebra $\mathfrak{g}$ coincides with the space $\text{Mat}_n(\mathbf{R})$ (see example 1 in 6.3).

Problem 5. Prove that the differential equation

$$x'(t) = X x(t)$$

with initial condition $x(0)=1$ has the function

$$x(t) = e^{tX} = \sum_{k=0}^{\infty} \frac{t^k X^k}{k!}. \tag{6}$$

as a solution.

Hint. See 4.2.

Corollary. *For matrix groups, the exponential mapping has the form*

$$\exp X = e^X = \sum_{k=0}^{\infty} \frac{X^k}{k!}.$$

Problem 6. Prove the equalities

1) $\exp \begin{pmatrix} \lambda & 1 & 0 & \ldots & 0 \\ 0 & \lambda & 1 & \ldots & 0 \\ \vdots & & & & \\ 0 & 0 & 0 & \ldots & \lambda \end{pmatrix} = \begin{pmatrix} e^\lambda & e^\lambda & \frac{e^\lambda}{2!} & \ldots & \frac{e^\lambda}{n!} \\ 0 & e^\lambda & e^\lambda & \ldots & \frac{e^\lambda}{(n-1)!} \\ \vdots & & & & \\ 0 & 0 & 0 & \ldots & e^\lambda \end{pmatrix},$

2) $\exp \begin{pmatrix} a & b \\ c & d \end{pmatrix} = e^{\frac{a+d}{2}} \left[\operatorname{ch}\rho \begin{pmatrix} 1 & 0 \\ 0 & 1 \end{pmatrix} + \frac{\operatorname{sh}\rho}{\rho} \begin{pmatrix} \frac{a-d}{2} & b \\ c & \frac{d-a}{2} \end{pmatrix} \right],$

where $\rho = \sqrt{\left(\frac{a-d}{2}\right)^2 + bc}$,

3) $\exp \begin{pmatrix} 0 & a & c \\ -a & b & b \\ -c & -b & 0 \end{pmatrix}$

$= \cos r \begin{pmatrix} 1 & 0 & 0 \\ 0 & 1 & 0 \\ 0 & 0 & 1 \end{pmatrix} + \frac{\sin r}{r} \begin{pmatrix} 0 & a & c \\ -a & 0 & b \\ -c & -b & 0 \end{pmatrix} + \frac{1-\cos r}{r} \begin{pmatrix} b^2 & -bc & ab \\ -bc & c^2 & -ac \\ ab & -ac & a^2 \end{pmatrix},$

where $r = \sqrt{a^2+b^2+c^2}$.

It follows from the general theory of ordinary differential equations (the theorem concerning the dependence of solutions on the initial conditions and the coefficients of the system) that exp is a smooth mapping. Let us show that the mapping $\exp_*(0): \mathfrak{g} \to \mathfrak{g}$ is the identity mapping.

In fact, the one-parameter subgroups corresponding to the vectors X and tX are obviously connected by the equality

$$x_{tX}(s) = x_X(ts).$$

From this we obtain the equalities $\exp tX = x_{tX}(1) = x_X(t)$, that is, the mapping exp carries the curve $\{tX\} \subset \mathfrak{g}$ into the curve $x_X(t) \subset G$. Passing to the derived

§ 6. Lie Groups and Lie Algebras

mapping, we obtain

$$\exp_*(0): X \mapsto X,$$

which we wished to prove.

Corollary. *The mapping* exp *is a diffeomorphism in a certain neighborhood U of the initial point* $X=0$.

We shall denote the mapping $\exp^{-1}$ by the symbol ln. In general, ln is defined only in a certain neighborhood of the identity of the group. The local system of coordinates $(\exp U, \ln)$ is called *canonical*.

A Lie group G is called *exponential* if the canonical system of coordinates covers the entire group G. The class of exponential groups contains all connected and simply connected nilpotent groups and is contained in the class of solvable groups.

Problem 7. Prove that the law of multiplication, written in canonical coordinates, has the form

$$\phi(x,y) = x + y + \tfrac{1}{2}[x,y] + \ldots, \tag{7}$$

where the sequence of dots represents terms of order ≥ 3.

Hint. Use formula (5) from 6.3 and the fact that $(\exp X)^{-1} = \exp(-X)$.

Formula (7) shows that the law of multiplication in the group can be reconstructed up to the second order from the law of commutation in the Lie algebra. It turns out that all following terms in the expansion of $\psi(x,y)$ as a Taylor series can also be expressed in terms of the commutation operation. Here we shall state only the final result. One can find a proof in the books [34] and [46].

Theorem 2 (the Campbell-Hausdorff-Dynkin formula). *In canonical coordinates, the law of multiplication has the form*

$$\psi(x,y) = \sum_{m=1}^{\infty} \frac{(-1)^{m-1}}{m} \sum_{\substack{k_i + l_i \geq 1 \\ k_i \geq 0, l_i \geq 0}} \frac{[x^{k_1} y^{l_1} \ldots x^{k_m} y^{l_m}]}{k_1! l_1! \ldots k_m! l_m!}, \tag{8}$$

where the symbol $[x_1 x_2 \ldots x_n]$ *denotes* $\frac{1}{n}[\ldots[[x_1, x_2], x_3], \ldots, x_n]$.

Remark 1. The formula (8) is obtained from the formal series for the function $\ln(e^x e^y)$ in the noncommuting variables x, y if we replace every monomial D by the expression $[D]$.

Remark 2. The series (8) is awkward for computations, since it contains many nonreduced similar terms.

Although it is bereft of practical significance, this series plays an important rôle in the theory, because it proves theorem 1 of 6.3.

In many questions of the theory of Lie groups, it is important to know not the function $\psi(x,y)$ itself, but only its derivative with respect to one of the variables at the origin of coordinates. For this derivative, we can point out the following explicit expression:

Theorem 3. We set $L_x(y) = R_y(x) = \psi(x, y)$. Then we have

$$(L_{\exp X})_*(0) = b(-\operatorname{ad} X), \quad (R_{\exp X})_*(0) = b(\operatorname{ad} X), \tag{9}$$

where $b(z) = \dfrac{z}{e^z - 1} = \sum\limits_{k=0}^{\infty} \dfrac{B_k}{k!} z^k$ (here the B_k are the Bernoulli numbers).

The preceding theorem is easily proved from the following formula, which also has independent interest.

Problem 8. Prove the equality

$$\exp(X + tY)\exp(-X) = \exp(t\alpha(\operatorname{ad} X) Y + o(t)), \tag{10}$$

where

$$\alpha(z) = \dfrac{e^z - 1}{z} = \sum\limits_{k=0}^{\infty} \dfrac{z^k}{(k+1)!}.$$

Hint. In view of Ado's theorem (see 6.2), it suffices to consider the case of a matrix algebra.

Corollary 1. *The mapping* exp *is a homeomorphism in a neighborhood of the point* X *if and only if the eigenvalues of the operator* $\operatorname{ad} X$ *are different from* $\pm 2\pi i, \pm 4\pi i, \pm 6\pi i, \ldots$.

Corollary 2. *In canonical coordinates, the left invariant (right invariant) measure on the group* G *has the form*

$$\rho_l(X) dX \quad (\text{or } \rho_r(X) dX),$$

where dX is the usual Euclidean measure in $\mathfrak{g}$, and the functions ρ_l and ρ_r are given by the formulas

$$\rho_l(X) = \det[\alpha(-\operatorname{ad} X)], \quad \rho_r(X) = \det[\alpha(\operatorname{ad} X)] \tag{11}$$

(the function α is as defined in problem 8).

Corollary 3. *There is a two-sided invariant measure on the group* G *if and only if the equalities*

$$\operatorname{tr} \operatorname{ad} X = 0 \tag{12}$$

hold for all $X \in \mathfrak{g}$.

In fact, we have $\rho_r(X)/\rho_l(X) = \det(e^{\operatorname{ad} X}) = e^{\operatorname{tr} \operatorname{ad} X}$.
One can derive formula (10) in another way, by considering the function

$$\phi(s, t) = \exp(-sc_t) \dfrac{\partial}{\partial t} \exp sc_t,$$

where c_t is a certain curve in $\mathfrak{g}$.

§ 6. Lie Groups and Lie Algebras

Problem 9. Prove the equality

$$\frac{\partial}{\partial s}\phi(s,t)=\exp(-sc_t)\left(\frac{d}{dt}c_t\right)\exp(sc_t). \tag{13}$$

Now setting $c_t = X + tY$ and integrating the relation (13) from $s=0$ to $s=1$, we obtain the equality (10).

Second Part. Basic Concepts and Methods of the Theory of Representations

§ 7. Representations of Groups

7.1. Linear Representations

We have already stated *supra* that the term "representation" in the wide sense means a homomorphism of the group G into the group of one-to-one mappings of a certain set X onto itself[1]. A representation T is called *linear* if X is a linear space and the mappings $T(g)$ are linear operators.

If X is an arbitrary G-space, then, in the linear space $L(X)$ of all functions on X, we obtain a linear representation T of G:

$$[T(g)f](x) = \begin{cases} f(xg) & \text{for a right } G\text{-space;} \\ f(g^{-1}x) & \text{for a left } G\text{-space.} \end{cases}$$

This observation allows us to use linear representations to study general representations.

The spectral theory of dynamical systems serves as an example of this sort of application.

A general dynamical system is a space X equipped with a measure μ and a group of one-to-one transformations of X onto itself that preserve the measure μ (or at least its equivalence class). The most important case is that in which X is the phase space of a mechanical system, μ is the Liouville measure, and G is the dynamical group that describes the evolution of the system. The spectral theory of dynamical systems concerns itself with the study and description of various properties of such a system in terms of the unitary representation T that arises in $L^2(X,\mu)$.

The class of what are called *projective* representations forms a close congener of the class of linear representations. These are representations T for which X is a projective space and $T(g)$ is a projective transformation for all $g \in G$. Since a projective space is the collection of lines in a linear space L, and since projective

---

[1] A. Yu. Geronimus has proposed a more general definition of a representation (as a sheaf on a certain Grothendieck topology; see [85]).

§ 7. Representations of Groups

transformations are induced by linear transformations in L, we can consider a projective representation as a mapping $T: G \to \operatorname{Aut} L$ that enjoys the property

$$T(g_1 g_2) = c(g_1, g_2) T(g_1) T(g_2),$$

where c is a certain numerical function on $G \times G$. In particular, every linear representation of G in L generates a projective representation.

Quantum mechanics is an important field of application of the theory of projective representations. As is well known, in the quantum theory, the rôle of phase space is assumed by the projective space $P(H)$ consisting of all complex lines in a complex infinite-dimensional Hilbert space. The admissible transformations of $P(H)$ are the projective transformations that correspond to unitary and anti-unitary operators in H.

In general we shall in the sequel consider only linear representations and for brevity will call them simply "representations".

The set of representations of a group G in linear spaces over the field (skew field) K forms a category $\Pi(G, K)$. The morphisms of this category are *intertwining operators*, that is, operators that commute with the action of the group. Isomorphic objects in $\Pi(G, K)$ are called *equivalent representations*.

We denote by the symbol $\mathscr{C}(T_1, T_2)$ or $\operatorname{Hom}_G(V_1, V_2)$ the set of intertwining operators for the representations T_1 and T_2, which act in spaces V_1 and V_2. We shall write $\mathscr{C}(T)$ or $\operatorname{End}_G V$ for $\mathscr{C}(T, T) = \operatorname{Hom}_G(V, V)$. The dimension of the space $\mathscr{C}(T_1, T_2)$ is denoted by $c(T_1, T_2)$ and is called the *intertwining number*. If $c(T_1, T_2) = c(T_2, T_1) = 0$, then the representations T_1 and T_2 are said to be *disjoint*.

Example. Let $G = \mathbf{R}$, and let the representation T_∞ act in the space of all polynomials in the variable x according to the formula

$$[T(t) f](x) = f(x + t).$$

The representation T_k is defined by exactly the same formula in the space P_k of polynomials of degree not exceeding k ($k = 0, 1, 2, \ldots$).

Problem 1. Prove that $c(T_\infty, T_k) = 0$, $c(T_k, T_\infty) = k + 1$.

Hint. The space $\mathscr{C}(T_k, T_\infty)$ is generated by the operators $1, d/dx, \ldots, d^k/dx^k$.

Problem 2. If representations T and S are finite-dimensional, then we have $c(T, S) = c(S, T)$.

Hint. Use the fact that the row rank and the column rank coincide for finite-dimensional matrices.

Let T be a representation of a group G in a space V. The correspondence $g \mapsto T(g^{-1})^*$, where $*$ denotes the adjoint operator in the conjugate space V^*, is plainly a representation of G in V^*. It is called *conjugate* or *contragredient* to the representation T. We will denote it by T^*.

To prevent confusion in the use of this notation, we emphasize that the operator $T^*(g)$ by the definition of T^* is equal to $T(g^{-1})^*$, and not to $T(g)^*$.

If the space V contains a subspace V_1 that is invariant under the representation T, then we say that T is *reducible*. The restriction of the operators $T(g)$ to V_1

defines a representation T_1 of the group G in the space V_1. It is called a *subrepresentation* of T. There is also a natural representation of G in the factor space $V_2 = V/V_1$. It is called a *factor representation* of T.

If an invariant subspace $V_1 \subset V$ admits an invariant complementary subspace (which is identified in a natural way with V_2), then the representation T is called *decomposable*, and we write $T = T_1 + T_2$.

Problem 3. Prove that the representation $T_1 + T_2$ is the sum of T_1 and T_2 in the category $\Pi(G, K)$.

A representation T that admits no nontrivial subrepresentations is called *algebraically irreducible*.

The representation T_k of the group $\mathbf{R}$ in the space P_k for $k > 0$ is an example of a representation that is reducible but not decomposable.

Representations for which every invariant subspace admits an invariant complement are called *completely reducible*.

In the category $\Pi(G, K)$ the operation of *tensor product* is defined. If T_i is a representation in a linear space V_i over K, $i = 1, 2$, then the representation $T_1 \otimes T_2$ acts in $V_1 \underset{K}{\otimes} V_2$ by the formula

$$(T_1 \otimes T_2)(g) : \xi_1 \otimes \xi_2 \mapsto T_1(g)\xi_1 \otimes T_2(g)\xi_2.$$

Problem 4. Prove that the operations of sum and tensor product are consistent with the relation of equivalence and consequently are carried over to the set of equivalence classes of representations.

Problem 5. Prove the equalities

$$c(T_1 + T_2, T_3) = c(T_1, T_3) + c(T_2, T_3),$$
$$c(T_1, T_2 + T_3) = c(T_1, T_2) + c(T_1, T_3),$$
$$c(T_1 \otimes T_2, T_3) = c(T_1, T_2^* \otimes T_3), \quad \text{if} \quad \dim T_2 < \infty.$$

Hint. One can interpret the space $\mathscr{C}(T_1, T_2)$ as the subspace of G-invariant elements in the space $\mathrm{Hom}(V_1, V_2)$ on which G acts by the formula

$$g : A \mapsto T_2(g) A T_1(g)^{-1}.$$

Let $\Pi' \subset \Pi(G, K)$ be an arbitrary subcategory closed under the operation of adding representations. In this subcategory, one can define the *Grothendieck group* $\Gamma(\Pi')$.

By definition, $\Gamma(\Pi')$ is the abelian group generated by equivalence classes $[T]$ of representations T in $\mathrm{Ob}\,\Pi'$ with the following relations.

$[T] = [T_1] + [T_2]$ if T_1 is a subrepresentation of T and T_2 is the corresponding factor representation.

If the category Π' is closed under the formation of tensor products, then $\Gamma(\Pi')$ will be a commutative ring under the operation

$$[T_1] \cdot [T_2] = [T_1 \otimes T_2].$$

§ 7. Representations of Groups

Suppose that Π' is closed under the formation of k-th exterior powers, for $k=2,3,\ldots$. Then we can define the structure of a λ-ring on $\Gamma(\Pi')$. This structure is given by a homomorphism λ_t of the additive group of the ring into the multiplicative group of power series of the form $1 + \sum_{k=1}^{\infty} a_k t^k$, where the a_k are elements of the ring. In the case of a Grothendieck ring, the homomorphism λ_t has the form

$$\lambda_t([T]) = 1 + \sum_{k=1}^{\infty} [\wedge^k T] t^k.$$

The mapping $G \mapsto \Gamma(\Pi(G,K))$ is a contravariant functor from the category of groups to the category of λ-rings. More details about this functor and its applications can be found in the book of D. Husemoller [32].

If K is a subfield of L, then there are naturally defined functors from $\Pi(G,K)$ to $\Pi(G,L)$ and back again, called *change of scalars*. The first functor consists in going from the space V over K to the space $V_L = V \otimes_K L$ over L. The second means that the space V over L is considered as a space over K. In this case, it is denoted by V_K. The very most important example is that in which $K = \mathbf{R}, L = \mathbf{C}$. The corresponding functors are ordinarily called *complexification* and *realification*.

7.2. Representations of Topological Groups in Linear Topological Spaces

Suppose that G is a topological group and that V is a linear topological space. We shall suppose as a rule, without explicit mention, that the representations T we consider are continuous, that is, that the mapping $(g,\zeta) \mapsto T(g)\zeta$ is continuous in both variables.

For many aims, a weaker condition of continuity suffices. This is the condition of separate continuity of the above mapping, which is nothing more than continuity of the homomorphism T of the group G into the group $\operatorname{Aut} V$, which is given the strong operator topology.

Problem 1. If G is locally compact and V is a Banach space, then separate continuity of a representation T is equivalent to continuity.

Hint. Use the Banach-Steinhaus theorem from 4.1.

The expression "intertwining operator" for continuous representations of topological groups will always mean "continuous intertwining operator", with similar restrictions on the expression "intertwining number".

The continuous representations of a topological group G into topological spaces over a field K form a category $T\Pi(G,K)$.

Suppose that there is a closed subspace V_1 of the space V of a representation T, invariant under the operators $T(g)$, $g \in G$. Then the restriction T_1 of the representation T to V_1 is called a *(topological) subrepresentation of T*. The representation T_2 in the factor space $V_2 = V/V_1$ is called a *(topological) factor representation of T*.

A representation T is called *topologically irreducible* if it admits no nontrivial subrepresentations, and *completely irreducible* if an arbitrary operator is the weak limit of a certain net consisting of linear combinations of operators $T(g)$, $g \in G$.

One says that a representation T is *(topologically) decomposable* if there exist closed invariant subspaces V_1 and V_2 in V such that V is isomorphic to the direct sum $V_1 \oplus V_2$. In this case, we write $T = T_1 + T_2$, where T_i denotes the restriction of T to V_i.

Problem 2. The representation $T_1 + T_2$ is the sum of T_1 and T_2 in the category $T\Pi(G,K)$.

One can reasonably define the tensor product only in certain subcategories of $T\Pi(G,K)$. Thus for representations in Banach spaces it is natural to use the operation $\hat{\otimes}$, and for unitary representations in Hilbert spaces, it is natural to use the operation of Hilbert product (see 4.1).

I. M. Gel'fand has proposed the following concept of *tensor irreducibility* of a representation.

A representation T of a group G in a space V is *tensorially irreducible* if for every trivial representation S of the group G in a nuclear space W, every closed G-invariant subspace in $V \hat{\otimes} W$ has the form $V \hat{\otimes} W_1$, where W_1 is a subspace in W.

Very little is known about the categories $T\Pi(G,K)$ for noncompact groups G. For example, for $G = \mathbf{Z}$ and $K = \mathbf{C}$, the classification of representations is equivalent to the classification of linear continuous operators in linear topological spaces up to similarity. This classical problem of functional analysis is very far from a solution. Not even the classification of topologically irreducible representations is known. The problem of existence of infinite-dimensional irreducible representations was posed by Banach about forty years ago, and remains unsolved up to the present day.

In the finite-dimensional case, the problem of classification of representations of the group $\mathbf{Z}$ is solved by means of the theorem about reduction of a matrix to Jordan normal form. For the group $\mathbf{Z} + \mathbf{Z}$, this problem is reduced to the problem of simultaneous reduction of a pair of commuting matrices to canonical form.

As I. M. Gel'fand and V. A. Ponomarev showed, this problem contains as a subproblem the classification of arbitrary finite sets of matrices (see the Journal of Functional Analysis, vol. 3, part 4 (1969), 81–82).

7.3. Unitary Representations

A representation T in a space V is called *unitary* if V is a Hilbert space and the operators $T(g)$ are unitary for all $g \in G$. Unitary representations are the most important and most studied class of representations. There are two reasons for this. First, unitary representations appear naturally in diverse applications (dynamical systems, quantum mechanics, the physical theory of fields). Second, unitary representations enjoy a series of remarkable properties, which greatly facilitate their study. We shall list some of these properties here. Others will be described in following sections.

§ 7. Representations of Groups

Problem 1. Every unitary representation is completely reducible.
Hint. The orthogonal complement of an invariant subspace is invariant.

We shall see *infra* that the study of unitary representations can in large measure be reduced to the study of topologically irreducible representations.

The set of equivalence classes of unitary irreducible representations of a group G is denoted by $\hat{G}$ and is called the *dual object* of the group G. If the group G is commutative, then all of its irreducible unitary representations are one-dimensional. In this case, the operation of tensor multiplication defines the structure of a commutative group in $\hat{G}$. The identity element is the trivial representation, and the inverse element is the complex adjoint representation.

We can introduce a *topology* in $\hat{G}$ in the following fashion. Let T be a irreducible representation of the group G in the space H. For every compact set $K \subset G$, every finite set of vectors $\xi_1, \ldots, \xi_n \in H$, and every number $\varepsilon > 0$, we define a subset $U(K, \xi_1, \ldots, \xi_n, \varepsilon)$ of $\hat{G}$. By definition, it consists of the classes $[S]$ of those irreducible representations S in the space of which there exist vectors $\eta_1, \ldots, \eta_n$ such that

$$|(T(g)\xi_i, \xi_j) - (S(g)\eta_i, \eta_j)| < \varepsilon \quad \text{for} \quad g \in K, \quad 1 \leq i, j \leq n.$$

The family of subsets $U(K, \xi_1, \ldots, \xi_n, \varepsilon)$ is taken as a basis of neighborhoods of the point $[T] \in \hat{G}$.

There is another method of defining a topology in $\hat{G}$. Let $\tilde{G}$ be the set of equivalence classes of arbitrary (not necessarily irreducible) unitary representations of the group G. To every subset $\mathscr{S} \subset \tilde{G}$ we assign a certain set $V(\mathscr{S})$ of function on G. By definition $V(\mathscr{S})$ contains all continuous bounded functions on G that can be uniformly approximated on every compact set by matrix elements of representations of the classes $[S] \subset \mathscr{S}$.

(A matrix element of a representation S in the space V is a function of the form $g \to (\eta, S(g)\xi)$, where $\xi \in V$, $\eta \in V'$.)

One says that the subset $\mathscr{S}_1$ is weakly contained in $\mathscr{S}_2$, if $V(\mathscr{S}_1) \subset V(\mathscr{S}_2)$, and that $\mathscr{S}_1$ is weakly equivalent to $\mathscr{S}_2$, if $V(\mathscr{S}_1) = V(\mathscr{S}_2)$.

Theorem 1. *Let T be an irreducible unitary representation of a locally compact group G and let $[T]$ be its equivalence class. The point $[T]$ lies in the closure of the set $\mathscr{S} \subset \hat{G}$ if and only if $[T]$ is weakly contained in $\mathscr{S}$.*

One of the assertions of the theorem is evident. We shall consider the other in § 10.

Unitary irreducible representations enjoy many special properties, which can conveniently be formulated with the aid of the concept of *k*-irreducibility.

Let L_k be a Hilbert space of dimension k, and S_k the trivial representation of the group G in the space L_k. We shall say that a representation T of the group G in a space H is *k-irreducible* if every closed subspace in $H \otimes L_k$ that is invariant under $T \otimes S_k$ has the form $H \otimes L_k^0$, where L_k^0 is a subspace of L_k.

We shall consider this notion in more detail for various values of k.

It is clear that 1-irreducibility coincides with topological irreducibility.

Problem 2. Prove that k-irreducibility for finite k is equivalent to the following condition. For arbitrary linearly independent vectors $\xi_1, ..., \xi_k \in H$, arbitrary vectors $\eta_1, ..., \eta_k \in H$ and arbitrary $\varepsilon > 0$, there exists an operator A of the form $A = \sum_{i=1}^{N} c_i T(g_i)$, such that

$$\|A\xi_i - \eta_i\| < \varepsilon, \quad i = 1, 2, ..., k.$$

Hint. In the space $H \otimes L_k \approx H \oplus ... \oplus H$, consider the smallest subspace that contains the vector $(x_1, ..., x_k)$.

Corollary. *If the representation T is k-irreducible for all finite k, then it is completely irreducible.*

Problem 3. Prove that ∞-irreducibility of T is equivalent to the following condition. For an arbitrary set of linearly independent vectors $\{\xi_i\}$ with the property that $\sum_{i=1}^{\infty} \|\xi_i\|^2 < \infty$, an arbitrary set of vectors $\{\eta_i\}$ such that $\sum_{i=1}^{\infty} \|\eta_i\|^2 < \infty$ and an arbitrary $\varepsilon > 0$, there exists an operator A of the form $A = \sum_{i=1}^{N} c_i T(g_i)$, such that

$$\sum_{i=1}^{\infty} \|A\xi_i - \eta_i\|^2 < \varepsilon.$$

A representation T is called *operator irreducible* if every closed operator that commutes with all of the operators $T(g)$, $g \in G$, is a scalar. (We recall that a linear operator A defined on a subspace $D(A) \subset H$ is said to be *closed* if its graph $\Gamma(A) = \{(x, Ax) \in H \oplus H; x \in D(A)\}$ is a closed subspace of $H \oplus H$.)

Problem 4. Prove that a unitary representation T is 2-irreducible if and only if it is operator irreducible.

Hint. An operator A commutes with $T(g)$ if and only if $\Gamma(A)$ is invariant under $(T+T)(g) = (T \otimes S_2)(g)$.

Theorem 2. *Let T be a unitary representation. All of the properties of k-irreducibility for $k = 1, 2, ..., \infty$ are equivalent and are equivalent to the condition $\mathscr{C}(T) = \mathbf{C} \cdot 1$.*

Proof. It is clear that if $k \geq l$, then k-irreducibility implies l-irreducibility. We shall show that 1-irreducibility implies the equality $\mathscr{C}(T) = \mathbf{C} \cdot 1$, and from this we infer ∞-irreducibility of T.

Let T be topologically irreducible and let $A \subset \mathscr{C}(T)$. Then $A + A^*$ and $i(A - A^*)$ are Hermitian and also belong to $\mathscr{C}(T)$. The spectral decompositions of these operators are invariant under the representation T. Therefore every projection that appears in this decomposition is equal either to 0 or 1. From this it follows (see theorem 2 of 4.5) that the operators $A + A^*$ and $i(A - A^*)$ are scalars. This means that A is also a scalar.

Suppose now that $\mathscr{C}(T) = \mathbf{C} \cdot 1$. The space $H \otimes L_\infty$ can be identified in a natural way with $H \oplus ... \oplus H \oplus ...$ (a countable set of summands). Let P_i be projection onto the i-th summand. Every operator $A \in \mathscr{C}(T \otimes L_\infty)$ can be written in the form of an infinite matrix $\|A_{ij}\|$, where $A_{ij} = P_i A P_j$. It is clear that $A_{ij} \in \mathscr{C}(T)$ and consequently, we have $A_{ij} = a_{ij} \cdot 1$, $a_{ij} \in \mathbf{C}$. It follows that $\mathscr{C}(T \otimes L_\infty) = 1 \otimes \mathfrak{B}(L_\infty)$.

If W is an invariant subspace in $H \otimes L_\infty$, then the orthogonal projection operator onto W lies in $\mathscr{C}(T \otimes L_\infty)$ and consequently has the form $1 \otimes P$, where P is a projection in L_∞. It follows that $W = H \otimes PL_\infty$, which proves that T is ∞-irreducible.

Problem 5. Every unitary topologically irreducible representation is tensorially irreducible (see 7.2).

Hint. Let T be a topologically irreducible unitary representation of G in a Hilbert space H, let W be a nuclear space with trivial action of G, and let V be a closed G-invariant subspace in $H \hat{\otimes} W$. Make use of the following fact. Since W is nuclear, there exists an antilinear isomorphism $\phi: H \hat{\otimes} W \to \mathrm{Hom}(H, W)$ that commutes with the action of G. Prove that $\phi(V) = \mathrm{Hom}(H, W_1)$, where W_1 is a subspace of W.

§ 8. Decomposition of Representations

One of the principal problems of the theory of representations is the problem of decomposing representations of a group G into the simplest possible components.

Many classical fields of analysis, the theory of functions, and mathematical physics are special cases of the general theory of decomposition of representations. As examples we may cite:

the spectral theory of unitary operators (decomposition of unitary representations of the group $\mathbf{Z}$);

the theory of reduction of matrices to normal form (decomposition of finite-dimensional representations of the group $\mathbf{Z}$);

the theory of Fourier series (decomposition of the regular representation of the group $\mathbf{T}$);

Fourier transforms (decomposition of the regular representation of the group $\mathbf{R}$);

the law of addition of moments in quantum mechanics (decomposition of the tensor product of two irreducible representations of the group $SO(3)$).

8.1. Decomposition of Finite Representations

Every representation T is either indecomposable or is the sum of two representations: $T = T_1 + T_2$. The same alternative applies to the representations T_1 and T_2. In the general case, this process can continue indefinitely. We say that a representation T of a group G in a space V is *finite* (or *topologically finite*) if every family of G-invariant subspaces of V (or closed G-invariant subspaces) that is strictly monotone with respect to inclusion is finite.

Every finite-dimensional representation is finite. The converse is evidently false, since irreducible representations are always finite but are not necessarily finite-dimensional.

Problem 1. Every (topologically) finite representation is the finite sum of (topologically) indecomposable finite representations.

Problem 2. Suppose that a representation T in a space V is (topologically) finite. Prove that there exists a strictly monotone finite collection of (closed) invariant subspaces

$$\{0\} = V_0 \subset V_1 \subset \ldots \subset V_{n-1} \subset V_n = V$$

such that the representations T_i, $i = 1, \ldots, n$ appearing in V_i/V_{i-1} are (topologically) irreducible.

Collections $\{V_i\}$ that satisfy the conditions of problem 2 are called *admissible*. For finite representations, we have

Theorem 1 (Jordan-Hölder). *The length of an admissible collection and the equivalence classes of the representations T_i (up to their order) are uniquely defined by the equivalence class of the representation T.*

We prove this theorem by induction on the length of the admissible collection. Suppose that the theorem has been proved for all representations having an admissible collection of length $n-1$, and let $\{V_i\}$, where $0 \leq i \leq n$, be an admissible collection for the representation T in the space V. We consider any other admissible collection

$$0 = W_0 \subset W_1 \subset \ldots \subset W_{m-1} \subset W_m = V,$$

and set $W_i' = W_i \cap V_{n-1}$. The sequence $\{W_i'\}$, $0 \leq i \leq m$, obviously enjoys all of the properties of an admissible collection except for strict monotonicity.

Problem 3. Prove that there is an index $j \in [1, m]$ such that $W_{j-1}' = W_j'$, and that the family $\{W_i'\}$, $i \neq j$, is strictly monotone.

Hint. Let W_i'' be the image of W_i in V/V_{n-1}. Since V/V_{n-1} is irreducible, the relations

$$W_0'' = W_1'' = \ldots = W_{j-1}'' = \{0\}, \quad W_j'' = W_{j+1}'' = \ldots = W_m'' = V/V_{n-1}$$

must hold for a certain $j \in [1, m]$. Verify that W_i'/W_{i-1}' is isomorphic to W_i/W_{i-1} for $i \neq j$ and that $W_j'/W_{j-1}' = \{0\}$.

By our inductive hypothesis, the equivalence classes that appear in W_i'/W_{i-1}' for $i \in [1, m]$, $i \neq j$, are exactly those of the representations $T_1, \ldots, T_{n-1}$. In particular, it follows from this that $m = n$. Furthermore, the representation that appears in W_j/W_{j-1} is obviously equivalent to T_n.

To prove the theorem, it remains to verify it for $n = 1$, which is trivial.

§ 8. Decomposition of Representations

Corollary. *The Grothendieck group of the category of finite representations is a free abelian group, generated by the equivalence classes of irreducible representations.*

The Jordan-Hölder theorem does not hold for all topologically finite representations.

We give an example. Let H be a Hilbert space with a basis $\{e_i\}$, $1 \leq i < \infty$. Let G_N be the group of invertible operators in H with the property that $(Ae_i, e_j) = (e_i, e_j)$, if $i > N$ or $j > N$. Let $G_\infty = \bigcup_{N=1}^{\infty} G_N$. We define the operator $\Lambda: e_k \mapsto \lambda_k e_k$, where $\{\lambda_k\}$ is a sequence of positive numbers with limit 0 as $k \to \infty$.

In the space $V = H \oplus H$ we consider the group G of operators defined by matrices of the form

$$g = \begin{pmatrix} A & \Lambda^{-1} B - A \Lambda^{-1} \\ 0 & B \end{pmatrix},$$

where $A, B \in G_\infty$.

Let V_1 be the set of vectors of the form $\begin{pmatrix} x \\ 0 \end{pmatrix}$, and V'_1 the set of vectors of the form $\begin{pmatrix} x \\ \Lambda x \end{pmatrix}$, $x \in H$. One can verify that $\{0\} = V_0 \subset V_1 \subset V_2 = V$ and $\{0\} = V'_0 \subset V'_1 \subset V'_2 = V$ are two admissible collections, leading to different sets of topologically irreducible representations

$$T_1(g) = A, \quad T_2(g) = B; \quad T'_1(g) = \Lambda^{-1} B \Lambda, \quad T'_2(g) = \Lambda A \Lambda^{-1}.$$

The proof given above does not carry over to the topological case, since the image of a closed subspace under a continuous mapping need not be closed.

In our example, the image V'_1 in V/V_1 is not closed (it is isomorphic to the image of the operator A). One can also show that the length of an admissible collection is not an invariant in the topological case. A relevant example is based on the fact that in a Hilbert space H, one can find three closed subspaces V_1, V_2, and W such that $V_1 \subset V_2$, $W \cap V_2 = 0$ and $(W \cup V_1)^\perp = 0$.

The result of problem 2 reduces the study of finite representations to the study of irreducible representations and certain complementary relations among them. We shall explain the nature of these relations for the case $n = 2$.

Let T be a reducible representation of a group G in a space V.

We shall consider to what extent the representation T is defined by its subrepresentations and the corresponding factor representations. Let V_1 be an invariant subspace and V_2 a certain subspace complementary to it.

In the topological case, we must suppose that V_1 is a closed complemented subspace, that is, that there exists a closed subspace V_2 such that V is isomorphic to $V_1 \oplus V_2$ (the sum being taken in the category of linear topological spaces).

In a Hilbert space V every closed subspace V_1 is complemented: it suffices to set $V_2 = V_1^\perp$.

In Banach spaces, this is no longer the case. For example, in the space $C(\mathbf{T})$, the subspace H of continuous functions that admit analytic extensions within the circle has no complement.

Let $T_1(g)$ denote the restriction of $T(g)$ to V_1, by $T_2(g)$ the restriction of $T(g)$ to V_2 followed by projection onto V_2 parallel to V_1, and by $T_{12}(g)$ the restriction of $T(g)$ to V_2 followed by projection onto V_1 parallel to V_2. The operator $T(g)$ is then written in the form of a matrix:

$$\begin{pmatrix} T_1(g) & T_{12}(g) \\ 0 & T_2(g) \end{pmatrix}. \qquad (1)$$

Evidently T_1 is a subrepresentation and T_2 a factor representation of T. The mapping $g \mapsto T_{12}(g) \in \mathrm{Hom}(V_2, V_1)$ is not a representation. It satisfies the following identity, which follows from the rule for multiplying matrices of the form (1):

$$T_{12}(g_1 g_2) = T_{12}(g_1) T_2(g_2) + T_1(g_1) T_{12}(g_2). \qquad (2)$$

We set $Z(g) = T_{12}(g) T_2(g)^{-1}$.

Problem 4. Prove that the identity (2) is equivalent to the assertion that $Z(g)$ is a one-dimensional cocycle on the group with values in the space $\mathrm{Hom}(V_2, V_1)$, on which G acts by the rule $g: A \mapsto T_1(g) A T_2(g)^{-1}$.

The cocycle just constructed depends not only on the original representation T in the space V, but also upon the choice of the complementary subspace V_2.

Problem 5. Prove that a change in the choice of V_2 leads to a change in $Z(g)$ to an arbitrary cohomological cocycle.

Thus the representation T defines a certain cohomology class $h \in H^1(G, \mathrm{Hom}(V_2, V_1))$ and is itself defined up to equivalence by this class. We will denote the equivalence class of the representation T by the symbol $[T_1, h, T_2]$. In particular we have $[T_1, 0, T_2] = [T_1 + T_2]$.

Thus the additional information needed to reconstruct the indecomposable representation T from the subrepresentation T_1 and the factor representation T_2 is a certain nonzero one-dimensional cohomology class of the group G.

The general case, where there are n irreducible representations T_i, can be studied by repeated application of the method described above. It would be of interest to find a direct method of reconstructing T from the (ordered) collection of irreducible components $\{T_i\}$ which were referred to in problem 2.

Another unsolved problem is that of carrying over the construction of one-dimensional cohomologies, as described above, to the topological case. For closed complemented subspaces, the theory has been worked out (see for example the article of C. C. Moore [119]).

The general case is significantly more complicated. It contains as a subproblem the theory of extensions of Banach spaces, which up to now is very little understood[1].

---

[1] In this connection, see the forthcoming paper by P.A. Kučment in "Funktionanalnyĭ Analiz i ego Priloženija", Vol. 10, No. 1, entitled "Reconstruction of a continuous representation from its subrepresentation and factor representation".

8.2. Irreducible Representations

The following property of algebraically irreducible representations is simple but important, and it plays a large rôle in the theory of representations.

Theorem 1 (Schur's lemma). *If representations T_1 and T_2 are algebraically irreducible, then every intertwining operator $A \in \mathscr{C}(T_1, T_2)$ is either zero or is invertible.*

The proof is immediate from the facts that $\ker A \subset V_1$ and $\operatorname{im} A \subset V_2$ are invariant subspaces, and consequently are either equal to $\{0\}$ or coincide with the entire space.

Corollary 1. *Two algebraically irreducible representations are either equivalent or disjoint.*

Corollary 2. *If T is an algebraically irreducible representation in a linear space over a field K, then the set $\mathscr{C}(T)$ of intertwining operators is a skew field over the field K.*

This assertion is sharpened in the following cases, which are important for applications.

Theorem 2. *Suppose that the dimension of an irreducible representation T in a space over the field K is no larger than countable. Then, if $K = \mathbf{C}$, we have $\mathscr{C}(T) \approx \mathbf{C}$; if $K = \mathbf{R}$, then $\mathscr{C}(T)$ is isomorphic to $\mathbf{R}$ or to $\mathbf{C}$ or to $\mathbf{H}$.*

Proof. Suppose that $A \in \mathscr{C}(T)$. We consider the expression $p(A)$, where p is a polynomial with coefficients in K. If all of the operators $p(A)$ are different from zero for $p \neq 0$, then the skew field $\mathscr{C}(T)$ contains the subring $K[A]$, which is isomorphic to the ring of polynomials. Therefore the skew field $C(T)$ contains the field $K(A)$, which is isomorphic to the field of rational functions.

Problem 1. Prove that the field $K(A)$ has dimension over K equal to the continuum.
Hint. All elements of the form $(A - \lambda)^{-1}$, $\lambda \in K$, are linearly independent over K.

Problem 2. If x is a nonzero vector in V, then the mapping $A \mapsto Ax$ is an embedding of $\mathscr{C}(T)$ in V.

The results of problems 1 and 2 show that the assumption $p(A) \neq 0$ for $p \neq 0$ contradicts the countability of the dimension of V.

Hence every element $A \in \mathscr{C}(T)$ satisfies a certain algebraic equation $p(A) = 0$. We now recall that every polynomial over the field $\mathbf{C}$ is the product of linear factors, and over the field $\mathbf{R}$ is the product of linear and quadratic factors.

For $K = \mathbf{C}$, it follows at once that $\mathscr{C}(T) = \mathbf{C} \cdot 1$. For $K = \mathbf{R}$, one can assert only that every element $A \in \mathscr{C}(T)$ generates a subfield of $\mathscr{C}(T)$ that is isomorphic to $\mathbf{R}$ or to $\mathbf{C}$.

Problem 3. Prove that any k elements $A_1, \ldots, A_k$ of the skew-field $\mathscr{C}(T)$ generate a sub skew field D of dimension $\leq 2^k$.

Hint. The space generated by all monomials $A_{i_1}A_{i_2}\ldots A_{i_l}$, $i_1 < \ldots < i_l$, is invariant under multiplication by generators, since $A_i^2 = a_i A_i + b_i 1$ and $A_i A_j + A_j A_i = (A_i + A_j)^2 - A_i^2 - A_j^2 = \alpha_{ij} A_i + \beta_{ij} A_j + \gamma_{ij} 1$.

Problem 4. Prove that any two elements $A, B \in \mathscr{C}(T)$ generate a skew field $D \subset \mathscr{C}(T)$ that is isomorphic to **R** or to **C** or to **H**.

Hint. It suffices to consider the case where D properly contains **C** as a subfield. From the result of the preceding problem, it follows that $\dim_{\mathbf{C}} D = 2$. Consider the mapping $I: D \to D: x \mapsto ixi^{-1}$, where i is the imaginary unit in **C**. Reducing I to diagonal form, prove that $D = \mathbf{C} + j\mathbf{C}$, where j is an element such that $j^2 = -1$, $ji = -ij$.

Problem 5. Prove that any three elements $A, B, C \in \mathscr{C}(T)$ generate a sub skew field D that is isomorphic to **R** or to **C** or to **H**.

Hint. It suffices to prove that D cannot be a two-dimensional left space over **H** with basis $1, l$, where $l^2 = -1$.

Problem 6. Every finite-dimensional skew field over **R** is isomorphic to **R** or **C** or **H**.

Hint. Use the result of problem 5 and use induction on the dimension.

Theorem 2 follows at once from problems 3 and 6.

Corollary. *For an irreducible representation T, the intertwining number $c(T)$ is equal to 1 if $K = \mathbf{C}$ and can assume the values 1, 2, 4 if $K = \mathbf{R}$.*

A real representation T is called a representation of *real, complex,* or *quaternion* type, depending upon the form of the skew field $\mathscr{C}(T)$.

It is clear that a representation of complex (or quaternion) type is obtained from a complex (or quaternion) representation by contracting the scalar field. For compact groups, we shall give in the sequel a criterion for the membership of a representation in one of these types.

It turns out that the type of a representation is determined by its behavior upon extension of the scalar field from **R** to **C**.

Theorem 3. *Let $T_{\mathbf{C}}$ be the complexification of a real representation T. If T is an irreducible representation of real, complex, or quaternion type, then $T_{\mathbf{C}}$ is irreducible or is the sum of two inequivalent irreducible representations or of two equivalent irreducible representations, depending upon the type of T.*

Corollary. *If T is an irreducible real representation of real, complex, or quaternion type, then the algebra $\mathscr{C}(T_{\mathbf{C}})$ is isomorphic to $\mathbf{C}$, $\mathbf{C} \oplus \mathbf{C}$, or to $\mathrm{Mat}_2(\mathbf{C})$ respectively.*

We can infer this corollary also from the following general fact.

Problem 7. Prove that $\mathscr{C}(T_{\mathbf{C}}) = \mathscr{C}(T) \underset{\mathbf{R}}{\otimes} \mathbf{C}$.

Hint. The space $V_{\mathbf{C}}$ of the representation $T_{\mathbf{C}}$ can be realized as the direct sum $V \oplus V$, where V is the space of the representation T. Then the algebra $\mathscr{C}(T_{\mathbf{C}})$ will be realized as the subalgebra of $\mathrm{Mat}_2(\mathscr{C}(T))$, consisting of the elements that commute with the operator $I = \begin{pmatrix} 0 & 1 \\ -1 & 0 \end{pmatrix}$, which gives multiplication by i in $V_{\mathbf{C}}$.

§ 8. Decomposition of Representations

We prove theorem 3 by considering the real invariant subspaces in $V_C \approx V \oplus V$.

By Schur's lemma, every such subspace W goes either into zero or into the entire space under projection on each summand. Hence W can be one of the following types:
1) $W = \{0\} \oplus \{0\}$,
2) $W = \{0\} \oplus V$,
3) $W = V \oplus \{0\}$,
4) $W = V \oplus V$,
5) $W = \{(x, Ax), x \in V\}$, where A is a nonzero element of $\mathscr{C}(T)$. It is natural to call the last space the *graph of the operator A*.

Problem 8. Nontrivial complex invariant spaces W in V_C are the graphs of operators $A \in \mathscr{C}(T)$ such that $A^2 = -1$.

Hint. A space W will be a complex subspace in V_C if and only if it is invariant under the operator $I = \begin{pmatrix} 0 & 1 \\ -1 & 0 \end{pmatrix}$, which gives multiplication by i in V_C.

If $\mathscr{C}(T) = \mathbf{R}$, then there are no such operators, and the space V_C is irreducible.

If $\mathscr{C}(T) = \mathbf{C}$, then there are exactly two such operators A (corresponding to the elements $i \in \mathbf{C}$). Thus there exist exactly two nontrivial invariant subspaces, in which there are realized inequivalent irreducible representations (compare 8.3 *infra*).

Finally, if $\mathscr{C}(T) = \mathbf{H}$, then there are an infinite number of operators A having the required property. All of them are carried into each other by inner automorphisms of $\mathbf{H}$. Thus V_C in this case is the sum of two subspaces, in which equivalent representations are realized.

8.3. Completely Reducible Representations

Suppose that the space V of a representation T is decomposed into the direct sum of invariant subspaces $V_i: V = \bigoplus_{i=1}^{n} V_i$. Let P_i be the projection operator onto V_i parallel to the remaining V_j.

Problem 1. The operators P_i enjoy the following properties:
a) $P_i^2 = P_i$;
b) $P_i P_j = P_j P_i = 0$ for $i \neq j$;
c) $\sum_{i=1}^{n} P_i = 1$;
d) $P_i \in \mathscr{C}(T)$.

(1)

The converse is also true (see problem 2).

Problem 2. Every collection of operators in the space V that enjoy the properties (1) produces a decomposition of V into a direct sum of invariant subspaces $V_i = P_i V$.

Hint. Property a) means that P_i is the projection operator onto the subspace $V_i = P_i V$ parallel to the subspace $W_i = (1 - P_i) V$. Property b) means that the spaces

V_i are linearly independent. Property c) means that the sum of all of the V_i coincides with V. Finally, property d) is equivalent to the invariance of the V_i.

Elements of a ring such that $x^2 = x$ are called *idempotent*. Two idempotents x and y are called *orthogonal* if $xy = yx = 0$. The results of problems 1 and 2 show that the problem of decomposing the space V into the sum of invariant subspaces is equivalent to the problem of decomposing the unit in the ring $\mathscr{C}(T)$ into a sum of orthogonal idempotents.

There is an order relation in the set of idempotents: we set $x < y$ if $xy = yx = x$.

Problem 3. Let P and Q be two idempotents in the ring of linear operators in the space V. Prove that the relation $P < Q$ is equivalent to the inclusion $PV \subset QV$.

Corollary. *Decompositions of the space V of a representation T into a sum of irreducible subspaces correspond to decompositions of the unit in $\mathscr{C}(T)$ into a sum of minimal idempotents.*

Suppose that the representation T is completely reducible and finite. Then it is the sum of a finite number of irreducible representations: $T = \sum_{i=1}^{n} T_i$. By the Jordan-Hölder theorem (see 8.1), the classes of the representations T_i are uniquely determined.

However, in general one can choose the subspaces V_i in which these representations act in a variety of ways.

To describe the situation arising here it is useful to introduce the following definition.

A representation T is called *primary* if it cannot be represented as the sum of two disjoint representations.

Problem 4. Prove that a finite completely reducible representation T is primary if and only if all of its irreducible components are equivalent.

Hint. Use the Jordan-Hölder theorem, Schur's lemma, and the additivity of the intertwining number.

Problem 5. Prove that a finite completely reducible representation is primary if and only if it has the form $T = U \otimes S$, where U is irreducible and S is a trivial representation (i.e., $S(g) \equiv 1$).

Hint. Use the isomorphism $V \otimes_K K^n = V \oplus \ldots \oplus V$ (n summands) and the result of problem 1.

Theorem 1. *Let T be a completely reducible finite representation in the space V. Then there exists a unique decomposition of V into a sum of invariant subspaces V_i for which the representations $T_i = T|_{V_i}$ are primary and pairwise disjoint. Every invariant subspace $W \subset V$ has the property that $W = \bigoplus_i W \cap V_i$.*

The proof of existence is simple. If T is not primary, then it is the sum of two disjoint summands T_1 and T_2. We apply this alternative again to T_1 and T_2. Continuing this process, we come to a decomposition of T into a sum of primary pairwise disjoint representations: $T = \sum_{i=1}^{n} T_i$.

§ 8. Decomposition of Representations

Let V_i be the space in which T_i acts, and let P_i be the projection operator onto V_i parallel to the sum of the remaining V_j.

Problem 6. The operators P_j belong to the center of the algebra $\mathscr{C}(T)$.

Hint. If $A \in \mathscr{C}(T)$, then we have $P_j A P_i \in \mathscr{C}(T_i, T_j)$. The pairwise disjointness of the T_i implies that $P_j A P_i = 0$ for $i \neq j$. Hence we have

$$A = \left(\sum_j P_j\right) \cdot A \cdot \left(\sum_i P_i\right) = \sum_{i,j} P_j A P_i = \sum_i P_i A P_i.$$

It follows that $A P_i = P_i A P_i$ and $P_i A = P_i A P_i$.

The result of problem 6 has a converse, in the following sense. Let $C(\mathscr{C}(T))$ be the center of the algebra $\mathscr{C}(T)$.

Problem 7. If the projection P belongs to $C(\mathscr{C}(T))$, then the spaces $W_1 = PV$ and $W_2 = (1-P)V$ are invariant, and the representations $S_1 = T|_{W_1}$ and $S_2 = T|_{W_2}$ to which they give rise are disjoint.

Hint. Invariance of W_1 and W_2 is equivalent to the condition $P \in \mathscr{C}(T)$. The disjointness of S_1 and S_2 follows from the relations $\mathscr{C}(S_1, S_2) = (1-P)\mathscr{C}(T)P$, $\mathscr{C}(S_2, S_1) = P\mathscr{C}(T)(1-P)$.

The results of problems 6 and 7 show that there is a one-to-one correspondence between decompositions of the representation T into the sum of disjoint subrepresentations and decompositions of the unit in $C(\mathscr{C}(T))$ into the sum of orthogonal idempotents.

A decomposition of T into primary disjoint representations obviously corresponds to a decomposition of the unit in $C(\mathscr{C}(T))$ into minimal idempotents.

Problem 8. Prove that in a commutative ring, there is at most one decomposition of the unit as a sum of minimal idempotents.

Hint. If x and y are two commuting idempotents, then xy is also an idempotent and has the properties that $xy < x$ and $xy < y$.

Let us prove the last assertion of theorem 1. Let W be an invariant subspace in V. Let P be the projection onto W parallel to the invariant complement of W (this complements exists, since T is completely reducible). Then we have $P \in \mathscr{C}(T)$. Furthermore, we may write $P = P \sum_{i=1}^{n} P_i = \sum_{i=1}^{n} P P_i$.

Since $P_i \in C(\mathscr{C}(T))$, the operators P and P_i commute with each other. Hence the operators PP_i are projections onto the subspaces $PP_i V = W \cap V_i$. Therefore

$$W = PV = \left(\sum_{i=1}^{n} PP_i\right) V = \bigoplus_{i=1}^{n} (W \cap V_i),$$

and the theorem is proved.

The decomposition of a primary reducible representation into irreducible components is never unique. To see this, suppose that the primary representation T has the form $U \otimes S$, where U is an irreducible representation in the space W' and S is the trivial representation in the space W'' (see problem 5).

Every decomposition of W'' as a sum of one-dimensional subspaces W''_i, $i = 1, 2, \ldots, \dim W''$, gives a decomposition of the space $V = W' \otimes W''$ as a sum of irreducible subspaces $V_i = W' \otimes W_i''$. If T is reducible, then we have $\dim W'' > 1$ and such a decomposition is not unique.

Problem 9. Suppose that $\mathscr{C}(U) = K$. Then every invariant subspace in V has the form $W' \otimes W_0''$, where W_0'' is a certain subspace of W''.

Hint. Use the isomorphism $\mathscr{C}(T) = \mathrm{Mat}_n(\mathscr{C}(U))$, where $n = \dim W''$.

We now state a useful sufficient condition for complete reducibility of a representation.

Theorem 2. *If the space V of a representation T is generated by its irreducible subspaces, then T is completely reducible.*

Proof. Let W be an invariant subspace of V. Consider a collection $\{V_\alpha\}$, $\alpha \in A$, of invariant irreducible subspaces having the following property: if $x_\alpha \in V_\alpha$, $y \in W$, and $y = \sum_\alpha x_\alpha$, then all x_α are equal to zero. We call such a collection of subspaces admissible. By Zorn's lemma, there exists a maximal admissible collection $\{V_\alpha\}$. Then $W' = \sum_{\alpha \in A} V_\alpha$ is an invariant complement to W. In fact, every irreducible subspace V_0 lies entirely within the sum $W + W'$ or is disjoint from this sum. The second possibility contradicts the maximality of the collection $\{V_\alpha\}$.

In conclusion, we show how to derive a certain classical theorem from the results of this section.

Theorem 3 (Burnside). *If T is an irreducible representation of a group G in a finite-dimensional linear space V over an algebraically closed field K, then the linear hull of the operators $T(g)$, $g \in G$ coincides with the space $\mathrm{End}\,V$ of all linear operators on V.*

Proof. We consider the representation $\tilde{T}$ of the group G in the space $\mathrm{End}\,V$ defined by the following formula:

$$\tilde{T}(g): A \mapsto T(g)A.$$

Identifying $\mathrm{End}\,V$ with $V \otimes V^*$ (see 3.3), we see that the representation $\tilde{T}$ is the tensor product of the representation T in V and the n-dimensional trivial representation in V^*, where $n = \dim V$. By theorem 2, $\tilde{T}$ is completely reducible. Since the field K is algebraically closed, the skew field $\mathscr{C}(T)$ coincides with K. By the result of problem 9, every invariant subspace in $\mathrm{End}\,V$ has the form $W = V \otimes V_1$, where V_1 is a certain subspace in V^*. We choose a basis in V such that V_1 is generated by the first k coordinates. Then W is realized in the form of a subspace of $\mathrm{Mat}_n(K) \approx \mathrm{End}\,V$, consisting of matrices which have zero entries except for the first k columns.

To prove the theorem, it remains to remark that the linear hull of the operators $T(g)$, $g \in G$ is an invariant subspace in $\mathrm{End}\,V$ and is contained in none of the subspaces W of the kind described above, for any $k < n$.

Corollary. *A finite group of order N cannot possess irreducible representations of dimension $> \sqrt{N}$ over an algebraically closed field.*

We shall see *infra* that this result can be considerably strengthened (see § 10).

8.4. Decomposition of Unitary Representations

Every unitary representation is completely reducible. Therefore the results of the preceding 8.3 are valid for finite unitary representations.

It turns out that also in the infinite case, there is a theory of decomposition for unitary representations. Here we use not decompositions into direct sums but decompositions into continuous sums (integrals) of Hilbert spaces. The necessity for such decompositions is evident even in the very simplest cases.

Consider for example the representation T of the group $\mathbf{R}$ in the space $H = L_2(\mathbf{R})$, given by the formula

$$[T(t)f](x) = e^{itx} f(x).$$

Let E be an arbitrary measurable subset of $\mathbf{R}$. It is evident that the subspace $H(E) \subset H$ consisting of all functions equal to zero outside of E is invariant under the operators of the representation.

Problem 1. Every closed invariant subspace of H has the form $H(E)$.

Hint. Every operator in $\mathscr{C}(T)$ is multiplication by some function of x (see problem 10 in 4.4).

Corollary. *The representation T has no irreducible subrepresentations.*

In fact, every subspace $H(E)$ is either zero (if E has measure zero) or has nontrivial invariant subspaces (if E has positive measure).

Let X be a space with a measure μ and suppose that to every point $x \in X$, there correspond a Hilbert space H_x and a unitary representation T_x of the group G in H_x. We will say that a unitary representation T of the group G is the *continuous sum* (or *integral*) of the representations T_x, and we will write $T = \int_X T_x d\mu(x)$, if

1) the space H of the representation T is the continuous sum of the spaces H_x with respect to the measure μ (see 4.5);
2) all of the operators $T(g)$ are decomposable and have the form

$$[T(g)f](x) = T_x(g) f(x).$$

It is clear that the representation T of the group $\mathbf{R}$ in $L^2(\mathbf{R})$ described above is the continuous sum of the one-dimensional representations

$$T_x(t) = e^{itx}.$$

It is natural to ask whether an arbitrary representation T can be decomposed into the continuous sum of simpler representations (for example, irreducible or primary). Suppose that we have a decomposition of the space H of the representation T into the continuous sum $\int_X H_x d\mu(x)$. By theorem 1 of 4.5, this decomposition is given by a commutative C^*-subalgebra $\mathscr{D}_0$ of continuously diagonal operators. The space X coincides with $\mathfrak{M}(\mathscr{D}_0)$ and the measure μ coincides with the basis measure of the identity representation of $\mathscr{D}_0$ in H. It is clear that the operators of the representation T are decomposable if and only if the algebra $\mathscr{D}_0$ is contained in $\mathscr{C}(T)$.

Nevertheless, this condition is in general not sufficient. The point is that the relations

$$T_x(g_1 g_2) = T_x(g_1) T_x(g_2)$$

for fixed g_1 and g_2 may fail on a set of measure zero that depends upon g_1 and g_2.

Furthermore, the correspondence $g \mapsto T_x(g)$ in general may fail to be continuous. We can remove these difficulties under certain additional assumptions about the group G and the space H.

Theorem 1. *Suppose that the group G is locally compact and has a countable basis and also that the space H is separable. Then an arbitrary commutative C*-subalgebra $\mathscr{D}_0$ of $\mathscr{C}(T)$ gives a decomposition of the representation T as a continuous sum*

$$T = \int_X T_x \, d\mu(x),$$

where $X = \mathfrak{M}(\mathscr{D}_0)$, and μ is the basis measure of the identity representation of $\mathscr{D}_0$.

We shall give the proof here only for the case of a countable discrete group. The general case will be investigated *infra* (see § 10), when we shall have the concept of group ring at our disposal.

Since all of the operators $T(g)$, $g \in G$ are decomposable, we can find for each $g \in G$ a measurable family of operators $T_x(g)$ such that

$$[T(g) f](x) = T_x(g) f(x). \tag{1}$$

For fixed g, g_1, g_2 the relations

$$T_x(g)^* = T_x(g^{-1}), \qquad T_x(g_1 g_2) = T_x(g_1) T_x(g_2) \tag{2}$$

hold almost everywhere in x. Since the group G is countable, we can alter the operators on a set of measure zero to obtain the equalities (2) for all $x \in X$, without disturbing the equalities (1).

Since the group G is discrete, the correspondence $g \mapsto T_x(g)$ is continuous, and in view of (2) it yields a unitary representation of G in the space H_x.

Theorem 2. *Suppose that a unitary representation T of a separable group G in a separable Hilbert space H is the continuous sum of representations T_x, $x \in X$, with respect to a measure μ. All of the representations T_x are irreducible if and only if the algebra $\mathscr{D}$ of diagonal operators is a maximal commutative subalgebra in $\mathscr{C}(T)$.*

Proof. Suppose that the algebra $\mathscr{D}$ is nonmaximal and that the operator A belongs to $\mathscr{D}' \cap \mathscr{C}(T)$. Since $\mathscr{D}' = \mathscr{R}$ consists of decomposable operators, the operator A has the form

$$[Af](x) = A_x f(x).$$

§ 8. Decomposition of Representations

Let $g_1, \ldots, g_n, \ldots$ be a countable everywhere dense subset of G. Then the equalities $T_x(g_i) A_x = A_x T_x(g_i)$ hold almost everywhere on X. It follows that $A_x \in \mathscr{C}(T_x)$ for almost all x.

Since $A \notin \mathscr{D}$, the operators A_x fail to be scalar operators on a set E of positive measure. Hence the representations T_x are reducible for $x \in E$.

Conversely, suppose that the representations T_x are reducible on a set E of positive measure. Then we have $\mathscr{C}(T_x) \neq \mathbf{C} \cdot 1$ for $x \in E$. Suppose that we have succeeded in finding, for each $x \in E$, an operator $A_x \in \mathscr{C}(T_x) \setminus \mathbf{C} \cdot 1$ such that the family $\{A_x\}$ is measurable. Then the operator A defined by this family would belong to $\mathscr{D}' \cap \mathscr{C}(T)$ and would not belong to $\mathscr{D}$, which would prove the nonmaximality of $\mathscr{D}$.

The possibility of a measurable selection of the operators A_x follows from the following general theorem of Luzin-Jankov. If X and Y are complete separable metric spaces, if Y is compact, and if Z is a Borel subset of $X \times Y$, whose projection onto X is equal to X, then there exists a mapping $f: X \to Y$ which is measurable with respect to any complete regular measure on X and whose graph lies in Z.

We omit the proof of this fact (it is given in Appendix III of the book of M. A. Naĭmark [42]).

Corollary. *Every unitary representation T of a locally compact group G with countable basis in a separable Hilbert space H can be decomposed into a continuous sum of irreducible representations*: $T = \int_X T_x d\mu(x)$.

One can single out a certain class of representations whose behavior under decomposition preserves the main features of the finite case (see 8.3).

We will say that a representation T:
1) has a *simple spectrum*, if the algebra $\mathscr{C}(T)$ is commutative;
2) has a *homogeneous spectrum of multiplicity n* (or is homogeneous of degree n), if it is equivalent to the tensor product of a representation with simple spectrum and an n-dimensional trivial representation;
3) belongs to *type I* if it is the sum of homogeneous representations.

Problem 2. A representation T belongs to type I (or is homogeneous of degree n) if and only if the algebra $\mathscr{C}(T)$ belongs to type I (or is homogeneous of degree n).

Hint. Use theorem 1 and the results of 4.5.

Thus every representation of type I has the form $T = \sum_{k=1}^{k=\infty} U_k \otimes S_k$, where the U_k's are representations with simple spectrum and S_k is the trivial k-dimensional representation. We call the group G *tame* if all of its unitary representations are of type I and *wild* in the contrary case.

In the author's opinion, this terminology, which has been borrowed from geometric topology (tame and wild knots), well expresses the nature of these groups and their representations. We shall see *infra* that tame groups enjoy many good properties—uniqueness of decomposition of representations, the existence of characters, T_1 separation of the dual space—which wild groups lack.

Problem 3. Prove that every primary representation of a tame group G is a multiple of an irreducible representation.

Hint. For a primary representation, the center of the algebra $\mathscr{C}(T)$ consists of scalar operators. Hence $\mathscr{C}(T)$ is a factor of type I.

One can show that this property is characteristic of tame groups—every wild group admits a primary representation that is not a multiple of an irreducible representation.

For tame groups, the theory of decomposition can be brought to a conclusion in the following sense.

Theorem 3. *Let G be a locally compact tame group with countable basis and let T be a unitary representation of G in a separable Hilbert space. Then the following assertions hold.*

1) There exists a collection of pairwise disjoint Borel measures μ_k on the space $\hat{G}$ dual to G such that the representation T can be written in the form

$$T = \sum_{k=1}^{k=\infty} U_k \otimes S_k,$$

where S_k is the trivial k-dimensional representation and U_k is a representation with simple spectrum, which can be written in the form of a continuous sum

$$U_k = \int_{\hat{G}} T_\lambda d\mu(\lambda).$$

(The representation T_λ belongs to the class $\lambda \in \hat{G}$.)

2) The measures μ_k are defined up to equivalence by the representation T.

3) An arbitrary collection $\{\mu_k\}$ of pairwise disjoint Borel measures arises in the manner just described from a certain representation T.

This result can be formulated in terms of decomposition into primary representations.

Every representation T can be written in the form

$$T = \int_{\hat{G}} W_\lambda d\mu(\lambda),$$

where W_λ is a primary representation of the form $T_\lambda \otimes S_\lambda$, T_λ belongs to the class $\lambda \in \hat{G}$, and S_λ is the trivial representation of dimension $n(\lambda)$. The measure μ is defined by the representation T up to equivalence and the measurable function $n(\lambda)$ is defined by T almost everywhere (for the measure μ).

Transition from one formulation to the other is carried out just as in 4.4.

The decomposition of a unitary representation T into primary representations can be carried out for an arbitrary locally compact group with a countable basis. To do this, it suffices to take the center of the algebra $\mathscr{C}(T)$ as the algebra $\mathscr{D}$ of diagonal operators. However, the further decomposition is nonunique, not only in the choice of subspaces of decomposition (or equivalently the choice of the algebra of diagonal operators) but also in the irreducible components themselves. In this connection, see § 19 in the third part of the book.

The class of tame groups is fairly large.

§ 9. Invariant Integration

Problem 4. Prove that every commutative group is tame.

Hint. If T is a representation of a commutative group, then the algebra $\mathscr{C}(T)'$ is commutative. Hence $\mathscr{C}(T)$ belongs to type I.

We shall see in the sequel that all compact groups, connected semisimple groups, and exponential Lie groups (in particular, nilpotent groups) are tame. It is also known (see the article [70]) that every linear algebraic group (that is, a subgroup of $GL(n, \mathbf{R})$ defined by algebraic equations) is tame.

On the other hand, discrete groups are as a rule wild. In fact, a countable discrete group is tame only if it contains a commutative subgroup of finite index (see the article [136]).

Among solvable Lie groups, we encounter both tame and wild groups. A criterion distinguishing these cases will be given in § 15.

§ 9. Invariant Integration

9.1. Means and Invariant Measures

One of the powerful tools for study of topological groups and their representations is invariant means. Let L be an arbitrary linear topological space consisting of functions on a G-space X and invariant under translations. An *invariant mean* is a positive[1] linear functional on L that is invariant under action of the group G. The most important case is that in which X is a topological group and G is the group of left, right, or two-sided translations on X. The corresponding means are also called *left, right,* or *two-sided*.

A group G is called *amenable* if there exists a left invariant mean on the space $B(G)$ of all continuous bounded functions on G with the norm $\|f\| = \sup |f(x)|$. In this case there is a two-sided mean on the subspace of $B(G)$ consisting of all right uniformly continuous functions.

von Neumann's criterion. *The group G fails to be amenable if and only if there exists functions $f_1, \ldots, f_N \in B(G)$ and elements $g_1, \ldots, g_N \in G$ such that*

$$\sum_{i=1}^{N} [f_i(g) - f_i(g_i g)] \geq 1 \quad \text{for all} \quad g \in G.$$

It is known that all solvable groups are amenable. As examples of nonamenable groups, we cite the free group with two generators or any noncompact semisimple Lie group.

In recent years, the theory of amenable groups has been vigorously pursued. It has been shown that for locally compact groups, amenability is equivalent to many other important properties: the existence of a fixed point for an arbitrary affine action of the group on a convex compact set; the existence of a two-sided

---

[1] That is, assuming nonnegative values on nonnegative functions.

mean on $B(G)$; the so-called R-property, which asserts that the trivial representation is weakly contained in the regular representation; and others.

The consideration of an invariant mean on $C_0(G)$ leads to the important concept of Haar measure.

When G is compact, the space $C_0(G)$ coincides with $C(G)$, and an invariant mean must have the form

$$I(f) = \int_G f(x) d\mu(x), \qquad (1)$$

where μ is a certain regular Borel measure (see 4.1).

If G is locally compact and has a countable basis, the space $C_0(G)$ is the union of a countable set of spaces of the form $C(K)$, where K is a compact subset of G. In this case the invariant mean (if it exists) can also be written in the form (1), where μ is a certain σ-finite measure.

Theorem 1 (Haar). *On every locally compact group with a countable basis, there exists a nonzero left-invariant σ-finite regular Borel measure. It is defined uniquely up to a numerical factor.*

We call this measure *left Haar measure*. We define *right Haar measure* analogously.

We will denote left Haar measure by the symbol $d_l g$, and right Haar measure by the symbol $d_r g$, omitting the indices l and r in those cases where a two-sided invariant measure exists.

By generalizing the notion of measure (dropping the condition of σ-finiteness), we obtain

Theorem 2 (A. Weil). *A complete topological group G admits a nonzero left-invariant regular Borel measure if and only if the group G is locally compact. If this condition holds, then the left invariant measure is defined uniquely up to a numerical factor.*

For specific groups, there is usually no difficulty in explicitly constructing a left invariant measure (see 6.4). Hence the principal value of the theorem is the assertion of uniqueness. We give several useful formulas, which are corollaries of the uniqueness property.

Problem 1. Let G be a locally compact group. There is a continuous homomorphism Δ_G of the group G into the multiplicative group of positive real numbers for which the following equalities hold:

$$d_r(xy) = \Delta_G(x) d_r y, \qquad d_l(xy) = \Delta_G(y)^{-1} d_l x, \qquad (2)$$

$$d_r x = \text{const} \cdot \Delta_G(x) d_l x, \qquad d_l(x) = \text{const} \cdot \Delta_G(x)^{-1} d_r x, \qquad (3)$$

$$d_r(x^{-1}) = \Delta_G(x)^{-1} d_r x = \text{const} \cdot d_l x, \qquad d_l(x^{-1}) = \Delta_G(x) d_l x. \qquad (4)$$

Hint. The left side of the first of the equalities (2) is right invariant. Hence there is a number $\Delta_G(x)$ such that this equality holds. It is clear that $x \mapsto \Delta_G(x)$ is a homo-

§ 9. Invariant Integration

morphism. Its continuity follows from the formula $\Delta_G(x) = \int_G f(x^{-1}y) d_r y$, where f is a continuous function on G with compact support such that $\int_G f(y) d_r y = 1$. The remaining relations are proved similarly.

Corollary. *The equivalence class of a left Haar measure is two-sided invariant.*

The function Δ_G is sometimes called the *modulus* of the group G. If $\Delta_G \equiv 1$, then one says that the group G is *unimodular*. In this case, there is a two-sided invariant measure.

Problem 2. Prove that if the group G is compact, then Haar measure is finite and two-sided invariant.

Hint. The group of positive numbers under multiplication has $\{1\}$ as its only compact subgroup.

Many questions of the theory of representations are most easily dealt with for unimodular groups.

For every locally compact group G, we define groups G_0 and G_1 as follows:

G_0 is the kernel of the homomorphism Δ_G;

G_1 is the semidirect product of G and $\mathbf{R}$ associated with the mapping $\Delta_G : G \to \operatorname{Aut} \mathbf{R}$.

Problem 3. Prove that the groups G_0 and G_1 are unimodular.

We now list some facts about measures on G-spaces.

Glimm's theorem[1]. *The following properties are equivalent for metrically complete separable G-spaces X.*

1. *Every G-orbit in X is locally closed* (that is, is the intersection of an open and a closed set).

2. *The factor-space X/G satisfies the T_0 separation axiom.*

3. *There exists a countable family of G-invariant Borel sets in X that distinguish any two G-orbits.*

4. *Every G-ergodic Borel measure on X is concentrated on a single orbit.* (We recall that a measure is called *G-ergodic* if every G-invariant measurable set either has measure zero or has complement of measure zero.)

If X is a homogeneous G-space and G is a locally compact group, then there exists a *quasi-invariant measure* on X, which is unique up to equivalence. (A measure is said to be quasi-invariant if it goes into an equivalent measure under all translations.) If X admits a countable basis, then such a measure μ is equivalent to a finite measure and can be defined by the equality

$$\int_X f(x) d\mu(x) = \int_G f(p(g)) dv(g),$$

where p is the natural projection of G onto X, and v is a finite measure on G that is equivalent to Haar measure.

Suppose for definiteness that $X = G/H$ is a left G-space and that s is a Borel mapping of X into G such that $p \circ s = 1$. (If G is a Lie group, such a mapping

---

[1] See J. Glimm, Locally compact transformation groups, Trans. Amer. Math. Soc. 101 (1961), 124–138.

can be chosen so as to be smooth almost everywhere.) Then every element of G can be written uniquely in the form

$$g = s(x)h, \quad x \in X, \quad h \in H, \tag{5}$$

and thus G can be identified with $X \times H$.

Under this identification, Haar measure on G goes into a measure equivalent to the product of a quasi-invariant measure on X and Haar measure on H. Therefore the equality

$$d_r g = \rho(x, h) d\mu(x) d_r h$$

holds. On each side of this equality, translate on the right by an element of H. Doing this, we see that the density $\rho(x, h)$ does not depend upon h. Replacing μ by an equivalent measure, we see that ρ can be an arbitrary positive function of x.

It is natural to choose ρ so that the corresponding measure μ undergoes the simplest possible transformations under translations of the G-space X. It turns out that for this we must set $\rho(x) = \Delta_G(s(x))$. We denote the corresponding measure by μ_s (it depends upon the choice of the mapping s).

Problem 4. Prove the validity of the equalities

$$d_l g = \frac{\Delta_H(h)}{\Delta_G(h)} d\mu_s(x) d_l h \tag{6}$$

and

$$\frac{d\mu_s(gx)}{d\mu_s(x)} = \frac{\Delta_G(h(g,x))}{\Delta_H(h(g,x))}, \tag{7}$$

where $h(g, x)$ is defined by the relation

$$g s(x) = s(g x) h(g, x). \tag{8}$$

Problem 5. The measure μ_s is defined up to a numerical factor by the relations (7) and (8).

Problem 6. There exists a G-invariant measure on the space $x = G/H$ if and only if we have

$$\Delta_G(h) = \Delta_H(h) \quad \text{for all} \quad h \in H. \tag{9}$$

Problem 7. If condition (9) is satisfied, then the measure μ_s is G-invariant and does not depend upon the choice of the mapping s.

Of course all of the above results apply as well to a right G-space $Y = H \backslash G$. We list the corresponding formulas

$$g = h s(y), \quad h \in H, \quad y \in Y, \tag{5'}$$

§ 9. Invariant Integration

$$d_r g = \frac{\Delta_G(h)}{\Delta_H(h)} dv_s(y) d_r h, \tag{6'}$$

$$\frac{dv_s(yg)}{dv_s(y)} = \frac{\Delta_H(h(y,g))}{\Delta_G(h(y,g))}, \tag{7'}$$

where $h(y,g)$ is defined by the relation

$$s(y)g = h(y,g)s(yg). \tag{8'}$$

Sometimes we can select a subgroup K of G that is complementary to H in the sense that almost every element of G can be uniquely written in the form

$$g = k \cdot h, \quad k \in K, \quad h \in H.$$

In this case, it is natural to identify $X = G/H$ with K and to choose s as the embedding of K in G. A simple verification shows that the measure μ_s will then be left Haar measure on K (since $h(k_1, k_2) = e$). Formula (6) in this case assumes the form

$$d_l g = \frac{\Delta_H(h)}{\Delta_G(h)} d_l k d_l h. \tag{10}$$

If G is unimodular, this equality can be rewritten as

$$dg = d_l k \cdot d_r h. \tag{11}$$

For a proof of Weil's theorem and further facts about integration on groups, see the books [8], [54]. A detailed exposition of the theory of amenable groups can be found in the book of F. P. Greenleaf, *Invariant means on topological groups*, Van Nostrand Math. Studies, 16.

9.2. Applications to Compact Groups

Suppose that G is a compact group and that dg is Haar measure on G, normalized by the condition $\int_G dg = 1$.

We consider a representation T of the group G in a complete locally convex topological space V. Let V_0 denote the subspace of V consisting of all G-invariant vectors.

Theorem 1. *The operator*

$$P = \int_G T(g) dg \tag{1}$$

commutes with the operators of the representation and is the operator of projection onto V_0.

Before we prove this theorem, we must explain the meaning of the integral (1).

Let X be a space with a measure and f a function on X with values in a locally convex space L. We call f *weakly integrable* if for every $\chi \in L'$, the scalar function $x \mapsto \langle \chi, f(x) \rangle$ is integrable with respect to the measure μ. The *weak integral* of the function f with respect to the measure μ is the element of the space $(L')^*$ (the algebraic dual to L') defined by the equality

$$\left\langle \chi, \int_X f(x) d\mu(x) \right\rangle = \int_X \langle \chi, f(x) \rangle d\mu(x) \quad \text{for all} \quad \chi \in L'. \tag{2}$$

We denote this integral by the symbol $\int_X f(x) d\mu(x)$.

One can show that the integral $\int_X f(x) d\mu(x)$ belongs to the closed convex hull of the set of values of the function f on X (or on any subset of full measure in X). We regard the space L as embedded in $(L')^*$: every element of L can be thought of as a linear functional on L'.

In particular, if X is compact and f is a continuous function, then the integral $\int_X f(x) d\mu(x)$ is an element of L. The expression (1) can be considered as the integral of a function on G whose values are in the space $\operatorname{End} V$.

Problem 1. Prove that the equality

$$P\xi = \int_G T(g)\xi \, dg$$

holds for all $\xi \in V$.

Hint. Prove the equality $\langle \chi, P\xi \rangle = \left\langle \chi, \int_G T(g)\xi \, dg \right\rangle$ for all $\chi \in V'$, comparing both sides with the expression $\int_G \langle \chi, T(g)\xi \rangle \, dg$.

From this it follows in particular that $P(\xi) = \xi$ for all $\xi \in V_0$.

Problem 2. If A is a bounded operator in V, then we have

$$AP = \int_G A T(g) \, dg, \quad PA = \int_G T(g) A \, dg.$$

Hint. For all $\xi \in V$ and $\chi \in V'$, we find

$$\langle \chi, AP\xi \rangle = \left\langle \chi, A \int_G T(g)\xi \, dg \right\rangle = \int_G \langle \chi, A T(g)\xi \rangle \, dg.$$

The assertion of theorem 1 now follows from the following relations.

$$T(g)P = \int_G T(g) T(g_1) dg_1 = \int_G T(gg_1) dg_1 = \int_G T(g_2) dg_2 = P,$$

$$P T(g) = \int_G T(g_1) T(g) dg_1 = \int_G T(g_1 g) dg_1 = \int_G T(g_2) dg_2 = P.$$

§ 9. Invariant Integration

Thus averaging on a group gives us a universal method of constructing invariants of the group: *for every vector $\xi \in V$, the vector $P\xi = \int_G T(g)\xi\, dg$ is invariant, and all invariants are obtained in this fashion.*

We shall draw a number of important consequences from this result, which comprise the foundation of the theory of representations of compact groups.

First of all we shall show that the study of an arbitrary representation can in large measure be reduced to the study of unitary representations. Let T be a representation in a linear topological space V. If χ is a linear continuous functional on V, the expression

$$(\xi,\eta)_\chi = \int_G \langle \chi, T(g)\xi \rangle \overline{\langle \chi, T(g)\eta \rangle}\, dg \tag{3}$$

is linear in ξ, conjugate linear in η, continuous in ξ and η jointly, and has the property that $(\xi,\xi)_\chi \geq 0$. Thus V is given the structure of a pre-Hilbert space (see 4.1). Let H_χ be the corresponding Hilbert space and ϕ_χ the natural mapping of V into H_χ. We define a representation T_χ in the space H_χ, obtained by factorization and extension by continuity of the representation T. The equality $(T(g)\xi, T(g)\eta)_\chi = (\xi,\eta)_\chi$ shows that T_χ is unitary.

Thus for every functional $\chi \in V'$ we have constructed a unitary representation T_χ and an intertwining operator $\phi_\chi \in \mathscr{C}(V, H_\chi)$. If the space V is locally convex, then the intersection of the kernels of the operators ϕ_χ for all $\chi \in V'$ consists of zero alone. In this case, all information about the representation V can be obtained by studying the unitary representations T_χ.

Problem 3. Let T be topologically irreducible and let $\chi \neq 0$. Prove that ϕ_χ is a continuous embedding of V in H_χ and that the representation T_χ is also topologically irreducible.

Problem 4. Every finite-dimensional representation of a compact group is equivalent to a unitary representation.

Now let T be a unitary representation of a compact group G in a Hilbert space H.

For arbitrary $\xi, \eta \in H$, we denote the function $g \mapsto (T(g)\xi, \eta)$ by the symbol $t_{\xi,\eta}(g)$. We call this function a *matrix element of the representation T*. (If ξ and η are elements of an orthonormal basis in H, then $t_{\xi,\eta}(g)$ is the corresponding coefficient of the matrix of the operator $T(g)$.)

Theorem 2. *Every topologically irreducible unitary representation T of a compact group G is finite-dimensional. Its matrix elements satisfy the conditions*

$$\int_G t_{\xi_1,\eta_1}(g)\overline{t_{\xi_2,\eta_2}(g)}\, dg = \frac{1}{\dim T}(\xi_1,\xi_2)\overline{(\eta_1,\eta_2)}. \tag{4}$$

If T_1 and T_2 are two inequivalent irreducible representations, then every matrix element of the representation T_1 is orthogonal in $L^2(G, dg)$ to every matrix element of the representation T_2.

Proof. The left side of the equality (4) is linear and continuous in ξ_1. Therefore it has the form (ξ_1, ζ), where ζ is a certain vector in H, which depends upon ξ_2, η_1 and η_2. For fixed η_1 and η_2, the vector ζ depends linearly and continuously on ξ_2. Therefore we may write $\zeta = A \xi_2$, where A is a certain operator in H that depends upon η_1 and η_2. The equality

$$t_{T(g)\xi,\eta}(g_1) = t_{\xi,\eta}(g_1 g)$$

shows that the operator A enjoys the property

$$(T(g)\xi_1, A T(g)\xi_2) = (\xi_1, A \xi_2)$$

for all $g \in G$. Therefore we have $A \in \mathscr{C}(T)$ and by theorem 1 of 7.3, it follows that $A = \lambda \cdot 1$. Obviously the number λ depends upon η_1 and η_2. Reasoning in the same fashion, we see that the dependence of λ on η_1 and η_2 has the form $\lambda = c \cdot \overline{(\eta_1, \eta_2)}$. Thus we have proved the equality

$$\int_G t_{\xi_1, \eta_1}(g) \overline{t_{\xi_2, \eta_2}(g)} \, dg = c(\xi_1, \xi_2) \overline{(\eta_1, \eta_2)}, \tag{5}$$

where the positive constant c depends solely on the representation T. Now let $e_1, \ldots, e_n$ be an arbitrary orthonormal collection of vectors in H. Then we have

$$\sum_{i=1}^{n} |(T(g)\xi, e_i)|^2 \leq \|T(g)\xi\|^2 = \|\xi\|^2.$$

From this we infer that

$$\sum_{i=1}^{n} |t_{\xi, e_i}(g)|^2 \leq \|\xi\|^2$$

for all g.

Integrating this inequality over the group G and using (5), we obtain the inequality $cn\|\xi\|^2 \leq \|\xi\|^2$, that is, $c \leq 1/n$. From this the finite dimensionality of T is obvious. If $\dim T = n$ and $e_1, \ldots, e_n$ is a basis in H, then the inequality considered above becomes an equality, and we obtain $c = 1/n = 1/\dim T$. The proof of the last assertion of the theorem is very similar to the proof of the relation (5). The only change is that instead of the equality $\mathscr{C}(T) = \mathbf{C} \cdot 1$ we must use the equality $\mathscr{C}(T_1, T_2) = 0$. The theorem is proved.

Combining the result just obtained with the result of problem 3, we see that the assertion about finite-dimensionality is true not only for unitary irreducible representations, but for arbitrary topologically irreducible representations in spaces admitting at least one nonzero linear continuous functional.

Corollary 1. *Every topologically irreducible representation of a compact group in a locally convex space is finite-dimensional and is equivalent to a unitary representation.*

§ 9. Invariant Integration 137

Corollary 2. *For a compact group G, the dual space $\hat{G}$ is discrete.*

Suppose that a topological group G acts continuously on an Abelian topological group M. Let $H_c^k(G,M)$ denote the k-dimensional *continuous cohomology* group of G with coefficients in M, that is, the factor group of the group $Z_c^k(G,M)$ of continuous cocycles with respect to the subgroup $B_c^k(G,M)$ of coboundaries of continuous cochains.

We note that the last-named group may not coincide with the group of continuous coboundaries: a discontinuous cocycle may have a continuous coboundary,

Theorem 3. *Let G be a compact group, let V be a linear space in which G acts by a certain representation T, and let V^G be the set of G-invariant elements in V. Then we have*

$$H_c^k(G,V) = \begin{cases} 0 & \text{for } k>0, \\ V^G & \text{for } k=0. \end{cases}$$

Proof. Let $\mathscr{F}^k$ be the set of all continuous functions on $G \times \cdots \times G$ (k factors) with values in V. We define the action of G in $\mathscr{F}^k$ by the formula

$$g: f(g_1,\ldots,g_k) \mapsto T(g) f(g^{-1}g_1,\ldots,g^{-1}g_k).$$

It is clear that the set of G-invariant elements in $\mathscr{F}^k$ coincides with the space $C_c^{k-1}(G,V)$ of continuous $(k-1)$-dimensional cochains of the group G with coefficients in V. The operator d can be naturally extended from $C_c^{k-1}(G,V)$ to $\mathscr{F}^k$, so that the diagram

$$\begin{array}{ccccccccc}
\mathscr{F}^1 & \to \cdots \to & \mathscr{F}^k & \to & \mathscr{F}^{k+1} & \to & \mathscr{F}^{k+2} & \to & \cdots \\
\uparrow i_1 & & \uparrow i_k & & \uparrow i_{k+1} & & \uparrow i_{k+2} & & \\
C_c^0(G,V) & \to \cdots \to & C_c^{k-1}(G,V) & \to & C_c^k(G,V) & \to & C_c^{k+1}(G,V) & \to & \cdots
\end{array} \quad (6)$$

is commutative. Here the vertical arrows i_k denote embeddings.

Problem 5. Prove that the upper line of the diagram (6) is an exact sequence in all terms, beginning with $k=2$.

Hint. Let $f \in \mathscr{F}^k$. The condition $df=0$ can be rewritten in the form

$$(-1)^{k+1} f(g_0,g_1,\ldots,g_{k-1}) = \sum_{i=0}^{k-1} (-1)^i f(g_0,\ldots,\hat{g}_i,\ldots,g_k),$$

from which it follows that $f=dh$, where $h(g_0,\ldots,g_{k-1}) = (-1)^{k+1} f(g_0,\ldots,g_k)$.

By theorem 1 the diagram (6) can be completed by projections p_k of the space $\mathscr{F}^k$ onto $C_c^{k-1}(G,V)$, in such a way that it remains commutative.

We shall show that this implies the exactness of the lower line beginning with $k=2$. In fact, if $z \in C_c^k(G,V)$ and $dz=0$, it follows that $i_{k+2} dz = 0$, and hence $d i_{k+1} z = 0$. Hence $i_{k+1} z = dw$ for a certain $w \in \mathscr{F}^k$. Then we have

$dp_k w = p_{k+1} dw = p_{k+1} i_{k+1} z = z$. The case $k=1$ can be analyzed similarly and leads to the equality $H^0(G,V) = V^G$.

The theorem is proved.

Corollary 1. *If a representation of a compact group is reducible and an invariant subspace is complemented, then this representation is decomposable.*

Corollary 2. *Every topological extension of a compact group with the aid of a linear space is chainable.*

We remark that the proof of theorem 3 used not the compactness of G but the consequence of compactness expressed in theorem 1. In the sequel we shall see that this fact holds for finite-dimensional representations of semisimple groups. From this will follow the validity of corollaries 1 and 2 for finite-dimensional representations of semisimple groups.

9.3. Applications to Noncompact Groups

A part of the results of the preceding section can be extended to a more general case.

Thus the "unitarization" of a representation T (that is, the construction of spaces H_χ, representations T_χ, and intertwining operators ϕ_χ) can be carried out for any bounded representation T of an amenable group G. For this purpose, $(\xi,\eta)_\chi$ becomes the invariant mean of the bounded function $g \mapsto \langle \chi, T(g)\xi \rangle \overline{\langle \chi, T(g)\eta \rangle}$.

Problem 1. Every bounded representation T of an amenable group G in a Hilbert space H is equivalent to a unitary representation.

Hint. Let A be the invariant mean of the operator function $g \mapsto T(g)T(g)^*$. Prove that A is a positive-definite invertible operator and that $T_1(g) = A^{1/2} T(g) A^{-1/2}$ is a unitary representation.

The relations of orthogonality of matrix elements (see theorem 2 of 9.2) can also be generalized. Let G be a locally compact unimodular group and let T be a unitary irreducible representation of G. We say that T is *square integrable* if all of its matrix elements belong to $L^2(G,dg)$. (It is not hard to verify that T has this property if at least one of its nonzero matrix elements lies in $L^2(G,dg)$.)

Problem 2. Prove that for a square integrable representation T, the relation

$$\int_G t_{\xi_1,\eta_1}(g) \overline{t_{\xi_2,\eta_2}(g)} dg = c(\xi_1,\xi_2) \overline{(\eta_1,\eta_2)} \tag{1}$$

holds. Prove that if S and T are inequivalent such representations, then

$$\int_G t_{\xi_1,\eta_1}(g) \overline{s_{\xi_2,\eta_2}(g)} dg = 0.$$

Corollary. *A separable locally compact group has no more than a countable set of pairwise inequivalent square integrable representations.*

We remark that the constant c in the equality (1) in the compact case is equal to the ratio of the volume of the group (in the sense of Haar measure) to the dimension of the representation. In the noncompact case, both of these quantities are infinite, but their ratio c remains finite. The quantity c^{-1} is called the *formal dimension of the representation T*. For real semisimple Lie groups that admit square integrable representations, one can normalize Haar measure in such a way that all formal dimensions are integers. In the general case this is not true, as has been proved by C.C. Moore.

The most essential application of invariant integration in the noncompact case is in the construction for every locally compact group of a sufficiently large number of unitary representations. We shall speak of this in more detail in § 13, and shall content ourselves with the following assertion.

Theorem 1 (Gel'fand-Raĭkov). *Every locally compact group G admits a complete system of irreducible unitary representations (that is, for every element $g \neq e$, there exists an irreducible unitary representation T such that $T(g) \neq 1$).*

This theorem will be proved in full detail in 10.3. Fór separable groups it follows from the decomposition into irreducible components of the regular representation of G, which acts in $L^2(G, d_l g)$ by the formula

$$[T(g)f](g_1) = f(g^{-1}g_1)$$

(see theorems 1 and 2 of 8.4).

§ 10. Group Algebras

10.1. The Group Ring of a Finite Group

Let G be a finite group. We denote by $\mathbf{Z}[G]$ the set of all formal linear combinations of elements of G with integer coefficients. In $\mathbf{Z}[G]$ we define the operations of addition and multiplication in a quite natural way:

$$\left. \begin{array}{l} \sum_i n_i g_i + \sum_i m_i g_i = \sum_i (n_i + m_i) g_i, \\ \sum_i n_i g_i \cdot \sum_j m_j g_j = \sum_{i,j} n_i m_j g_i g_j. \end{array} \right\} \quad (1)$$

The ring obtained in this way is called the *group ring* of the group G. It is clear that every linear representation T of the group G can be extended to a representation of its group ring if we set

$$T\left(\sum_i n_i g_i\right) = \sum_i n_i T(g_i). \quad (2)$$

Conversely, every nondegenerate representation of the group ring (that is, a representation in which the unit of the ring goes into the unit operator) is obtained by formula 2 from a representation of G.

Thus the theory of representations for the group G is completely equivalent to the theory of $\mathbf{Z}[G]$-modules. If we are interested in representations of G in linear spaces over a field K, it is natural to define the *group algebra* $K[G]$ of the group G over the field K, setting $K[G] = \mathbf{Z}[G] \underset{\mathbf{Z}}{\otimes} K$. The elements of $K[G]$ can be considered either as formal linear combinations of elements of G with coefficients from K or as functions on G with values in K. Under the second interpretation, addition in $K[G]$ is ordinary addition of functions, and multiplication is given by the formula

$$f_1 * f_2(g) = \sum_{g_1 g_2 = g} f_1(g_1) f_2(g_2) = \sum_{h \in G} f_1(h) f_2(h^{-1} g). \tag{3}$$

The function $f_1 * f_2$ is called the *convolution* of the functions f_1 and f_2.

It is clear that the category $\Pi(G, K)$ of representations of G in linear spaces over the field K is equivalent to the category of $K[G]$-modules.

The principal properties of representations of finite groups are contained in

Maschke's theorem. *If the characteristic of the field K does not divide the order of the group G, then the algebra $K[G]$ is semisimple.*

Proof. We define an invariant mean in the space $K[G]$ by setting

$$I(f) = \frac{1}{|G|} \sum_{g \in G} f(g).$$

Therefore theorems 1 and 3 from 9.2 hold for representations of the group G over the field K. In particular, every finite-dimensional representation of the group G (and hence of the algebra $K[G]$) is the sum of irreducible representations. From this and from the result of problem 2 of 3.3, it follows that the radical of $K[G]$ is contained in the kernel of an arbitrary finite-dimensional representation. But the representation of $K[G]$ in the algebra $K[G]$ itself (left translations) has kernel zero. Hence $K[G]$ is semisimple.

Corollary. *The algebra $K[G]$ is isomorphic to the product of a finite number of algebras of the form $\mathrm{Mat}_m(D)$, where D is a finite-dimensional skew field over K.*

This follows from the general theory of finite-dimensional semisimple algebras (see 4.1). For an algebraically closed field K, a sharper result holds.

Theorem 2 (Burnside). *Let G be a finite group, and k the number of conjugacy classes of elements of G. Let K be an algebraically closed field. Then there are exactly k equivalence classes of irreducible representations of G. Their dimensions n_i, $i = 1, \ldots, k$, are connected by the relation*

$$\sum_{i=1}^{k} n_i^2 = |G|. \tag{4}$$

§ 10. Group Algebras

The algebra $K[G]$ is isomorphic to the product $\prod_{i=1}^{k} \text{Mat}_{n_i}(K)$.

Proof. Let $T_1, \ldots, T_r$ be a complete set of pairwise inequivalent irreducible representations of G. We extend the representations T_i to representations of the algebra $K[G]$, that is, we set

$$T_i(f) = \sum_{g \in G} f(g) T_i(g).$$

Problem 1. The mapping $f \mapsto \{T_i(f)\}_{1 \leq i \leq r}$ is an epimorphism of $K[G]$ onto the algebra $\prod_{i=1}^{k} \text{End } V_i$, where V_i is the space of the representation T_i.

Hint. Use the orthogonality relations (see 9.2, Theorem 2) for matrix elements of the representations T_i. Their validity follows from the equalities $\mathscr{C}(T_i) = K$.

Let us show that the mapping of problem 1 is in fact an isomorphism.

Indeed, the kernel of this mapping coincides with the radical of $K[G]$ (see problem 2 of 3.3) and hence consists of zero alone.

Since the algebra $\text{End } V_i$ is isomorphic to $\text{Mat}_{n_i}(K)$, we have proved the last assertion of the theorem.

The equality (4) follows from comparison of the dimensions of $K[G]$ and $\prod_{i=1}^{r} \text{End } V_i$.

It remains to show that $r = k$.

Problem 2. Prove that a function $f \in K[G]$ belongs to the center of $K[G]$ if and only if it is constant on every conjugacy class of elements of G.

Thus the dimension of the center of the algebra $K[G]$ coincides with the number k of conjugacy classes of elements of the group G. On the other hand, the center of the algebra $\prod_{i=1}^{r} \text{Mat}_{n_i}(K)$, plainly has dimension r (it consists of sequences of scalar matrices).

Thus theorem 2 is completely proved.

The structure of group algebras $K[G]$ for a field that is not algebraically closed has been well studied in the case where the characteristic of the field does not divide the order of the group. For fields that do not satisfy this condition, the study of the algebra $K[G]$ is considerably more complicated. The same is true of the ring $\mathbf{Z}[G]$ and its analogues $R[G]$, where R is the ring of integers of a certain number field.

One may acquaint one's self with the principal methods and results of this domain of the theory of representations, and their applications to the theory of finite groups, in the books [13], [48].

10.2. Group Algebras of Topological Groups

The natural attempt to generalize the concept of the group algebra $K[G]$ to infinite groups leads to a number of inequivalent definitions.

First of all, just as in the finite case, one can consider formal linear combinations of a finite number of elements of the group G with coefficients in K. The algebra obtained in this way, which we denote as before by $K[G]$, is the simplest and most direct generalization of the concept of a group algebra (although it is not the most convenient and useful one).

For the case $K = \mathbf{C}$, we give a number of more frequently used variants of this definition.

1. The algebra $l^1(G)$ consists of all complex functions on G that satisfy the condition

$$\|f\| = \sum_{g \in G} |f(g)| < \infty.$$

The law of multiplication is given by the formula

$$f_1 * f_2(g) = \sum_{h \in G} f_1(h) f_2(h^{-1} g).$$

Problem 1. Prove that $l^1(G)$ is a Banach algebra with an involution, defined by the formula

$$f^*(g) = \overline{f(g^{-1})}.$$

Like $K[G]$, this definition has meaning for an arbitrary group G. One may regard as a deficiency of this definition the fact that the object it gives rise to reflects only the algebraic structure of the group G and is in no way connected with its topology.

2. The algebra $M(G)$ consists of all complex Borel measures on G. The norm in $M(G)$ has the form

$$\|\mu\| = |\mu|(G),$$

and the operation of convolution, which plays the role of multiplication, is defined by the formula

$$\int_G f(g) d(\mu_1 * \mu_2)(g) = \int_G \int_G f(g_1 g_2) d\mu_1(g_1) d\mu_2(g_2).$$

Here f is an arbitrary measurable bounded function.

Problem 2. Prove that $M(G)$ is a Banach algebra with involution $\mu \mapsto \mu^*$ defined by the equality

$$\int_G f(g) d\mu^*(g) = \int_G \overline{f(g^{-1})} d\mu(g).$$

Let $M_0(G)$ denote the subalgebra of $M(G)$ consisting of all regular Borel measures with compact support.

§ 10. Group Algebras

Problem 3. Prove that the algebra $l^1(G)$ is isomorphic to the subalgebra of $M(G)$ consisting of all measures concentrated on a countable subset.

3. For a locally compact group G, one can define an algebra structure in the space $L^1(G, d_l g)$, where $d_l g$ is left Haar measure. The convolution of two functions in $L^1(G, d_l g)$ is given by the formula

$$(f_1 * f_2)(g) = \int_G f_1(h) f_2(h^{-1} g) d_l h,$$

and involution has the form

$$f^*(g) = \Delta_G(g) \overline{f(g^{-1})}.$$

Problem 4. Prove that the measures absolutely continuous with respect to Haar measure form a two-sided ideal J in $M(G)$ and that the mapping

$$f \to f(g) d_l g$$

is an isomorphism of the algebra $L^1(G, d_l g)$ onto J.

Hint. Use the results of § 9, para. 1.

4. The algebra $C^*(G)$ (called the C^*-algebra of the group G) is defined as $C^*(L^1(G, d_l g)$ (see 4.3). It is obtained from the algebra $L^1(G, d_l g)$ by completion in the norm

$$\|f\|_1 = \sup_T \|T(f)\|,$$

where the supremum is taken over all *-representations of the algebra $L^1(G, d_l g)$ (and subsequent adjunction of a unit, if the group G is nondiscrete).

We now consider the connection between representations of the group G and representations of its group algebras. It is clear that any representation T of the group G in a linear space over a field K can be extended to a representation of $K[G]$ and that every nondegenerate representation of $K[G]$ is obtained in this fashion. Hence, just as for finite groups, the category $\Pi(G, K)$ is equivalent to the category of nondegenerate $K[G]$-modules. However, in view of the greater complexity of the structure of $K[G]$, this fact does not play the rôle that it does for the finite case.

If the representation T is continuous and acts in a complete locally convex space, then T can be extended to a representation of the algebra $M_0(G)$. We have only to set

$$T(\mu) = \int_G T(g) d\mu(g).$$

For the definition of operator integrals, see 9.2. We use the fact that for a locally convex space V, the space $\text{End } V$ of continuous operators carrying V into V is locally convex in the strong topology, and that the mapping $g \mapsto T(g)$ of the group G into $\text{End } V$ is continuous.

Let T be a uniformly bounded representation of a group G in a reflexive Banach space V. Then T can be extended to a representation of the algebra $M(G)$ in the following way. If $\xi \in V$ and $\chi \in V'$, the function $g \mapsto \langle \chi, T(g)\xi \rangle$ is continuous and bounded. Therefore the integral

$$I_\mu(\xi, \chi) = \int_G \langle \chi, T(g)\xi \rangle d\mu(g) \tag{1}$$

exists.

It is clear that $|I_\mu(\xi, \chi)| \leq C \cdot \|\xi\| \cdot \|\chi\| \cdot \|\mu\|$, where $C = \sup_{g \in G} \|T(g)\|$. From this and the bilinearity of $I_\mu(\xi, \chi)$ it is easy to show that

$$I_\mu(\xi, \chi) = \langle \chi, T(\mu)\xi \rangle, \tag{2}$$

where $T(\mu)$ is a certain linear operator on V with norm $\leq C\|\mu\|$.

Problem 5. Prove that the correspondence $\mu \mapsto T(\mu)$ is a representation (in fact a *-representation if T is unitary) of the algebra $M(G)$ in the space V.
Hint. Prove the equalities

$$I_{\mu_1 * \mu_2}(\xi, \chi) = I_{\mu_1}(T(\mu_2)\xi, \chi), \quad I_{\mu^*}(\xi, \chi) = \overline{I_\mu(\chi, \xi)}.$$

Thus every unitary representation of the group G can be extended by formulas (1), (2) to a *-representation of the algebra $L^1(G, d_l g)$. This representation in turn can be extended to a representation of the algebra $C^*(G)$, which we also denote by T. The algebras $L^1(G, d_l g)$ and $C^*(G)$, in contrast to $K[G]$, $l^1(G)$, and $M(G)$, do not contain the group G. Hence the following questions arise. Can we reconstruct the original representation of the group G from the representation of the group algebra, and does every representation of the group algebra correspond to a representation of the group?

A *-representation T of the algebra $L^1(G, d_l g)$ or of $C^*(G)$ in a Hilbert space H will be called *nondegenerate* if vectors of the form $T(f)\xi$, $f \in L^1(G, d_l g)$, $\xi \in H$, are dense in H.

Theorem 1. *A representation T of the algebra $L^1(G, d_l g)$ (or of $C^*(G)$) is generated by a unitary representation of the group G if and only if it is nondegenerate. If this condition holds, then the representation of the group is uniquely determined by the representation of the group algebra which it engenders.*

Proof. Let T be a unitary representation of the group G in the space H and let $\xi \in H$. Since the correspondence $g \mapsto T(g)\xi$ is continuous, there exists for each $\varepsilon > 0$ a neighborhood of the identity U_ε such that $\|T(g)\xi - \xi\| < \varepsilon$ for $g \in U_\varepsilon$. If f is a function on the group with support contained in U_ε and satisfying the condition $\int_G f(g) d_l g = 1$, then we have $\|T(f)\xi - \xi\| < \varepsilon$, which proves that T is nondegenerate.

Suppose now that T is nondegenerate. Let $\{U_\alpha\}$, $\alpha \in A$, be a basis of neighborhoods of the point $g \in G$. We consider a family of functions $\{f_\alpha\}$, $\alpha \in A$, defined on G and satisfying the following conditions:
1) $f_\alpha(g) \geq 0$ and $\int_G f_\alpha(g) d_l g = 1$;
2) $f_\alpha(g) = 0$ outside of U_α.

We order the set A by setting $\alpha > \beta$ if $U_\alpha \subset U_\beta$. We shall show that the limit $\lim_{\alpha \in A} T(f_\alpha)\xi$ exists for every $\xi \in H$. Since $\|T(f_\alpha)\| \leq \|f_\alpha\| = 1$, it suffices to prove this for the dense set of vectors of the form $\xi = T(f)\eta$, where f is a continuous function on G with compact support and η is a certain vector from H. The uniform continuity of f implies that for every $\varepsilon > 0$, the inequality $\|f_\alpha * f - L_g f\| < \varepsilon$ holds for all α beginning with a certain index α. Here $L_g f$ means the left translate of the function f by the element g^{-1}: $L_g f(g_1) = f(g^{-1} g_1)$. This implies that the net $T(f_\alpha)\xi = T(f_\alpha * f)\eta$ converges to the vector $T(L_g f)\eta$. We define the operator $T(g)$ first on H_0 by the formula

$$T(g)T(f)\eta = T(L_g f)\eta.$$

From what was said above it follows that $\|T(g)\| \leq 1$ and consequently the operator $T(g)$ can be extended by continuity over the entire space H.

Problem 6. Prove that the operators $T(g)$ constructed as above enjoy the following properties:
1) $\quad T(g_1 g_2) = T(g_1) T(g_2),$
2) $\quad T(g^{-1}) = T(g)^*,$
3) $\quad T(e) = 1.$

Hint. If $\{U_\alpha\}$ is a basis of neighborhoods of g_1 and $\{V_\alpha\}$ is a basis of neighborhoods of g_2, then $\{U_\alpha V_\alpha\}$ is a basis of neighborhoods of $g_1 g_2$.

To complete the proof of the theorem, it remains to show that the construction of a group representation from a nondegenerate representation of the algebra, which we have just given, is the inverse of the construction given by formula (1). We leave this to the reader.

10.3. Application of Group C^*-Algebras

If G is a separable locally compact group, then the space $L^1(G, d_l g)$ is separable. Therefore the algebra $C^*(G)$ is also separable. This permits us to prove the theorem on decomposition of representation which we stated in 8.4.

Let T be a unitary representation of G in a separable Hilbert space $H = \int_X H_x d\mu(x)$, and suppose that all of the operators $T(g)$, $g \in G$, are decomposable. Then the operators $T(a)$, $a \in C^*(G)$, are also decomposable. Let $\mathfrak{A}$ be a countable dense subset of $C^*(G)$. We may obviously suppose that $\mathfrak{A}$ is an algebra over the field $\mathbf{Q}(i)$ of all rational complex numbers (adding to $\mathfrak{A}$ if need be all polynomials in elements of $\mathfrak{A}$ with rational coefficients).

The operators $T(a)$, $a \in \mathfrak{A}$, are decomposable and have the form

$$[T(a)f](x) = T_x(a)f(x).$$

Each of the following relations $(a, a_1, a_2 \in \mathfrak{A}, \lambda \in \mathbf{Q}(i))$ holds almost everywhere on X:

$$\left.\begin{array}{c} T_x(a_1 + a_2) = T_x(a_1) + T_x(a_2), \\ T_x(a_1 a_2) = T_x(a_1) T_x(a_2), \\ T_x(\lambda a) = \lambda T_x(a), \\ T_x(a^*) = T_x(a)^*. \end{array}\right\} \qquad (1)$$

Hence, changing the operators $T_x(a)$ on a set of measure zero, we may suppose that (1) holds everywhere. Then for every x, the correspondence $a \to T_x(a)$ can be extended by continuity to a $*$-representation T_x of the algebra $C^*(G)$.

Problem 1. Prove that almost all representations T_x of the algebra $C^*(G)$ are nondegenerate.

Hint. Use the nondegeneracy of the representation T and the separability of H.

Changing the operators once more on a set of measure zero, we may suppose that all of the representations T_x are nondegenerate. Then by theorem 1 of 10.2 they are generated by unitary representations of the group G in H_x, which we also denote by T_x.

Problem 2. Prove that the representation T of the group G is a continuous sum of the representations T_x with respect to the measure μ on X.

Hint. It suffices to verify the equalities $[T(g)f](x) = T_x(g)f(x)$ for vectors f of the form $f = T(a)f_1$, where $a \in \mathfrak{A}$.

For nonseparable groups, the decomposability of a representation into a continuous sum and the decomposability of each operator of the representation are obviously inequivalent properties.

However, for arbitrary locally compact groups, we have

Theorem 1 (Gel'fand-Raĭkov). *Every unitary representation T of a locally compact group G can be decomposed into the continuous sum of irreducible representations.*

Proof. Without loss of generality, we may suppose that T is a cyclic representation. Let ξ be a source of T. Then ξ is also a source for the corresponding representation of the algebra $C^*(G)$. We set $\phi_\xi(a) = (T(a)\xi, \xi)$. This is a positive functional on $C^*(G)$. The set of all positive functionals on $C^*(G)$ that satisfy the condition $\phi(1) = \|\xi\|^2$ forms a compact convex set K in $C^*(G)'$. By Choquet's theorem (see 4.1) the functional ϕ_ξ can be written in the form

$$\phi_\xi(a) = \int_X \phi_x(a) d\mu(x),$$

where X is the set of extreme points of the compact set K and μ is a certain Borel measure on X. Let T_x be the representation of $C^*(G)$ corresponding to the functional ϕ_x, and let H_x be the space of the representation T_x.

Problem 3. Prove that the representation T of the algebra $C^*(G)$ is the continuous sum of the representations T_x with respect to the measure μ.

Hint. Let ξ_x be a source in H_x. The desired isomorphism between the space H of the representation T and $\int_X H_x d\mu(x)$ maps the vector $T(a)\xi \in H$ into the vector function $\xi(x) = T_x(a)\xi_x$ (ξ_x is a source in H_x).

Each of the functionals ϕ_x either corresponds to a nondegenerate representation of $C^*(G)$ or has the form $\phi_0(\lambda \cdot 1 + a) = \lambda \|\xi\|^2$, where a belongs to the closure of $L^1(G, d_l g)$ in $C^*(G)$. It is easy to show that the point ϕ_0 has measure zero with respect to μ [in the contrary case, the representation T of the algebra $C^*(G)$ would be degenerate].

§ 10. Group Algebras

In view of theorem 1 of 10.2, the representation T_x of the algebra $C^*(G)$ is generated by a certain representation of the group G (which we shall also denote by T_x).

Problem 4. Prove that the representation T of the group G is the continuous sum of the representations T_x with respect to the measure μ.

Hint. It suffices to verify the equality $[T(g)f](x) = T_x(g)f(x)$ on the dense subset of H consisting of vectors of the form $T(a)f$, where a is a continuous function on G with compact support.

We now turn to applications connected with the topology of the set $\hat{G}$. For every C^*-algebra $\mathfrak{A}$, we can introduce a topology in the set $\hat{\mathfrak{A}}$ of equivalence classes of irreducible *-representations of $\mathfrak{A}$.

Two methods of topologizing $\hat{\mathfrak{A}}$ are completely analogous to the methods of topologizing the space $\hat{G}$, as described in § 7. According to the first method, we define a neighborhood of a point $[T] \in \hat{\mathfrak{A}}$ by means of a number $\varepsilon > 0$, a finite set of vectors $\xi_1, \ldots, \xi_n$ in the space of the representation T, and a finite set $a_1, \ldots, a_m$ of elements of $\mathfrak{A}$. This neighborhood consists of the equivalence classes of those representations S in the spaces of which there exist vectors $\eta_1, \ldots, \eta_n$ such that

$$|(T(a_i)\xi_k, \xi_l) - (S(a_i)\eta_k, \eta_l)| < \varepsilon$$

for all $i \in [1, m]$, $k \in [1, n]$, $l \in [1, n]$.

The second method is based on the concept of "weak containment". For every representation T of the algebra $\mathfrak{A}$ in the space H (this includes also reducible representations) we construct the closed subspace $V(T) \subset \mathfrak{A}'$ generated by all functionals of the form $a \mapsto (T(a)\xi, \eta)$, $\xi, \eta \in H$. We shall say that the representation T is *weakly contained* in a family of representations $\mathscr{S}$ if $V(T)$ is contained in the closure of the subspace generated by all $V(S)$, $S \in \mathscr{S}$. The point $[T] \in \hat{\mathfrak{A}}$ is in the closure of a set $X \subset \hat{\mathfrak{A}}$ if the representation T is weakly contained in the family of all representations the classes of which belong to X.

We remark that we get the same definition if we replace the space $V(T)$ by the cone $V_+(T)$ of positive functionals contained in $V(T)$.

A third topology for $\hat{\mathfrak{A}}$ is based upon the *Jacobson topology*, which can be defined in the set of all prime two-sided ideals of an arbitrary algebra. The closure of a set M of ideals consists of all ideals J that contain the ideal $J_M = \bigcap_{J \in M} J$.

We recall that an ideal J is called *prime* if the inclusion $J \supset J_1 J_2$ implies that either $J \supset J_1$ or $J \supset J_2$. If $\mathfrak{A}$ is a C^*-algebra and T is an irreducible representation of $\mathfrak{A}$, then the kernel of this representation is a prime two-sided ideal in $\mathfrak{A}$.

It turns out that all three methods lead to the same topology in the set $\hat{\mathfrak{A}}$. In the case where $\mathfrak{A} = C^*(G)$ and G is nondiscrete, the set $\hat{\mathfrak{A}}$ can be naturally identified with $\hat{G} \cup [T_0]$ where T_0 is the degenerate representation of $C^*(G)$ that annihilates $L^1(G, d_l g)$ (see 10.2, theorem 1).

This identification gives a homeomorphism of the set of all nondegenerate representations of $\mathfrak{A}$ onto the set $\hat{G}$.

The proof of all the above assertions can be found in the book [15].

10.4. Group Algebras of Lie Groups

When G is a Lie group, we can single out various subalgebras of $L^1(G, d_l g)$ by imposing on the functions f various requirements of smoothness (up to analyticity) and of rapidity of decrease at infinity (up to having compact support). If G is compact, then there is a smallest algebra among these subalgebras, namely, the algebra generated by matrix elements of irreducible representations. For noncompact groups, one uses most often the algebra $C_0^\infty(G)$ of infinitely differentiable functions with compact supports.

On the other hand, one can define for Lie groups an algebra larger than $M_0(G)$. Let $C^\infty(G)$ be the space of all infinitely differentiable functions on G with the topology of uniform convergence on compact sets. It is known that $C^\infty(G)$ is a countably normed nuclear space and that the topological tensor product $C^\infty(G) \hat{\otimes} C^\infty(G)$ can be identified in a natural way with $C^\infty(G \times G)$. The conjugate space $R(G)$ of $C^\infty(G)$ consists of all generalized functions with compact supports.

For every $\chi \in R(G)$, there exists a smallest compact set K such that $\langle \chi, \phi \rangle = 0$ for all $\phi \in C^\infty(G)$ that vanish together with all of their derivatives on K. This compact set K is called the *support* of χ and is denoted by $\operatorname{supp} \chi$. The space $M_0(G)$ is a subspace of $R(G)$: to each measure μ there corresponds the functional $\phi \mapsto \int_G \phi(g) d\mu(g)$.

Problem 1. Verify that $\operatorname{supp} \mu$ denotes one and the same set, independent of whether μ is regarded as an element of $M_0(G)$ or as an element of $R(G)$.

We now define *convolution* in $R(G)$, setting $\langle \chi_1 * \chi_2, \phi \rangle = \langle \chi_1, \psi \rangle$, where $\psi(g) = \langle \chi_2, L_g^{-1} \phi \rangle$, and the operator L_g has the form $[L_g \phi](g_1) = \phi(g^{-1} g_1)$.

Problem 2. Prove that the operation of convolution on $M_0(G) \subset R(G)$ coincides with the operation introduced in 10.2.

Problem 3. Prove the inclusion

$$\operatorname{supp}(\chi_1 * \chi_2) \subset \operatorname{supp} \chi_1 \cdot \operatorname{supp} \chi_2.$$

Hint. If the function ϕ vanishes together with all of its derivatives on the set $\operatorname{supp} \chi_1 \cdot \operatorname{supp} \chi_2$, then the function $\psi(g) = \langle \chi_2, L_g^{-1} \phi \rangle$ vanishes with all of its derivatives on the set $\operatorname{supp} \chi_1$.

As a corollary of this, we see that the set $R(G, K)$ of all generalized functions on G with compact supports contained in a subgroup K forms a subalgebra of $R(G)$. The structure of algebras $R(G, K)$ is as yet little studied, with the exception of the case $K = \{e\}$ (see *infra*).

The case is which G is a semisimple Lie group and K is a maximal compact subgroup is of particular interest. Many of the deep results obtained in recent years in the theory of representations of semisimple groups are closely connected with the algebraic structure of the algebra $R(G, K)$.

The algebra $R(G, \{e\})$ is completely defined by arbitrarily small neighborhoods of the identity in the group G. Hence it can be reconstructed from the Lie algebra $\mathfrak{g}$ of this group (see § 6).

§ 10. Group Algebras

Let $U(\mathfrak{g})$ denote the *enveloping algebra* of the algebra $\mathfrak{g}$. By definition, $U(\mathfrak{g})$ is the factor algebra of the tensor algebra $T(\mathfrak{g})$ (see 3.5) by the two-sided ideal J which is generated by elements of the form

$$X \otimes Y - Y \otimes X - [X, Y], \quad X, Y \in \mathfrak{g}. \tag{1}$$

Problem 4. Prove that the natural embedding $\phi: \mathfrak{g} \to U(\mathfrak{g})$ is a universal object in the category of all mappings ψ of the Lie algebra $\mathfrak{g}$ into associative algebras satisfying the identity

$$\psi([X, Y]) = \psi(X)\psi(Y) - \psi(Y)\psi(X). \tag{2}$$

(Morphisms of this category are commutative diagrams of the form

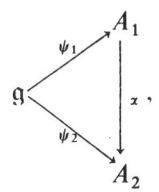

where α is a homomorphism of A_1 into A_2.)

Theorem 1 (L. Schwartz). *Let $\mathfrak{g}$ be the Lie algebra of the group G. The algebra $R(G, \{e\})$ is isomorphic to $U(\mathfrak{g})$.*

Proof. We realize the Lie algebra $\mathfrak{g}$ by left invariant vector fields on G and consider the mapping $\psi: \mathfrak{g} \to R(G)$, defined by the formula

$$\langle \psi(X), f \rangle = (Xf)(e). \tag{3}$$

It is clear that $\psi(X)$ belongs to $R(G, \{e\})$.

Problem 5. Prove that the mapping ψ enjoys property (2) of problem 4.

Hint. Verify that $\langle \psi(X) * \psi(Y), f \rangle = (XYf)(e)$ and use the fourth definition of the commutator $[X, Y]$ in 6.3.

In view of problem 4, there exists a homomorphism $\bar{\psi}$ of the algebra $U(\mathfrak{g})$ into $R(G, \{e\})$, which extends the mapping ψ. It is easy to establish that this homomorphism $\bar{\psi}$ carries the element $X_1 \otimes \cdots \otimes X_n \mod J \in U(\mathfrak{g})$ into the generalized function

$$f \mapsto (X_1 \ldots X_n f)(e).$$

Let us show that $\bar{\psi}$ is an isomorphism. For this we introduce a filtration in $U(\mathfrak{g})$ and $R(G, \{e\})$ and show that $\bar{\psi}$ preserves the filtration and generates an isomorphism of the corresponding graded algebras $\operatorname{gr} U(\mathfrak{g})$ and $\operatorname{gr} R(G, \{e\})$. From this we infer without difficulty that $\ker \bar{\psi} = 0$ and $\operatorname{im} \bar{\psi} = R(G, \{e\})$, that is, $\bar{\psi}$ is an isomorphism.

The filtration in $U(\mathfrak{g})$ arises from the grading in $T(\mathfrak{g})$, defined by
$$U_k(\mathfrak{g}) = \bigoplus_{t=0}^{k} T^t(\mathfrak{g}) \bmod J.$$

Problem 6. Prove that the algebra $\operatorname{gr} U(\mathfrak{g})$ is isomorphic to the symmetric algebra $S(\mathfrak{g})$ (see 3.5).

We define a filtration in $R(G, \{e\})$ in the following way. Let $X_1, \ldots, X_n$ be a basis in $\mathfrak{g}$. We introduce the corresponding canonical system of coordinates in a neighborhood of the identity of the group G (see 6.4). The point $\exp(x_1 X_1 + \ldots + x_n X_n)$ has coordinates $x_1, \ldots, x_n$.

Every element $\chi \in R(G, \{e\})$ has the form

$$f \to \sum c_{k_1, \ldots, k_n} \frac{\partial^{k_1 + \cdots + k_n} f}{\partial x_1^{k_1} \ldots \partial x_n^{k_n}}(0) \tag{4}$$

(the sum is finite). We denote by $R_k(G, \{e\})$ the subspace of $R(G, \{e\})$ consisting of those elements χ for which in the notation of formula (4), all coefficients $c_{k_1, \ldots, k_n}$ are zero for $k_1 + \cdots + k_n > k$.

Problem 7. Prove that $\{R_k(G, \{e\})\}$ is a filtration in $R(G, \{e\})$ and that the algebra $\operatorname{gr} R(G, \{e\})$ is isomorphic to the algebra $\mathbf{C}\left[\dfrac{\partial}{\partial x_1}, \ldots, \dfrac{\partial}{\partial x_n}\right]$.

Hint. With each element $\chi \in R_k(G, \{e\})$ having the form (4), we associate the expression

$$[\chi] = \sum_{k_1 + \ldots + k_n = k} c_{k_1, \ldots, k_n} \frac{\partial^k}{\partial x_1^{k_1} \ldots \partial x_n^{k_n}}.$$

Prove that $[\chi_1 * \chi_2] = [\chi_1][\chi_2]$.

The filtrations in $U(\mathfrak{g})$ and $R(G, \{e\})$ have the property that $\tilde{\psi}(U_k(\mathfrak{g})) \subset R_k(G, \{e\})$. In fact, for $k = 1$, this is evident, and for other k it follows from the fact that $\tilde{\psi}$ is a homomorphism. Hence $\tilde{\psi}$ generates a homomorphism $\operatorname{gr} \tilde{\psi} : \operatorname{gr} U(\mathfrak{g}) \to \operatorname{gr} R(G, \{e\})$. It is easy to prove from the definition of a canonical system of coordinates that

$$\tilde{\psi}(X_k): f \mapsto \frac{\partial f}{\partial x_k}(0).$$

Hence we have $\operatorname{gr} \tilde{\psi}(X_k) = \partial / \partial x_k$, and consequently $\operatorname{gr} \tilde{\psi}$ is an isomorphism. The theorem is proved.

The algebraic structure of $U(\mathfrak{g})$ plays a great rôle in the theory of representations of the corresponding group.

We can obtain much information about this structure by comparing $U(\mathfrak{g})$ with the commutative algebra $\operatorname{gr} U(\mathfrak{g}) = S(\mathfrak{g})$. For each subspace $L \subset U(\mathfrak{g})$, we denote by $\operatorname{gr} L$ the subspace of $\operatorname{gr} U(\mathfrak{g})$ generated by elements of the form $\operatorname{gr}^k X$, $X \in L \cap U_k(\mathfrak{g})$ (see 3.1).

It is clear that if L is a subalgebra in $U(\mathfrak{g})$, then $\operatorname{gr} L$ is a subalgebra of $\operatorname{gr} U(\mathfrak{g})$. We apply this method to describe the center $Z(\mathfrak{g})$ of the algebra $U(\mathfrak{g})$.

§ 10. Group Algebras

We extend the adjoint representation Ad of the group G in $\mathfrak{g}$ (see 6.3) to a representation $\widetilde{\mathrm{Ad}}$ in the space $U(\mathfrak{g})$. [This extension is uniquely defined by the condition that $\widetilde{\mathrm{Ad}}\,g$ be an automorphism of the algebra $U(\mathfrak{g})$.]

Theorem 2 (I. M. Gel'fand). *Let G be a connected Lie group. Then the subspace $\operatorname{gr} Z(\mathfrak{g}) \subset \operatorname{gr} U(\mathfrak{g}) = S(\mathfrak{g})$ coincides with the subspace of G-invariant elements in $S(\mathfrak{g})$.*

Proof. We consider the linear mapping $\sigma : S(\mathfrak{g}) \to U(\mathfrak{g})$ which is called symmetrization, and which carries $X_1 \ldots X_k \in S^k(\mathfrak{g})$ into

$$\frac{1}{k!} \sum_s X_{s(1)} \ldots X_{s(k)} \in U(\mathfrak{g})$$

(the sum is over all permutations s of the indices $1, 2, \ldots, k$).

We shall prove the theorem from the following fact, which also has other applications.

Problem 8. The mapping σ commutes with the action of the group G and is an isomorphism of the linear spaces $S(\mathfrak{g})$ and $U(\mathfrak{g})$.

Hint. The space $S(\mathfrak{g})$ is generated by elements of the form X^k, $X \in \mathfrak{g}$. The second assertion follows from problem 6.

We shall now show that for a connected Lie group, the set of G-invariant elements in $U(\mathfrak{g})$ coincides with $Z(\mathfrak{g})$. In fact, since G is generated by an arbitrary neighborhood of the identity, to prove that an element is G-invariant, it suffices to prove its invariance relative to one-parameter subgroups $\exp tX$, $X \in \mathfrak{g}$. But this in turn is equivalent to the annihilation of this element by all operators of the corresponding representation of the Lie algebra.

Problem 9. The representation $\widetilde{\mathrm{ad}}$ of the algebra $\mathfrak{g}$ in $U(\mathfrak{g})$ that corresponds to the representation $\widetilde{\mathrm{Ad}}$ of the group G has the form

$$\widetilde{\mathrm{ad}}\, X : Y \mapsto XY - YX, \quad X \in \mathfrak{g}, \quad Y \in U(\mathfrak{g}). \tag{5}$$

Hint. Use the fact that $\widetilde{\mathrm{ad}}\, X$ is a differentiation of the algebra $U(\mathfrak{g})$.

The condition $(\widetilde{\mathrm{ad}}\, X) Y = 0$ for all $X \in \mathfrak{g}$ is easily seen to be equivalent to the condition that $Y \in Z(\mathfrak{g})$.

Gel'fand's theorem follows from problems 8 and 9 and the easily verified identity

$$\operatorname{gr}^k(\sigma(X)) = X \quad \text{for} \quad X \in S^k(\mathfrak{g}).$$

The skew field of quotients $D(\mathfrak{g})$ of the algebra $U(\mathfrak{g})$ (which is called the Lie skew field) is a very interesting and as yet little studied object. Since the algebra $U(\mathfrak{g})$ is noncommutative, even the definition of $D(\mathfrak{g})$ is not obvious.

To formulate this definition, we need some preparation.

Problem 10. Prove that $U(\mathfrak{g})$ admits no divisors of zero: if $X \neq 0$ and $Y \neq 0$, then $XY \neq 0$.

Hint. Use the analogous property of the algebra $\operatorname{gr} U(\mathfrak{g}) = S(\mathfrak{g})$.

Problem 11. Every strictly increasing sequence of left (or right) ideals in $U(\mathfrak{g})$ is finite.

Hint. Prove that for every ideal $J \subset U(\mathfrak{g})$, the set $\operatorname{gr} J$ is an ideal in $\operatorname{gr} U(\mathfrak{g})$. The algebra $\operatorname{gr} U(\mathfrak{g})$ is isomorphic to the algebra of polynomials in $n = \dim \mathfrak{g}$ variables, and for this algebra, as is well known, every strictly increasing chain of ideals has finite length.

Problem 12. Any two elements $X \neq 0, Y \neq 0$ in $U(\mathfrak{g})$ have a nonzero common left (right) multiple.

Hint. Consider the chain $\{J_k\}$, where J_k is the left ideal generated by the elements $X, XY, \ldots, XY^k$, and use problems 10 and 11.

We are now in a position to define the Lie skew field $D(\mathfrak{g})$. We consider the set of all expressions of the form XY^{-1} and $Y^{-1}X$, where $X, Y \in U(\mathfrak{g})$ and $Y \neq 0$. We will call these expressions *right* and *left fractions*, respectively. We will say that the left fraction $Y_1^{-1}X_1$ is equivalent to the right fraction $X_2 Y_2^{-1}$, if $X_1 Y_2 = Y_1 X_2$. Two left (right) fractions are called equivalent if they are equivalent to one and the same right (left) fraction.

We leave it to the reader to convince himself that the equivalence relation just introduced is symmetric, reflexive, and transitive.

We define the *Lie skew field* $D(\mathfrak{g})$ as the set of equivalence classes of equivalent fractions. From problem 12, it follows that each such class contains both right and left fractions.

We define the operations in $D(\mathfrak{g})$ in the following way.

Let $X_1 Y_1^{-1}$ and $X_2 Y_2^{-1}$ be representatives of the classes a_1 and a_2 in $D(\mathfrak{g})$. The elements Y_1 and Y_2 have a nonzero common right multiple: $Z = Y_1 Z_1 = Y_2 Z_2$. Evidently the fraction $X_1 Y_1^{-1}$ is equivalent to $X_1 Z_1 Z^{-1}$, and the fraction $X_2 Y_2^{-1}$ is equivalent to $X_1 Z_1 Z^{-1}$. We define the sum $a_1 + a_2$ as the equivalence class of the fraction $(X_1 Z_1 + X_2 Z_2) Z^{-1}$ and the quotient $a_1 a_2^{-1}$ as the equivalence class of the fraction $X_1 Z_1 (X_2 Z_2)^{-1}$. The remaining arithmetical operations are defined by means of those already defined: $a_1 a_2 = a_1 (a_2^{-1})^{-1}, a_1 - a_2 = a_1 + (-1) a_2$. It is not hard to check that these definitions are consistent, and we omit this.

It has been conjectured that *for every algebraic group G, the skew field $D(\mathfrak{g})$ is isomorphic to the skew field of quotients $D_{n,k}$ of the algebra $R_{n,k}$* defined in 3.5. This means that generators of the algebra $\mathfrak{g}$ are connected by one-to-one birational transformations with the canonical generators $p_1, \ldots, p_n, q_1, \ldots, q_n, r_1, \ldots, r_k$ of the algebra $R_{n,k}$. The numbers n and k have a simple interpretation in the language of the theory of representations. A typical irreducible representation of the group G depends upon k parameters and is realized in a space of functions of n variables.

Proofs of the conjecture stated above for some special cases and other facts about Lie skew fields can be found in the articles of I. M. Gel'fand and the author [81], [82].

10.5. Representations of Lie Groups and their Group Algebras

Every representation T of a Lie group G in a complete locally convex space V can be extended to a representation of the algebra $M_0(G)$.

§ 10. Group Algebras

If V is a Banach space and T is a locally bounded representation (that is, the norms of the operators $T(g)$ are bounded on every compact subset of G), then the representation of the group G can be uniquely reconstructed from the representation of the subalgebra $C_0^\infty(G) \subset M_0(G)$. This assertion follows easily from theorem 1 of para. 2 and its proof.

If the representation T is finite-dimensional, then the mapping $g \mapsto T(g)$ is an analytic operator function on G (see 6.4). Hence the representation T can be extended to a representation of the algebra $R(G)$ by the formula

$$\langle \eta, T(\chi)\xi \rangle = \langle \chi, t_{\xi,\eta} \rangle, \tag{1}$$

where $t_{\xi,\eta}(g) = \langle \eta, T(g)\xi \rangle$ is a matrix element of the representation T, corresponding to the vectors $\xi \in V$ and $\eta \in V^*$.

The situation is more complex for infinite-dimensional representations. Matrix elements of such representations are continuous (by the definition of a representation) but are not necessarily smooth. For example, for the regular representation T of the group $\mathbf{R}$, acting in the space $L^2(\mathbf{R}, dx)$ by the formula

$$[T(s)\xi](x) = \xi(x+s),$$

the matrix element $t_{\xi,\eta}$ has the form

$$t_{\xi,\eta}(s) = \int_{-\infty}^{\infty} \xi(x+s)\overline{\eta(x)}dx,$$

where ξ, η are arbitrary functions in $L^2(\mathbf{R}, dx)$.

Problem 1. Prove that $t_{\xi,\eta}$ can be an arbitrary piecewise linear continuous function with compact support.

Hint. Consider the case in which the functions ξ and η are piecewise constant and have compact supports.

From this same example it is seen that the definition of the operator $T(\chi)$ by formula (1) has meaning in a smaller space, for the vectors of which the corresponding matrix elements belong to $C^\infty(G)$.

Consider a representation T of a Lie group G acting in a complete locally convex space V. Consider also the subspace $V^\infty \subset V$ consisting of all vectors $\xi \in V$ for which the matrix element $t_{\xi,\eta}(g) = \langle \eta, T(g)\xi \rangle$ belongs to $C^\infty(G)$ for all $\eta \in V'$. This is called the *Gårding space* of the representation T.

Problem 2. Prove that the Gårding space is isomorphic to $\mathrm{Hom}_G(V', C^\infty(G))$, if we define the action of G in V' consistently with the representation T' adjoint to T and in $C^\infty(G)$ with the aid of left translations $L(g)$.

Hint. To every $\xi \in V^\infty$, there corresponds a mapping $\eta \mapsto t_{\xi,\eta}$ of the space V' into $C^\infty(G)$.

Since the space $C^\infty(G)$ is nuclear, one can identify $\mathrm{Hom}(V', C^\infty(G))$ with the topological tensor product $V \hat{\otimes} C^\infty(G)$, and the space $\mathrm{Hom}_G(V', C^\infty(G))$ with $V_G \hat{\otimes} C^\infty(G)$. Thus one can consider the Gårding space V^∞ as the subspace of G-invariant elements in the space $V \hat{\otimes} C^\infty(G)$ of infinitely differentiable V-

valued functions on G. We give V^∞ the *topology* which it inherits from $V \hat{\otimes} C^\infty(G)$. In other words, we shall consider that the net ξ_α converges to ξ if the net of vector-functions $g \mapsto T(g)\xi_\alpha$ converges to the vector function $g \mapsto T(g)\xi$ together with all derivatives uniformly on every compact subset of G.

Every representation T of a Lie group G in a complete locally convex space V generates a representation of the algebra $R(G)$ in the Gårding space V^∞ by formula (1). We denote this representation by T^∞. Its restriction to $R(G, \{e\}) = U(\mathfrak{g})$ is a representation of the algebra $U(\mathfrak{g})$.

In the example given above, the Gårding space consists of all infinitely differentiable functions on the line for which all derivatives belong to $L^2(\mathbf{R}, dx)$.

Theorem 1 (Gel'fand-Gårding). *For an arbitrary representation T of a Lie group G in a complete locally convex space V, the Gårding space V^∞ is dense in V.*

The proof is based on the very useful and frequently used technique of *smoothing*. A *smoothing operator* in the space V is an operator of the form

$$T(\phi) = \int_G \phi(g) T(g) d_r g ,$$

where $\phi \in C_0^\infty(G)$, $\phi \geq 0$ and $\int_G \phi(g) d_r g = 1$.

We shall show that every smoothing operator carries the space V into V^∞. This follows from the following fact, which has many applications.

Problem 3. Let ψ be a generalized function on the group G (that is, a linear continuous functional on the space C_0^∞) and let $\phi \in C_0^\infty(G)$. Then the function

$$\psi * \phi(g) = \langle \psi, L(g^{-1})\phi \rangle$$

belongs to $C^\infty(G)$.

Hint. Prove the identity

$$L(g)(\psi * \phi) = \psi * L(g)\phi ,\qquad(2)$$

where $[L(g)\phi](g_1) = \phi(g_1 g)$. Infer from this that the function $\psi * \phi$ is differentiable and that its derivative along the left invariant field X is given by the identity $X(\psi * \phi) = \psi * X\phi$ (see 6.3, problem 5).

To finish the proof of theorem 1, it remains to note that the set of "smoothed" vectors of the form $T(\phi)\xi$ is dense in V. In fact, if U is an arbitrary convex neighborhood of the point ξ in V, then the continuity of T implies that $T(g)\xi \in U$ for all g in a certain neighborhood W of the identity in the group G. Then we have $T(\phi)\xi \in U$ provided that ϕ vanishes outside of W, $\phi \geq 0$, and $\int_G \phi(g) d_r g = 1$. The theorem is proved.

Corollary. *If V is finite-dimensional, then $V^\infty = V$.*

Problem 4. Prove that $(V^\infty)^\infty = V^\infty$.

§ 10. Group Algebras

Hint. Use the isomorphism

$$C^\infty(G) \hat{\otimes} C^\infty(G) \approx C^\infty(G \times G)$$

and the isomorphism

$$C^\infty(G) \hat{\underset{G}{\otimes}} C^\infty(G) \approx C^\infty(G)$$

which follows from it.

Problem 5. Let T_1 and T_2 be two representations of the group G in complete locally convex spaces V_1 and V_2. Prove that if $A \in \mathscr{C}(T_1, T_2)$, then $AV_1^\infty \subset V_2^\infty$.

Hint. Use the relation $t^{(2)}_{A\xi,\eta} = t^{(1)}_{\xi,A'\eta}$ for $\xi \in V_1$ and $\eta \in V_2$.

We remark that theorem 1 can be proved from problem 5 and fact that $C_0^\infty(G)$ is contained in the Gårding space of the left regular representation of G.

Sometimes we replace the Gårding space V^∞ by the smaller space V^ω of *analytic vectors* in the space V.

A vector $\xi \in V$ belongs to V^ω, by definition, if the vector-function $g \mapsto T(g)\xi$ is analytic on G. For representations in Banach spaces, one can prove that V^ω is dense in V. As in the case of Gårding spaces, the proof is based on the technique of smoothing operators $T(\phi)$, where ϕ is an analytic function on G which is "almost concentrated" in a small neighborhood of the identity. The existence of such functions can be obtained from the theory of parabolic equations (by considering the fundamental solution of the heat equation on the group G).

Let G be a connected group. A representation T of G can be uniquely reconstructed from the representation T^ω of its Lie algebra $\mathfrak{g}$ in the space V^ω. In fact, a representation of the covering algebra $U(\mathfrak{g})$ can be reconstructed from the representation of the Lie algebra $\mathfrak{g}$. Knowing this representation, we can compute all derivatives of the analytic function $g \mapsto T(g)\xi$ at the point e, and consequently we can reconstruct this function. The representation of G in V^ω obtained in this way can be extended by continuity to the desired representation T in the space V.

Suppose now that we have a representation T^0 of the Lie algebra $\mathfrak{g}$ of the group G in a certain dense subspace $V^0 \subset V$. The following question arises naturally. When does there exist a representation T of the group G in the space V such that $V^0 \subset V^\infty(T)$ and T^0 coincides with the restriction of T^∞ to V^0?

Even for $G = \mathbf{R}$, this question is nontrivial. In this case (for a Banach space V) an answer is provided by the well-known theorem of I. M. Gel'fand on the structure of the generating operator of a strongly continuous one-parameter group. Not long ago, S. G. Kreĭn and A. M. Šihvatov [111] extended this result to arbitrary Lie groups. We give here the definitive formulation in terms that are convenient for our purposes.

Theorem 2. *Let T^0 be a representation of the algebra $\mathfrak{g}$ in a dense subspace V^0 of a Banach space V. This representation has the form $T^\infty|_{V^0}$ for a certain representation T of the corresponding Lie group G, which leaves V^0 invariant, if and only if the following conditions hold:*

1) *each of the operators $T^0(X)$, $X \in \mathfrak{g}$, admits a closure;*

2) *there exist positive constants C and ε such that the resolvents of all of the operators* $T^0(X)$, *where X runs through a bounded neighborhood of zero in* $\mathfrak{g}$, *are defined outside of the strip* $|\operatorname{Re}\lambda|\leqslant\varepsilon$ *and satisfy the condition*

$$\|R_\lambda(T^0(X))^n\| \leqslant \frac{C}{|\cdot|\operatorname{Re}\lambda|-\varepsilon|^n}$$

for $n=1,2,\ldots$;

3) *the space* V^0 *is invariant under the resolvents* $R_\lambda(T^0(X))$, $X\in\mathfrak{g}$.

The first two conditions of this theorem are a natural generalization of Gel'fand's conditions, which are satisfied by the generator A of a one-parameter group of operators.

We remark that the third condition is satisfied automatically in the one-dimensional case, if we take as V^0 the intersection of the domains of definition of the closures of the operators A^n.

For the case in which V is a Hilbert space and T is a unitary representation, a more precise result is known.

Theorem 3 (E. Nelson). *Let* T^0 *be a representation of the Lie algebra* $\mathfrak{g}$ *of a connected and simply connected Lie group G, acting in a certain dense subspace* H^0 *of a Hilbert space H. This representation is generated by a unitary representation T of the group G in the space H if and only if the operators*

$$iT^0(X_1),\ldots,iT^0(X_n) \quad \text{and} \quad \varDelta = \sum_{k=1}^{n} [T^0(X_k)]^2,$$

where $X_1,\ldots,X_n$ *is a basis in* $\mathfrak{g}$, *admit self-adjoint closures.*

A proof, as well as many interesting examples, can be found in the article [122] of E. Nelson.

§ 11. Characters

11.1. Characters of Finite-Dimensional Representations

By the *character* of a finite-dimensional representation T of a group G, we mean the function

$$\chi_T(g) = \operatorname{tr} T(g).$$

From the definition and from properties of the trace of an operator (see § 3, para. 3) it follows that the function χ_T is constant on every conjugacy class of elements in G and also depends only upon the equivalence class of the representation T. If T_1 is a subrepresentation of T and T_2 is the corresponding factor

§ 11. Characters

representation, then we have $\chi_T = \chi_{T_1} + \chi_{T_2}$. Furthermore, it is easy to check (see 3.3) that

$$\chi_{T_1 \otimes T_2} = \chi_{T_1} \chi_{T_2}.$$

Thus the correspondence $T \to \chi_T$ gives a homomorphism of the Grothendieck ring $\Gamma(G)$ (see 7.1) into the ring of functions on G that are constant on all conjugacy classes.

Theorem 1. *An irreducible representation T of a group G in a finite-dimensional linear space is defined up to equivalence by its character χ_T.*

Proof. We consider the space $V(T)$ of functions on G generated by matrix elements of the representation T. Consider next the representation S of the group G by left translations of functions in the space $V(T)$. It is clear that S is the sum of $n = \dim T$ irreducible representation each equivalent to T. By the Jordan-Hölder theorem (see 8.1) every irreducible subrepresentation of S is equivalent to T. This shows that to reconstruct T, it suffices to choose any irreducible subspace of the space $V \subset V(T)$ where V is the subspace generated by translates of the function χ_T. (One can show that $V = V(T)$, but we need only the inclusion $V \subset V(T)$.) The theorem is proved.

For unitary representation of compact groups, we have

Theorem 2. *Characters of unitary irreducible representations of a compact group G form an orthonormal basis in the space $H \subset L^2(G, dg)$ consisting of functions that are constant on conjugacy classes of elements of G.*

Proof. The relations

$$(\chi_{T_1}, \chi_{T_2}) = 0 \quad \text{for inequivalent} \quad T_1 \quad \text{and} \quad T_2 \tag{1}$$

and

$$(\chi_T, \chi_T) = 1 \tag{2}$$

follow at once from the orthogonality relations proved in 9.2. It remains to verify that if $f \in H$ and $(f, \chi_T) = 0$ for all irreducible representations T, then $f = 0$.

Problem 1. If $f \in H$, then for an arbitrary irreducible representation T, we have the equality $T(f) = \dfrac{(f, \chi_T)}{\dim T} \cdot 1$.

Hint. Prove that $T(f) \in \mathscr{C}(T)$ and that $\operatorname{tr} T(f) = (f, \chi_T)$.

Thus, if $f \in H$ and $(f, \chi_T) = 0$ for all irreducible representations T, then the operators $T(f)$ are equal to zero for all unitary representations, since every unitary representation is the continuous sum of irreducible representations[1]. Applying this fact to the regular representation of G in $L^2(G, dg)$, we see that $f = 0$. Theorem 2 is proved.

---

[1] For compact groups, it suffices to consider discrete sums.

Corollary. *Let S be a finite-dimensional representation of a compact group G. The decomposition of this representation into irreducible components has the form*

$$S = \sum_{|T| \in \hat{G}} (\chi_S, \chi_T) T. \tag{3}$$

Problem 2. Prove that for irreducible representations of a compact group G in real linear spaces, the equalities (1) continue to hold, while the equality (2) assumes the form

$$(\chi_T, \chi_T) = \dim_R \mathscr{C}(T). \tag{2'}$$

Hint. Use theorem 3 from 8.2.

The theory of characters is one of the most effective means for studying the categories $\Pi(G, K)$ and the connections among these categories for various fields and rings K. The book [48] of J.-P. Serre is an excellent introduction to this part of the theory of representations.

We limit ourselves here to two results that have many applications.

Let T be an irreducible unitary representation of a compact group G. As we have seen in 8.2, there are three possible cases:

1) $\chi_T \neq \bar{\chi}_T$; the representation T is not equivalent to $\bar{T}$; the real representation T_R is irreducible and is of complex type;

2) $\chi_T = \bar{\chi}_T$; the representation T is equivalent to $\bar{T}$ and to a certain real representation of real type;

3) $\chi_T = \bar{\chi}_T$; the representation T is equivalent to $\bar{T}$ but is equivalent to no real representation; the representation T_R is of quaternion type.

Theorem 3 (Schur's criterion). *The cases listed above are distinguished by the quantity $\int_G \chi_T(g^2) dg$, which assumes the values $0, +1, -1$ in the three cases, respectively. (We normalize Haar measure so that $\int_G dg = 1$.)*

The proof is based on the following useful identity:

$$\chi(g^2) = \chi_{S^2 T}(g) - \chi_{\wedge^2 T}(g), \tag{4}$$

where $S^2 T$ is the symmetric power and $\wedge^2 T$ is the exterior power of the representation T. In fact, if $\mu_1, \ldots, \mu_n$ are the eigenvalues of the operator $T(g)$, then the operator $S^2 T(g)$ has eigenvalues $\mu_i \mu_j$, $i \leq j$, and the operator $\wedge^2 T$ has eigenvalues $\mu_i \mu_j$, $i < j$. Hence we have

$$\operatorname{tr} S^2 T(g) - \operatorname{tr} \wedge^2 T(g) = \sum_{i \leq j} \mu_i \mu_j - \sum_{i < j} \mu_i \mu_j = \sum_i \mu_i^2 = \operatorname{tr} T(g)^2 = \chi(g^2).$$

We now remark that by formula (3), the quantity $\int_G \chi_T(g) dg$ is equal to the multiplicity of the identity representation in the decomposition of the representation T.

§ 11. Characters

It follows that $\int_G \chi_T(g^2)\,dg$ is the difference between the number of independent invariant vectors in $S^2 V$ and in $\wedge^2 V$, where V is the space of the representation T. But we also have

$$S^2 V \oplus \wedge^2 V = V \otimes V = \mathrm{Hom}(V, V^*).$$

The number of independent invariants in the last space is the intertwining number $c(T, \overline{T})$. By Schur's lemma, it is equal to zero in the first case (where T is not equivalent to $\overline{T}$) and is equal to unity in the second and third cases (where T and $\overline{T}$ are equivalent). To prove the theorem, it remains to verify that an invariant element $A \in \mathscr{C}(T, \overline{T})$ (defined by Schur's lemma uniquely up to a numerical factor) lies in $S^2(V)$ if T is equivalent to a real representation and in $\wedge^2 T$ in the opposite case. Let T be real and hence orthogonal. Then $S^2(V)$ contains an invariant element $\sum_i v_i \otimes v_i$, where $\{v_i\}$ is an orthonormal basis in V. Conversely, if $S^2(V)$ contains an invariant element, then it has the form $\sum v_i \otimes v_i$ in an appropriate basis (since every nonsingular quadratic form in a complex space can be reduced to a sum of squares). Consequently the image of the group G under the representation T is contained in the group $O(n, \mathbf{C})$, $n = \dim V$. It is known that every maximal compact subgroup of $O(n, \mathbf{C})$ is conjugate to $O(n, \mathbf{R})$. Hence the representation T is equivalent to a real representation.

There are also other methods of finishing the proof of the theorem.

Problem 3. Prove that in case (3) the invariant element in $\wedge^2 V$ is the image of a quaternion $k \in \mathbf{H}$ under the natural identification of $\mathbf{H}$ with $\mathscr{C}(T_\mathbf{R})$.

Problem 4. Prove theorem 3 by applying the construction of intertwining operators given in § 9 to the representations T and $\overline{T}$.

The next result relates to the theory of representations of finite groups.

Theorem 4. *If T is an irreducible unitary representation of a finite group G, then $\dim T$ is a divisor of the number $|G|$.*

Proof. The orthogonality relations for matrix elements (see 9.2) imply that

$$\chi_T * \chi_T = \frac{|G|}{\dim T} \chi_T. \tag{5}$$

Iterating this relation n times, we come to the equality

$$\chi_T * \ldots * \chi_T(e) = \frac{|G|^n}{(\dim T)^{n-1}}. \tag{6}$$

We now use the concept and the simplest properties of algebraic integers. A number $\lambda \in \mathbf{C}$ is called an *algebraic integer* if it satisfies an equation of the form

$$\lambda^n + a_1 \lambda + \ldots + a_n = 0, \tag{7}$$

where $a_i \in \mathbf{Z}$.

Problem 5. Prove that the set of all algebraic integers is a subring of **C**.

Hint. Use the formulas of Vieta and show that if $\lambda_1,\ldots,\lambda_n$ and $\mu_1,\ldots,\mu_m$ are sets of roots of two equations of the form (7), then $\{\lambda_i \pm \mu_j\}$ and $\{\lambda_i \mu_j\}$ are also the sets of roots of certain equations of the form (7).

Problem 6. Prove that a rational number is an algebraic integer if and only if it is an integer.

Hint. Suppose that $\lambda = p/q$ is a fraction in irreducible form, and multiply the left side of the equality (7) by q^n. Prove that one obtains in this way an integer not divisible by q.

Problem 7. Prove that a character of a representation of a finite group G assumes algebraic integers as values at every point $g \in G$.

Hint. If $g \in G$, then we have $g^N = e$ for a certain integer N. Hence the eigenvalues of the operator $T(g)$ satisfy the equation $\lambda^N - 1 = 0$.

The theorem now follows from the following fact.

Problem 8. Prove that both sides of the equality (6) are integers.

Hint. Write the left side in the form

$$\sum_{g_1 g_2 \cdots g_{n+1} = e} \chi_T(g_1) \cdots \chi_T(g_{n+1})$$

and use the results of problems 5–7.

We can give a different proof of theorem 4, also using the concept of an algebraic integer, by comparing two bases in the center of the algebra **C**[G]. One basis consists of characters of irreducible representations, and the other of characteristic functions of conjugacy classes of elements in G (see for example [48]).

The result of theorem 4 can be sharpened as follows.

Problem 9. Let Z be the center of the group G. Then the dimension of an arbitrary irreducible representation T of the group G is a divisor of the number $|G|/|Z|$.

Hint (Tate). Prove that the representation $T \times \cdots \times T$ of the group $G \times \cdots \times G$ (n factors) is irreducible and trivial on the subgroup of $Z \times \cdots \times Z$ consisting of all elements $(z_1,\ldots,z_n)$ such that $\prod_{i=1}^{n} z_i = e$.

The above property of dimensions in conjunction with the equalities of 10.1 allow us for many specific groups to "predict" the dimensions of irreducible representations and even to construct them in detail.

11.2. Characters of Infinite-Dimensional Representations

The definition of character given in 11.1 is inappropriate for infinite-dimensional representations, since the operators $T(g)$ are practically never of trace class and hence do not admit a reasonable trace. However it is often possible to define the character of an infinite-dimensional representation as a generalized function

§ 11. Characters

on the group G. This definition, which is due to I. M. Gel'fand, can be stated as follows.

Let $D_T(G)$ be a certain subalgebra of the group algebra $M(G)$ having the following properties:

1) the space $D_T(G)$ is invariant under right and left translations on G;
2) the representation T of the group G generates a representation $\tilde{T}$ of the algebra $D_T(G)$ and can be reconstructed uniquely from this representation;
3) the operators $T(a)$, $a \in D_T(G)$, are of trace class (that is, they belong to $V' \hat{\otimes} V$) and the mapping $\tilde{T}: D_T(G) \to V' \hat{\otimes} V$ is continuous.

The *character of the representation* T is defined as the linear functional $\chi \in D_T(G)'$ defined by the equality

$$\langle \chi, a \rangle = \operatorname{tr} \tilde{T}(a). \tag{1}$$

If G is a Lie group and $D_T(G) = C_0^\infty(G)$, then the character χ is a generalized function on G in the ordinary sense of this word.

For a finite-dimensional representation T, one may take $D_T(G)$ to be the subalgebra $M_0(G)$ (or the whole algebra $M(G)$ if T is bounded). In this case the character χ has the form

$$\langle \chi, \mu \rangle = \int_G \chi_T(g) \, d\mu(g). \tag{2}$$

where χ_T is the usual character of the representation T, defined in 11.1.

For semisimple Lie groups, Harish-Chandra has established the following remarkable fact (see [92]).

If T is a continuous unitary representation of G, then the character of T is defined on the subalgebra $C_0^\infty(G)$ and has the form

$$\langle \chi, f \rangle = \int_G \chi_T(g) f(g) \, dg. \tag{3}$$

where χ_T is a measurable and locally integrable function on G.

We remark that for specific representations T, it is usually not hard to compute the function χ_T explicitly. We shall do this in § 13 for a wide class of representations. The main difficulty lies in proving a theorem without knowing the explicit form of the representations.

For nilpotent Lie groups, the characters of irreducible representations are also defined on $C_0^\infty(G)$, but can be generalized functions of a more complicated sort (for example, they may contain the δ-function in several variables).

The characters of irreducible unitary representations of solvable Lie groups are even more complicated. As a rule, their domain of definition is much smaller than $C_0^\infty(G)$. In this case, smoothness and compact support of f do not suffice for the operator $T(f)$ to have a trace. Additional restrictions on f are needed, which single out in $C_0^\infty(G)$ a subspace of infinite codimension. We shall go into this in more detail in §§ 12 and 15.

Finally, the following case can occur (even for Lie groups). It is possible that the operator $\tilde{T}(a)$ is of trace class only for the elements of the group algebra that belong to the kernel of the representation $\tilde{T}$.

Theorem 1. *Let G be a locally compact group. The following conditions are equivalent:*

(1) for every irreducible representation T in a space H, the character of T is defined on a certain subalgebra $D_T(G) \subset C^*(G)$ for which $\tilde{T}(D_T(G)) \neq 0$;

(2) G is a tame group.

Thus, tame groups are singled out among all locally compact groups from the point of view of the theory of characters.

Theorem 1 follows from the following theorem for C^*-algebras.

Theorem 2 (Dixmier-Glimm-Sakai). *The following properties of a C^*-algebra $\mathfrak{A}$ are equivalent.*

1) For every irreducible $*$-representation T of the algebra $\mathfrak{A}$, the image $T(\mathfrak{A})$ contains at least one completely continuous operator.

2) For every irreducible $*$-representation T of the algebra $\mathfrak{A}$, the image $T(\mathfrak{A})$ contains all completely continuous operators.

3) Every $*$-representation of $\mathfrak{A}$ is of type I.

A proof of this theorem for separable C^*-algebras $\mathfrak{A}$ can be found in the book [15] of Dixmier. There one will find a citation of Sakai's paper in which the nonseparable case is taken up.

Characters as defined above retain many properties of ordinary characters of finite-dimensional representations. Thus, they are invariant under inner automorphisms of G:

$$\langle \chi, L_g R_g f \rangle = \operatorname{tr} T(L_g R_g f) = \operatorname{tr}[T(g) T(f) T(g^{-1})] = \operatorname{tr} T(f) = \langle \chi, f \rangle.$$

(This property implies that if χ is an ordinary function, then it is constant on conjugacy classes of elements of G).

Theorem 3. *An irreducible representation T is defined by its character up to equivalence.*

Proof. The set of elements $a \in C^*(G)$ for which $a^*a \in D_T(G)$ is a pre-Hilbert space with respect to the scalar product $(a,b)_\chi = \langle \chi, b^*a \rangle$. Let H_χ be the corresponding Hilbert space. The correspondence $a \to \tilde{T}(a)$ yields an isomorphism of H_χ onto a space of Hilbert-Schmidt operators in the space H of the representation T, under which left and right translations by $g \in G$ go into the operators of left and right multiplication by $T(g)$. Hence the representation of G in H_χ by left translations is equivalent to $(\dim T) \cdot T$.

Thus we have a representation equivalent to T in an arbitrary irreducible subspace of H_χ.

For tame groups, Theorems 1 and 3 reduce the problem of classifying irreducible unitary representations of the group to the identification of all of its characters. We shall see in the following paragraph that characters of Lie groups satisfy a certain system of differential equations. In many important cases, this permits us to write them explicitly.

Working with characters is made difficult by the fact that the domain of definition of a character is not known in advance. In particular it is not clear

§ 11. Characters

whether or not this domain contains at least one "genuine" function on the group. (We recall that the C^*-algebra of the group is obtained by completing $L^1(G, d_l g)$ in a certain norm. To compute this norm, we need to know a large class of representations of G. Hence the structure of the "ideal elements" to be added to the L^1 algebra is not known *a priori*.) In all examples that have been studied, the intersection $D_T(G) \cap L^1(G, d_l g)$ is dense in $D_T(G)$. However, no general theorem of this sort has been proved.

11.3. Infinitesimal Characters

For irreducible representations of Lie groups, one can give yet another variant of the definition of a character, which has meaning for infinite-dimensional representations.

Let G be a connected Lie group, $\mathfrak{g}$ its Lie algebra, $U(\mathfrak{g})$ the enveloping algebra, and $Z(\mathfrak{g})$ its center.

Problem 1. Prove that $Z(\mathfrak{g})$ lies in the center of $R(G)$ (under the natural embedding $U(\mathfrak{g}) \approx R(G, \{e\})$ in $R(G)$: see 10.4).

Hint. Linear combinations of delta-functions δ_g form a dense set in $R(G)$ (by definition, $\langle \delta_g, f \rangle = f(g)$ for $f \in C^\infty(G)$). The assertion of the problem follows from the fact that a connected group is generated by the canonical neighborhood of the identity and from the identity given below.

Problem 2. Let $X \in U(\mathfrak{g})$, and let $Y \in \mathfrak{g}$. Then we have

$$\delta_{\exp Y} * X * \delta_{\exp(-Y)} = e^{\widetilde{\mathrm{ad}} Y} X = \widetilde{\mathrm{Ad}}(\exp Y) X.$$

Here $\widetilde{\mathrm{Ad}}$ ($\widetilde{\mathrm{ad}}$) denotes the representation of $G(\mathfrak{g})$ by automorphisms (differentiations) of $U(\mathfrak{g})$, generating the adjoint (regular) representation.

Hint. First work out the case $X \in \mathfrak{g}$.

Suppose that T is a finite-dimensional irreducible representation of a connected Lie group G. Then by Schur's lemma, elements of the center of $R(G)$ go into scalar operators. In particular, for $z \in Z(\mathfrak{g})$ we have

$$T(z) = \lambda_T(z) \cdot 1. \tag{1}$$

It is clear that the correspondence $z \mapsto \lambda(z)$ is a homomorphism of the algebra $Z(\mathfrak{g})$ in the field $\mathbf{C}$. This homomorphism is called the *infinitesimal character* of the representation T. We explain its connection with the ordinary character.

Problem 3. Prove the identity

$$\lambda_T(z) = \frac{\langle z, \chi_T \rangle}{\dim T}. \tag{2}$$

Hint. Substitute in the left side of (1) the definition of $T(z)$ (see 10.5) and take the trace of both sides.

Thus we can find the infinitesimal character λ_T from the ordinary character χ_T.

The question of whether or not the converse is true is more complicated, and its answer depends upon the group under consideration.

Problem 4. Let T be a finite-dimensional irreducible representation of a group G and let λ_T be its infinitesimal character. Prove that an arbitrary matrix element $t_{\xi,\eta}$ of the representation T satisfies the system of differential equations

$$z * t_{\xi,\eta} = \lambda_T(z) t_{\xi,\eta}, \quad z \in Z(\mathfrak{g}). \tag{3}$$

Hint. Use the identity

$$T(z * \delta_g) = T(z) T(\delta_g) = \lambda_T(z) T(g).$$

Thus, to reconstruct the character χ_T from the infinitesimal character λ_T, it is necessary to solve the system of differential equations (3) in the class of functions that are invariant under inner automorphisms. For semisimple Lie groups, this problem can be completely solved (it reduces to the solution of a system of differential equations with constant coefficients: see [62]). The general case can as a practical matter be reduced to the semisimple case. The point is that every irreducible linear group is either semisimple or is the product of a semisimple subgroup and a one-dimensional center. Thus, if T is a faithful irreducible representation of the group G, then its character χ_T (and hence the representation itself) can be reconstructed from the infinitesimal character λ_T.

We note that the requirement of faithfulness for T is essential. For example, the group of affine transformations of the line $x \to ax + b$ admits a series of one-dimensional representations $T_\rho(a, b) = a^\rho$, which are not defined by their infinitesimal characters, since for this group, $Z(\mathfrak{g})$ consists only of scalars.

We turn to the infinite-dimensional case. Let T be an irreducible representation of G in a complete locally convex space V and T^∞ the corresponding representation of $R(G)$ in the Gårding space $V^\infty \subset V$. Then the operators $T^\infty(z)$, $z \in Z(\mathfrak{g})$, commute with all operators of the representation T. Generally speaking, one cannot infer from this that they are scalar operators. The standard way of circumventing this difficulty is to consider only representations for which (1) holds. Such representations are called *simple*.

Theorem 1. *Every irreducible unitary representation T of a Lie group G is simple and consequently has an infinitesimal character λ_T.*

Proof. We use the fact that the representation T is 2-irreducible (see 7.3) and consequently every closed operator commuting with the operators of the representation must be a scalar operator. Let $z \in Z(\mathfrak{g})$. We represent z as a sum of homogeneous components: $z = \sum z_k$, $z_k \in \sigma(S^*(\mathfrak{g}))$. (See 10.4.) It follows from theorem 2 and problem 8 of 10.4 that every component z_k also lies in $Z(\mathfrak{g})$. Let us show that the operator $T^\infty(z_k)$ admits a closure. In fact, the operators $T^\infty(X)$, $X \in \mathfrak{g}$, have the property that $T^\infty(X) \subset T^\infty(-X)^*$ (see 10.5). Hence we have $T^\infty(z_k) \subset (-1)^k T^\infty(z_k)^*$, which we wished to show. This completes the proof.

Just as in the finite-dimensional case, we prove that the matrix elements of the representation T satisfy the system of equations (3). Suppose that the

§ 11. Characters

representation T has a generalized character (in the sense stated in 11.2), which is defined on a certain subspace $D(G) \subset C_0^\infty(G)$. Let us show that this character also satisfies the system of equation

$$z * \chi = \lambda_T(z) \chi, \quad z \in Z(\mathfrak{g}). \tag{4}$$

By definition, this means that for an arbitrary function $f \in D(G)$, the equality

$$\langle z * \chi, f \rangle = \lambda_T(z) \langle \chi, f \rangle$$

must hold. That is, we must have

$$\langle z, \operatorname{tr} T(g) T(f) \rangle = \lambda_T(z) \operatorname{tr} T(f),$$

which follows from the fact that $T(f)$ is of trace class and from the relations (3) for matrix elements of the representation T.

This result discloses the possibility of finding generalized characters of irreducible unitary representations as solutions of the system of differential equations (4) in the class of generalized functions on G that are invariant under inner automorphisms of G.

For complex semisimple Lie groups, this method leads to a complete classification of a wide class of simple representations containing all unitary representations and representations in Banach spaces.

Unfortunately, fundamental difficulties appear when one tries to single out the unitary representations in this class, and up to the present time, the problem has not been completely solved.

More details about this direction of the theory can be found in the works [62], [139], [120].

The theory of infinitesimal characters depends in an essential way on the structure of the algebra $Z(\mathfrak{g})$. We saw in 10.4 that the algebra $\operatorname{gr} Z(\mathfrak{g})$ coincides with the algebra $S(\mathfrak{g})^G$ of invariant elements of the algebra $S(\mathfrak{g})$. In all known examples, the algebra $Z(\mathfrak{g})$ is isomorphic with $\operatorname{gr} Z(\mathfrak{g})$. In particular, this has been proved for semisimple and solvable Lie groups.

We shall give in § 15 an interpretation of this fact from the point of view of the method of orbits[1].

A very interesting problem is presented by the construction and investigation of *generalized infinitesimal characters*, which are connected with the Lie skew field $D(\mathfrak{g})$ (see 10.4). Let X be an element of the center of $D(\mathfrak{g})$. This element can be written in the form PQ^{-1}, $P, Q \in U(\mathfrak{g})$.

If T is an irreducible unitary representation of the group G and T^∞ is the corresponding representation of $U(\mathfrak{g})$, then the operators $T^\infty(P)$ and $T^\infty(Q)$ differ only by a scalar multiple: $T^\infty(P) = \lambda T^\infty(G)$.

One can show that the number λ depends only on X (and not on the way in which we write $X = PQ^{-1}$), and we denote this number by $\lambda(X)$. The corre-

---

[1] M. Duflo has recently proved this fact in the the general case, making use of the method of orbits: see [14].

spondence $X \mapsto \lambda(X)$ gives a mapping of the center of the skew field $D(\mathfrak{g})$ into the extended complex plane $\bar{C}$ and has the properties

$$\lambda(X+Y) = \lambda(X) + \lambda(Y), \quad \lambda(XY) = \lambda(X)\lambda(Y)$$

in all cases when the equalities have a meaning. This is the generalized infinitesimal character of the representation T.

§ 12. Fourier Transforms and Duality

One of the directions of the theory of representations of groups, which has been given the name of harmonic analysis, consists in the study of function spaces on groups and homogeneous spaces in terms of the representations that appear in them. Here a great rôle is played by the generalized Fourier transform. As is known, the classical Fourier transform on the real line carries translations into multiplication by functions and thus gives the spectral decomposition of an arbitrary operator that commutes with translations (for example, a differential operator with constant coefficients). The generalized Fourier transform is designed to play exactly this rôle for an arbitrary group. This transform is defined as follows. Let G be a topological group and $\hat{G}$ the set of equivalence classes of irreducible unitary representations of G. For each class $\lambda \in \hat{G}$, we choose a representation T_λ belonging to this class and acting in a Hilbert space H_λ.

The representation T_λ can be extended to a representation of the group ring $M(G)$ in the same space H_λ. The *Fourier transform* of an element $\mu \in M(G)$ is the operator function $\tilde{\mu}$ on $\hat{G}$ which assumes at the point $\lambda \in G$ the value in $\mathrm{End}\, H_\lambda$ given by the formula

$$\tilde{\mu}(\lambda) = \int_G T_\lambda(g)\, d\mu(g). \tag{1}$$

In particular, if the group G is locally compact and $f \in L^1(G, d_l g)$, we will write

$$\tilde{f}(\lambda) = \int_G f(g)\, T_\lambda(g)\, d_l g. \tag{2}$$

If G is a Lie group, then the definition of the Fourier transform can be generalized to elements of the ring $R(G)$, by setting

$$\tilde{x}(\lambda) = T_\lambda^\infty(x) \in \mathrm{End}\, H_\lambda^\infty. \tag{3}$$

It is evident that the transform just defined carries left (right) translation by g into the operator of multiplication by the operator function $\lambda \mapsto T_\lambda(g)$. We shall see *infra* that many of the usual properties of the classical Fourier transform are preserved even in the most general situation. For certain classical groups (com-

§ 12. Fourier Transforms and Duality

mutative, compact, finite) the Fourier transform enjoys special properties and has been studied in more detail. We shall consider these classes of groups separately, and then we shall turn to the general case.

12.1. Commutative Groups

The basic distinguishing feature of commutative groups from the point of view of harmonic analysis is the fact that the dual space $\hat{G}$ is itself a topological group. The crux of the matter is that for a commutative group G, all unitary irreducible representations are one-dimensional (see theorem 2 of 7.3 or 4.4). Hence the tensor product of two irreducible representations is again irreducible. Every such representation can be regarded as an ordinary numerical function on G that satisfies the conditions

$$f(g_1 g_2) = f(g_1) f(g_2), \quad |f(g)| \equiv 1. \tag{1}$$

The operation of tensor multiplication reduces to ordinary multiplication of functions. It is clear that $\hat{G}$ is a group under this operation. The identity in $\hat{G}$ is the trivial representation $f_0(g) \equiv 1$, and the inverse element to f is the complex conjugate representation $\bar{f}$. The topology defined in 7.3 becomes in this case the topology of uniform convergence on compact sets.

Problem 1. Verify that $\hat{G}$ is a topological group and that if G is locally compact (resp. discrete, compact) that $\hat{G}$ is locally compact (resp. compact, discrete).

The group $\hat{G}$ arising in this way is called the *Pontrjagin dual of the group* G, or the *dual group*, or the *character group of* G. (We remark that one-dimensional representations coincide with their characters.)

We can construct the dual group $(\hat{G})\hat{\,} = \hat{\hat{G}}$ of the group $\hat{G}$, and so on. There is a natural mapping of G into $\hat{\hat{G}}$, which maps each element $g \in G$ into the character (that is, one-dimensional representation) of the group $\hat{G}$ which assumes the value $\chi(g)$ at every point $\chi \in \hat{G}$.

For the most important examples, where $G = \mathbf{R}$, $\mathbf{Z}$, or $\mathbf{T}$, the dual groups (without topology) were in fact computed in § 6 (see problem 4 from 6.1).

Problem 2. Prove that the following are isomorphisms of topological groups:

$$\hat{\mathbf{R}} = \mathbf{R}, \quad \hat{\mathbf{T}} = \mathbf{Z}, \quad \hat{\mathbf{Z}} = \mathbf{T}, \quad \hat{\mathbf{Z}}_n = \mathbf{Z}_n.$$

The corresponding Fourier transforms have been used for a long time in various branches of mathematics and its applications. If $G = \mathbf{R}$, the Fourier transform has the form

$$\tilde{f}(\lambda) = \int_{-\infty}^{\infty} e^{i\lambda t} f(t) dt.$$

that is, it coincides with the classical Fourier transform.

For $G=\mathbf{T}$, we obtain

$$\tilde{f}(n) = \frac{1}{2\pi} \int_0^{2\pi} e^{in\phi} f(\phi) d\phi, \quad n \in \mathbf{Z},$$

where ϕ is the classical parameter (angle) on the circle $\mathbf{T}$. Thus, in this case the Fourier transform of the function f is the collection of coefficients of its Fourier series.

For $G=\mathbf{Z}$, the general formula has the form

$$\tilde{f}(z) = \sum_{-\infty}^{\infty} f(n) z^n, \quad z \in \mathbf{T},$$

that is, the Fourier transform of $\{f(n)\}$ is its *generating function*.

Finally, for $G=\mathbf{Z}_n$, the Fourier transform has the form

$$\tilde{f}(k) = \sum_{l=0}^{n-1} f(l) e^{2\pi k l i/n}, \quad k, l \in \mathbf{Z}_n.$$

In all of the above examples, the group $\hat{\hat{G}}$ is isomorphic to G, and furthermore, this isomorphism can be defined in such a way that the Fourier transforms from G to $\hat{G}$ and from $\hat{G}$ to $\hat{\hat{G}} \approx G$ will be inverses of each other. This fact is an illustration of the general duality principle, which is formulated in the following way.

Theorem 1 (L. S. Pontrjagin). *For an arbitrary locally compact commutative group G, the canonical mapping of G into $\hat{\hat{G}}$ is an isomorphism of topological groups. Haar measures on G and $\hat{G}$ can be normalized so that the Fourier transforms from G to $\hat{G}$ and from $\hat{G}$ to $\hat{\hat{G}} = G$ will be connected by the relations*

$$\left. \begin{aligned} \tilde{f}(\chi) &= \int_G f(g) \chi(g) dg, \\ f(g) &= \int_{\hat{G}} \tilde{f}(\chi) \overline{\chi(g)} d\chi, \\ \int_G |f(g)|^2 dg &= \int_{\hat{G}} |\tilde{f}(\chi)|^2 d\chi. \end{aligned} \right\} \quad (2)$$

If the group G is compact and the measure dg is normalized so that $\int_G dg = 1$, then the group $\hat{G}$ is discrete and each of its points has measure 1. Conversely, if G is discrete and each of its points has measure 1, then $\hat{G}$ is compact and $\int_{\hat{G}} d\chi = 1$.

There are two types of proofs of this theorem. The first is based on the study of the structure of locally compact commutative groups. It is the case that an arbitrary locally compact group is obtained by the operations of sum, product, and projective and inductive limits from the "elementary" groups $\mathbf{R}, \mathbf{Z}, \mathbf{T}$, and $\mathbf{Z}_n$. For elementary groups, the truth of the theorem is checked immediately.

§ 12. Fourier Transforms and Duality

Then one proves that the operations of sum, product, and inductive and projective limit are replaced, when one goes to the dual group, by the operations of product, sum, and projective and inductive limit, respectively. A detailed exposition can be found in the book [44] of L. S. Pontrjagin and in the book [54] of A. Weil.

A different type of proof is based on the theory of Banach algebras. It requires no facts about the structure of the group, but also does not yield any information about this structure. A proof of this sort was first given by D. A. Raĭkov. Variants of it are set forth in the books [9], [22], and [42].

We now give the principal components of this proof. We begin with the fact that if G is nondiscrete, then by theorem 1 of 10.2, the set $\hat{G}$ is obtained from $\mathfrak{M}(C^*(G))$, that is, the spectrum of the algebra $C^*(G)$, by removing the point corresponding to the degenerate representation of $C^*(G)$. For a discrete group G, the set $\hat{G}$ coincides with $\mathfrak{M}(C^*(G))$. The Fourier transform on the group G obviously coincides with the Gel'fand transform of the C^*-algebra $C^*(G)$.

By the theorem of Gel'fand in 4.3, the image of $C^*(G)$ under this transform is all of $C(\mathfrak{M}(C^*(G)))$. Hence the image of $L^1(G,dg)$ under the Fourier transform is dense in the set $C_1(\hat{G})$ of all continuous functions on $\hat{G}$ that go to zero at infinity.

We now consider the set of Fourier transforms of functions in $L^1(G,dg) \cap C(G)$. On the space of all such functions, we define the linear functional

$$F(\tilde{f}) = f(e). \qquad (3)$$

This functional is invariant under translations on $\hat{G}$, since the translation of a function $\tilde{f}$ by $\chi \in \hat{G}$ is equivalent to multiplication of f by χ, which does not alter the value $f(e)$. If we knew in advance that the domain of definition of F contains the space $C_0(\hat{G})$ of continuous functions on $\hat{G}$ with compact supports, then we could write the functional F in the form

$$F(\phi) = \int_{\hat{G}} \phi(\chi) d\chi, \qquad \phi \in C_0(\hat{G}), \qquad (4)$$

where $d\chi$ is some measure on $\hat{G}$. From the invariance of F it follows that $d\chi$ is Haar measure on $\hat{G}$. Now let f be the convolution of functions f_1 and f_2^*. We would then obtain from (3) and (4) the equality

$$(f_1 * f_2^*)(e) = \int_{\hat{G}} (f_1 * f_2^*)\tilde{\ }(\chi) d\chi.$$

From the definitions of convolution, involution, and Fourier transform, we could rewrite the last equality in the form

$$\int_G f_1(g) \overline{f_2(g)} dg = \int_G \tilde{f}_1(\chi) \overline{\tilde{f}_2(\chi)} d\chi. \qquad (5)$$

Thus the Fourier transform is extended to an isomorphism of the Hilbert spaces $L^2(G,dg)$ and $L^2(\hat{G},d\chi)$. This proves the last of the equalities (2) in the statement of the theorem.

Next, in equality (5) let f_2 run through a sequence converging to the δ-function at e. Also suppose that f_1 is continuous and that $\tilde{f}_1$ is summable. We thus arrive at the second of the equalities (2).

Problem 3. Deduce from the properties of the Fourier transform obtained above the fact that the groups G and $\hat{\hat{G}}$ are topologically isomorphic.

Hint. Prove in turn the following. The canonical mapping of G into $\hat{\hat{G}}$ is an embedding. The topology of G coincides with the topology induced on G by its embedding in $\hat{\hat{G}}$. Finally, G is closed in $\hat{\hat{G}}$.

Unfortunately, up to this time there exists no direct proof that the functional F is defined and continuous on $C_0(\hat{G})$. Hence one usually employs the following circuitous route to prove the equality (4).

Problem 4. Suppose that f_1 and f_2 belong to $L^1(G) \cap L^2(G)$. Then the function $f_1 * f_2$ is a continuous function on G.

Hint. Verify that

$$|(f_1 * f_2)(x)| \leq \|f_1\|_{L^2(G)} \|f_2\|_{L^2(G)}$$

and use the fact that continuous functions are dense in $L^1(G) \cap L^2(G)$.

Problem 5. If $\phi \in C^*(G)$, $f \in L^2(G)$, then we have $\phi * f \in L^2(G)$, and furthermore, $\|\phi * f\|_{L^2(G)} \leq \|\phi\|_{C^*(G)} \|f\|_{L^2(G)}$.

Hint. Use the result of problem 2 of 4.3.

Let $A(G)$ denote the set of linear combinations of functions of the form $f_1 * f_2$, $f_i \in L^1(G) \cap L^2(G)$. It follows from what has been said above that for $f \in A(G)$, the mapping

$$\phi \to (\phi * f)(e)$$

is a continuous functional on $C^*(G)$. Hence for every $f \in A(G)$, there exists a measure μ_f on $\hat{G}$ such that

$$(\phi * f)(e) = \int_{\hat{G}} \tilde{\phi}(\chi) d\mu_f(\chi) \tag{6}$$

for all $\phi \in L^1(G)$.

Problem 6. Prove the equality

$$\tilde{f}\mu_h = \tilde{h}\mu_f$$

for all $f, h \in A(G)$.

Hint. Use the equalities

$$\phi * (f * h) = (\phi * f) * h = (\phi * h) * f.$$

§ 12. Fourier Transforms and Duality

Problem 7. Prove that there exists a measure μ on $\hat{G}$ such that

$$\mu_f = \tilde{f}\mu$$

for all $f \in A(G)$.

Hint. Use the result of problem 6 and the fact that the Fourier transform of the space $A(G)$ is dense in $C_0(\hat{G})$.

Problem 8. Prove that the measure μ described in problem 7 is Haar measure on $\hat{G}$.

Hint. Use the equality

$$[(\phi \cdot \chi) * (f \cdot \bar{\chi})](e) = (\phi * f)(e).$$

Thus we come to the relation

$$(\phi * f)(e) = \int_{\hat{G}} (\phi * f)\check{\;} d\chi, \quad \phi \in C^*(G), \quad f \in A(G),$$

which replaces (and refines) the inadequate equality (4).

Much work has been devoted to the study of the Fourier transform on locally compact commutative groups. A quite detailed exposition of this part of harmonic analysis is contained in the book of E. Hewitt and K. Ross [30]. We give here some of the most important facts.

First of all, it is appropriate to point out the functorial properties of the mapping $G \to \hat{G}$. To every morphism $\phi: G_1 \to G_2$, there corresponds the dual morphism $\hat{\phi}: \hat{G}_2 \to \hat{G}_1$, which carries $\chi \in \hat{G}_2$ into $\chi \circ \phi$.

Theorem 2. *Let G_0 be a closed subgroup in a locally compact commutative group G, and let G_1 be the corresponding factor group. Then $\hat{G}_1$ coincides with the subgroup $G_0^{\perp} \subset \hat{G}$, which consists of all characters that are trivial on G_0. The character group $\hat{G}_0$ coincides with the corresponding factor group.*

Thus we have the exact sequence

$$1 \to G_0 \xrightarrow{i} G \xrightarrow{p} G_1 \to 1,$$

and to it there corresponds the exact sequence

$$1 \leftarrow \hat{G}_0 \xleftarrow{\hat{i}} \hat{G} \xleftarrow{\hat{p}} \hat{G}_1 \leftarrow 1.$$

Suppose that the Haar measures dg, dg_0, and dg_1 on G, G_0, and G_1 are normalized so that the equality $dg = dg_0 \cdot dg_1$ obtains.

We consider the linear functional F on $C_0(G)$ which maps the function f into its integral on the subgroup $G_0 \subset G$. That is, we have

$$F(f) = \int_{G_0} f(g_0) dg_0.$$

Plainly this functional is invariant under translations by elements of G_0 and also under multiplication by characters of the group G that are trivial on G_0. We shall consider F as a functional of $\tilde{f} \in C(\hat{G})$. It is invariant under multiplication by characters of the group G corresponding to elements of G_0 (that is, trivial on $\hat{G}_1 = G_0^\perp$). The functional F is also invariant under translation by elements $\chi_1 \in \hat{G}_1$. This leads us to the following relation.

Problem 9. Prove that F can be written in the form

$$F(\tilde{f}) = \int_{\hat{G}_1} \tilde{f}(\chi_1) d\chi_1,$$

where $d\chi_1$ is the Haar measure on $\hat{G}_1 = G_0^\perp$ that corresponds to the measure dg_1 on G_1.

Hint. Apply the inversion formula to the function ϕ on G_1 that is defined by the equality

$$\phi(gG_0) = \int_{G_0} f(gg_0) dg_0.$$

Corollary. *Define the Fourier transform on $\mathbf{R}$ by the formula $\tilde{f}(y) = \int_{-\infty}^{\infty} e^{ixy} f(x) dx$. Then for every $\alpha > 0$, we have the following equality* (Poisson's formula):

$$\sum_{k \in \mathbf{Z}} f(k\alpha) = \frac{\sqrt{2\pi}}{\alpha} \sum_{k \in \mathbf{Z}} \tilde{f}\left(\frac{2\pi k}{\alpha}\right).$$

There are a multitude of results about the effect of taking the Fourier transform of one or another class of functions. From these, we cite the following fact.

Theorem 3. *Consider the restriction of the Fourier transform to the space $L^1(G) \cap L^p(G)$, $1 < p < 2$. This mapping can be extended to a linear operator with norm ≤ 1 that carries $L^p(G)$ into $L^q(\hat{G})$, where $q = \dfrac{p}{p-1}$.*

The proof of this theorem (and of other analogous assertions) is based on the elegant idea of interpolation. Specifically, one proves that the logarithm of the $L^q(\hat{G})$ norm of the function $\tilde{f}$ (where $f \in L^1(G) \cap L^p(G)$) is a convex function of $1/p$ in the closed interval $\tfrac{1}{2} \leq 1/p \leq 1$.

There are many results which both sharpen and generalize the following principle:

the properties of smoothness and rapid decrease to zero at infinity go into each other under the Fourier transform.

For the case $G = \mathbf{R}^n$, one of these precise formulations is the following. For $f \in C_0^\infty(\mathbf{R}^n)$, we set

$$\|f\|_{k,l} = \int_{\mathbf{R}^n} |(1 + \|x\|^{2k} + \Delta^l) f|^2 dx,$$

where $\|x\|^2 = x_1^2 + \ldots + x_n^2$, $\Delta = -\dfrac{\partial^2}{\partial x_1^2} - \ldots - \dfrac{\partial^2}{\partial x_k^2}$.

§ 12. Fourier Transforms and Duality

The closure of $C_0^\infty(\mathbf{R}^n)$ in the norm $\|\cdot\|_{k,l}$ consists, roughly speaking, of the functions that have $2l$ derivatives that decrease at infinity like $\dfrac{1}{\|x\|^{2k-n-2}}$. The completion of $C_0^\infty(\mathbf{R}^n)$ in the collection of all the norms $\|\cdot\|_{k,l}$ coincides with the Schwartz space $S(\mathbf{R}^n)$.

Theorem 4. *The Fourier transform carries the Schwartz space into itself, and furthermore we have*

$$\|\tilde{f}\|_{k,l} = \|f\|_{l,k}.$$

The proof follows at once from the following assertion.

Problem 10. Suppose that we choose coordinates $x_1, \ldots, x_n$ in $G = \mathbf{R}^n$ and $y_1, \ldots, y_n$ in $\hat{G} \approx \mathbf{R}^n$ in such a way that

$$\langle \chi, g \rangle = e^{i(x_1 y_1 + \ldots + x_n y_n)}.$$

Prove that the operators of multiplying by x_k and differentiating with respect to x_k go into the operators $i\dfrac{\partial}{\partial y_k}$ and multiplication by $-iy_k$, respectively, under the Fourier transform.

We note that analogues of the spaces C^∞, C_0^∞ and S can be defined for an arbitrary locally compact group. This definition is based on the fact that every locally compact group is a projective limit of Lie groups.

More precisely, every neighborhood of the identity $e \in G$ contains a compact normal subgroup K such that the factor group G/K is a Lie group. As the spaces $C^\infty(G)$ and $C_0^\infty(G)$ we take the inductive limit of the spaces $C^\infty(G/K)$ and $C_0^\infty(G/K)$ respectively. Details can be found in the article of G. I. Kac [95] and the work [68] of F. Bruhat. The definition of $S(G)$ is more complicated. For semisimple Lie group, see [93].

12.2. Compact Groups

Let G be a noncommutative compact group. In this case, the set $\hat{G}$ is discrete. The Fourier transform of a function f on G is an operator function $\tilde{f}$ on $\hat{G}$. Since all irreducible representations of the group G are finite-dimensional, we may suppose that $\tilde{f}$ assumes the value $\tilde{f}(\lambda) \in \mathrm{Mat}_{n(\lambda)}(\mathbf{C})$, at each point $\lambda \in \hat{G}$. Here $n(\lambda)$ is the dimension of a representation T_λ of the class λ. The question of identifying the Fourier transform of one or another class of functions on G arises quite naturally. (One may ask the same question for an arbitrary version of the group algebra.)

The answer is much the simplest for the function space $L^2(G, dg)$.

Let $L^2(\hat{G})$ denote the set of all matrix functions on $\hat{G}$ that satisfy the following conditions:

1) $\phi(\lambda) \in \mathrm{Mat}_{n(\lambda)}(\mathbf{C})$ for all $\lambda \in \hat{G}$,

2) $\sum\limits_{\lambda \in \hat{G}} n(\lambda) \mathrm{tr}[\phi(\lambda)^* \phi(\lambda)] < \infty.$

It is clear that $L^2(\hat{G})$ is a Hilbert space under the scalar product

$$(\phi_1, \phi_2) = \sum_{\lambda \in \hat{G}} n(\lambda) \text{tr}[\phi_1(\lambda)\phi_2(\lambda)^*].$$

Theorem 1. *The Fourier transform can be extended to an isometry of the space $L^2(G, dg)$ onto the space $L^2(\hat{G})$.*

The proof follows without difficulty from the orthogonality relations for matrix elements of irreducible representations of a compact group (see 9.2).

It is also not hard to identify the image of the C^*-algebra of the group G. We recall that by the definition of the algebra $C^*(G)$, the mapping $f \to \tilde{f}(\lambda)$ for every $\lambda \in \hat{G}$ can be extended to a $*$-representation of the algebra $C^*(G)$, the image of which is the full matrix ring $\text{Mat}_{n(\lambda)}(\mathbf{C})$. Furthermore, the mapping that arises in this way of $C^*(G)$ into $\prod_{\lambda \in \hat{G}} \text{Mat}_{n(\lambda)}(\mathbf{C})$ is an isometric embedding of $C^*(G)$ into the C^*-algebra of bounded matrix functions on $\hat{G}$.

Theorem 2. *The image of $C^*(G)$ in $\prod_{\lambda \in \hat{G}} \text{Mat}_{n(\lambda)}(\mathbf{C})$ consists of all bounded matrix functions ϕ on $\hat{G}$ that satisfy the following condition: there exists a constant $c = c(\phi)$ such that for an arbitrary $\varepsilon > 0$, the inequality*

$$\|\phi(\lambda) - c \cdot \mathbf{1}\| < \varepsilon$$

holds for all but a finite number of elements $\lambda \in \hat{G}$.

In other words, the image of $C^*(G)$ is generated by the function $\varepsilon(\lambda) = \mathbf{1}_{n(\lambda)}$ and by functions that go to zero at infinity.

Proof. Let $C^*(\hat{G})$ denote the algebra of bounded matrix functions that satisfy the condition of theorem 2.

Problem 1. Prove that $C^*(\hat{G})$ is a C^*-algebra under the norm

$$\|\phi\| = \sup_{\lambda \in \hat{G}} \|\phi(\lambda)\|$$

and coordinatewise operations of addition, multiplication, and involution.

Since the algebra $C^*(G)$ is generated by functions in $C(G)$ and the adjoined unit e, it suffices to prove that the images of $C(G)$ and e lie in $C^*(\hat{G})$ and generate a dense subset of $C^*(\hat{G})$. The image of e is the function $\varepsilon(\lambda) = \mathbf{1}_{n(\lambda)}$. It obviously belongs to $C^*(\hat{G})$ (it suffices to set $c(\varepsilon) = 1$). The space $C(G)$ is contained in $L^2(G)$. By theorem 1, the image $\tilde{f}$ of an arbitrary function $f \in L^2(G)$ lies in $L^2(\hat{G})$ and consequently satisfies the condition of theorem 2, if we set $c(\tilde{f}) = 0$.

It remains to show that the image of $C^*(G)$ is dense in $C^*(\hat{G})$.

Problem 2. Let χ_λ be the character of a representation T_λ of the class λ and let λ^* be the class of the representation T_λ^*. Prove that

$$\hat{\chi}_\lambda(\mu) = \begin{cases} 0 & \text{if } \mu \neq \lambda^*, \\ n(\lambda)^{-1} \cdot \mathbf{1}_{n(\lambda)} & \text{if } \mu = \lambda^*. \end{cases}$$

§ 12. Fourier Transforms and Duality

Problem 3. The image of $L^1(G)$ in $C^*(\hat{G})$ contains all functions that are different from zero only at a finite number of points.

Hint. It suffices to consider the case of a single point.

The assertion about density that we need evidently follows from problem 3. This completes the proof.

With regard to other properties of the Fourier transform connected with the spaces $L^p(G,dg)$, the Schwartz space $S(G)$, and so on, we refer the reader to the book [30] of Hewitt and Ross, cited in 12.1. We note only that the theory has been much less worked out than in the commutative case.

For noncommutative groups G, the set $\hat{G}$ is no longer a group. The operation of tensor multiplication in general takes one out of the class of irreducible representations. Nevertheless, one can define a certain additional structure on $\hat{G}$, with respect to which one can reconstruct the original group G.

Let $\Pi(G)$ be the category of all finite-dimensional representations of the compact group G. Besides the usual categorial operations, $\Pi(G)$ admits the operation of tensor multiplication of any two objects. It also admits the operation of involution, which associates with the object $T \in \mathrm{Ob}\,\Pi(G)$ the adjoint representation T^*.

We define the dual object of the group G as the category $\Pi(G)$ together with the operations of tensor multiplication and involution. When G is commutative, the dual object in this sense is completely determined by the group structure on $\hat{G}$. In the general case, we can obtain $\Pi(G)$ from the set $\hat{G}$ only by specifying the involution $\tau: [T] \to [T^*]$ on $\hat{G}$ and for every $\lambda, \mu \in \hat{G}$ the isomorphism of the space

$$L(T_\mu) \otimes L(T_\nu) \approx L(T_\mu \otimes T_\nu)$$

onto a direct sum of the form

$$\oplus N^\nu_{\lambda,\mu} L(T_\nu).$$

Here $L(T)$ denotes the space of the representation T and $N^\nu_{\lambda,\mu}$ denotes the multiplicity of occurrence of the irreducible component T_ν in the decomposition of the product $T_\lambda \otimes T_\mu$.

We shall now show how to reconstruct the original group G from the dual object.

First we define a *representation of the category* $\Pi(G)$. This is a nonzero function ϕ on $\mathrm{Ob}\,\Pi(G)$ assuming values in $\mathrm{End}\,L(T)$ at the point $T \in \mathrm{Ob}\,\Pi(G)$ and satisfying the following conditions:

1) $\quad A\phi(T_1) = \phi(T_2)A \quad$ for all $\quad A \in \mathscr{C}(T_1, T_2)$,
2) $\quad \phi(T_1 \otimes T_2) = \phi(T_1) \otimes \phi(T_2)$.

Problem 4. Prove that every representation ϕ of the category $\Pi(G)$ also has the properties

3) $\quad \phi(T_1 + T_2) = \phi(T_1) \oplus \phi(T_2)$,
4) $\quad \phi(T^*) = [\phi(T)^*]^{-1}$.

Hint. To obtain 3), replace A in 1) by the projection from $L(T_1+T_2) \approx L(T_1) \oplus L(T_2)$ onto $L(T_1)$ parallel to $L(T_2)$. To derive 4), make use of the isomorphism $L(T)^* \otimes L(T) \approx \operatorname{End} L(T)$.

The set $\Gamma(\Pi(G))$ of all representations of the category $\Pi(G)$ can also be given an operation of multiplication $(\phi_1 \phi_2(T) = \phi_1(T) \cdot \phi_2(T))$ and a topology $(\phi_\alpha \to \phi$, if $\phi_\alpha(T) \to \phi(T)$ for all $T \in \operatorname{Ob} \Pi(G))$.

Problem 5. Prove that the set $\Gamma(\Pi(G))$ is a compact topological group under the operations described above.

Hint. Prove that $\Gamma(\Pi(G))$ is a closed subgroup of the compact group of all functions on $\operatorname{Ob} \Pi(G)$ which assume values in the group of unitary operators in the space $L(T)$ for all points T. (We recall that every finite-dimensional representation of the group G is unitary.)

Theorem 1 (Tannaka). *Let G be a compact group and ϕ_g the representation of the category $\Pi(G)$ given by the formula*

$$\phi_g(T) = T(g).$$

Then the mapping $g \mapsto \phi_g$ is an isomorphism of the topological groups G and $\Gamma(\Pi(G))$.

Proof. Suppose that $T \in \operatorname{Ob} \Pi(G)$ and that $\xi \in L(T)$ is a vector invariant under the operators $T(g)$, for all $g \in G$. We will show that this vector is also invariant under the operator $\phi(T)$, $\phi \in \Gamma(\Pi(G))$. In fact, let T_0 be the trivial one-dimensional representation of G. The operator A that embeds $L(T_0)$ in the subspace generated by the vector $\xi \in L(T)$ obviously belongs to $\mathscr{C}(T_0, T)$. Hence condition 1) and the definition of a representation of the category $\Pi(G)$ imply that $\phi(T)\xi = \phi(T) A \xi_0 = A \phi(T_0) \xi_0$, where $\xi_0 \in L(T_0)$.

Problem 6. Prove that $\phi(T_0) = 1$.

Hint. Use condition 2) and the relation $T_0 \otimes T_0 \approx T_0$.

It follows that $\phi(T)\xi = A\xi_0 = \xi$, and thus our assertion is proved.

We now assume that the image of G in $\Gamma(\Pi(G))$ does not coincide with all of the latter group. This means that for a certain $T \in \operatorname{Ob} \Pi(G)$ and a certain $\phi \in \Gamma(\Pi(G))$, the operator $\phi(T)$ does not belong to the group $T(G)$. Then there is a continuous function f on $\operatorname{End} L(T)$ that vanishes on the set $T(G)$ and is different from zero at the point $\phi(T)$. Approximating this function uniformly by a polynomial on the compact set $T(G) \cup \{\phi(T)\}$ and then averaging this polynomial over the group G, which acts in a natural way on $L(T)$, we arrive at the following conclusion.

There exists a polynomial P on $\operatorname{End} L(T)$ that is invariant under the group G but is not invariant under the group $\Gamma(\Pi(G))$.

Let N be the degree of the polynomial P and S the representation of the group G in the finite-dimensional space of all polynomials of degree $\leqslant N$ on $\operatorname{End} L(T)$. Then the vector $P \in L(S)$ is an example contradicting the assertion proved above.

Thus the image of G coincides with $\Gamma(\Pi(G))$. It is also obvious that the mapping of G into $\Gamma(\Pi(G))$ is continuous. Since G is compact, it follows that G and $\Gamma(\Pi(G))$ are homeomorphic. The proof is complete.

§ 12. Fourier Transforms and Duality

We remark that compactness of the group G in this proof can be replaced by the hypothesis that the groups $T(G)$ be algebraic for a sufficiently large set of representations T.[1]

It is natural to ask what categories can appear as the dual object to a compact group G.

The following theorem answers this question.

Theorem 2 (M. G. Kreĭn). *Let Π be a certain category of finite-dimensional linear spaces, equipped with the operations of tensor product and involution. The following conditions are necessary and sufficient for there to be a compact group G for which Π is the dual object.*

1. There exists exactly one (up to isomorphism) object $L_0 \in \Pi$ having the property

$$L_0 \otimes L \approx L \quad \text{for all} \quad L \in \operatorname{Ob}\Pi.$$

2. Every object $L \in \operatorname{Ob}\Pi$ can be decomposed into the sum of minimal objects.

3. If L_1 and L_2 are two minimal objects, then the space $\operatorname{Hom}(L_1, L_2)$ is either one-dimensional (if L_1 and L_2 are isomorphic) or consists of zero alone.

If these conditions are satisfied, then $\Pi = \Pi(G)$, where G is the group of all representations of the category Π.

We recommend to the reader that he prove theorem 2 and that he also establish the equivalence of this theorem and the classical duality theorem of M. G. Kreĭn, which is stated in terms of block-algebras (see [42], § 32, para. 4)[2].

12.3. Ring Groups and Duality for Finite Groups

If we compare the Tannaka-Kreĭn duality theorem for compact groups with Pontrjagin duality for locally compact commutative groups, the following basic difference springs to one's attention. While the Pontrjagin dual of a locally compact commutative group G is again a locally compact commutative group, the Tannaka-Kreĭn dual of a compact group G is a certain category of linear spaces (or block-algebras). The following question all but asks itself: does there exist a concept that would include as a special case both groups and their dual objects, and for which one can define the operation of going to a dual object of the same type?

The sought-for concept has indeed been found, and with its aid there has been constructed a definitive theory of duality which includes the results of Pontrjagin and Tannaka-Kreĭn *for finite groups*. We shall consider possible generalizations of this theory to infinite groups in the following paragraph.

---

[1] See A. L. Rosenberg, Theorems of duality for Lie groups and algebras. Uspehi mat. Nauk XXVI, 6 (1971), 253–254.

[2] Translator's note. A complete proof of theorem 2, different from Kreĭn's original proof, appears in [30], § 30.

Let L be a linear space over a field K. We shall say that L has the structure of a *coalgebra* over K if there is given a K-linear mapping

$$m: L \to L \underset{K}{\otimes} L$$

(comultiplication) for which the following diagram is commutative:

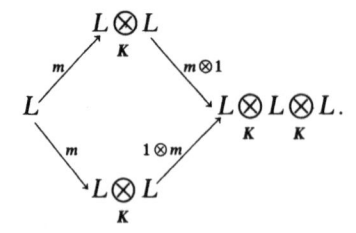

$$\begin{array}{ccc}
& L \underset{K}{\otimes} L & \\
{}^m\nearrow & & \searrow{}^{m \otimes 1} \\
L & & L \underset{K}{\otimes} L \underset{K}{\otimes} L. \\
{}_m\searrow & & \nearrow{}_{1 \otimes m} \\
& L \underset{K}{\otimes} L &
\end{array}$$

A coalgebra (L,m) is called commutative if $m(x)$ belongs to $S^2(L)$, the symmetric part of the product $L \otimes L$.

Problem 1. Let $m^*: L^* \underset{K}{\otimes} L^* \to L^*$ be the mapping adjoint to m. Prove that (L,m) is a (commutative) coalgebra over K if and only if L^* is an associative (and commutative) K-algebra under the operation $x \cdot y = m^*(x \otimes y)$, $x, y \in L^*$.

A linear space L over K is called a *ring group* (other terms are *Hopf algebra* and *bigebra*) if it has the structures of an algebra and a coalgebra, and if comultiplication $m: A \to A \otimes A$ is an algebra homomorphism and multiplication defines a coalgebra homomorphism: $A \otimes A \to A$.

Example 1. Let G be a finite group. In the space $\mathbf{C}[G]$ we define multiplication by

$$f_1 f_2(g) = f_1(g) f_2(g)$$

and comultiplication by

$$m(f)(g_1, g_2) = f(g_1 g_2).$$

One verifies without difficulty that $\mathbf{C}[G]$ is a ring group under these operations.

We note that the algebra in question is commutative, while the coalgebra is commutative if and only if the group G is commutative.

Example 2. We consider the same space $\mathbf{C}[G]$ and define multiplication as convolution:

$$f_1 * f_2(g) = \sum_{h \in G} f_1(h) f_2(h^{-1} g).$$

§ 12. Fourier Transforms and Duality

We define comultiplication by the formula

$$m(f)(g_1, g_2) = \begin{cases} f(g_1) & \text{if } g_1 = g_2, \\ 0 & \text{if } g_1 \neq g_2. \end{cases}$$

Here it is also easy to check that these operations make $C[G]$ into a ring group. In this example, the coalgebra is commutative, and the algebra is commutative if and only if G is commutative.

One can show that the ring group structure in this example defines a certain block-algebra in the sense of M. G. Kreĭn and that the ring group is itself uniquely defined by this block-algebra.

Let L be a ring group over K. The dual ring group to L is the space $L^* = \operatorname{Hom}_K(L, K)$, equipped with the operations of multiplication and comultiplication adjoint to the corresponding operations in L.

Problem 2. Show that the ring groups from examples 1 and 2 are duals of each other.

We will say that a ring group corresponds to an ordinary group G if it is obtained from G by the construction of example 1. We will say that it is dual to the ordinary group G if it is obtained by the construction of example 2. It is easy to give examples of ring groups belonging to neither of these types. It suffices to take the sum (in the category of ring groups) of two ring groups one of which corresponds to an ordinary noncommutative group and the other of which is dual to an ordinary noncommutative group.

One can construct more interesting examples by considering extensions of ring groups.

The study of such extensions was begun by G. I. Kac (see [98], [97]).

Problem 3. Let L_1 be the ring group corresponding to the ordinary group $\mathbb{Z}_2$, and let L_2 be the ring group corresponding to the ordinary commutative group $\mathbb{Z}_2 + \mathbb{Z}_2$. Show that there is an extension of L_1 with the aid of L_2 that corresponds to no ordinary group.

12.4. Other Results

For groups that are neither commutative nor compact, the Fourier transform has as yet been studied only inadequately. One of the few general results here is the analogue of Plancherel's formula for unimodular locally compact groups. For complex semisimple Lie groups, this result was obtained by M. A. Naĭmark (under the name "continuous analogue of Schur's lemma"). We will state this theorem here only for tame groups.

A general statement and proof can be found in the works of I. Segal [130] and F. Mautner [117].

Theorem 1. *Let G be a locally compact unimodular tame group with a countable basis.*

There exists a measure μ (called the Plancherel measure) on the space $\hat{G}$ such that

$$\int_G |f(g)|^2 \, dg = \int_G \mathrm{tr}[\tilde{f}(\lambda)\tilde{f}(\lambda)^*] \, d\mu(\lambda)$$

for all $f \in L^1(G, dg) \cap L^2(G, dg)$.

The Fourier transform can be extended to an isometry of $L^2(G, dg)$ onto the space $L^2(\hat{G}, \mu)$ of square μ-integrable operator functions on $\hat{G}$ that assume at each point $\lambda \in \hat{G}$ values in the space of Hilbert-Schmidt operators on H_λ.

A proof can be constructed from the general theorem on the decomposition of unitary representations of tame groups (see 8.4, theorem 3). We apply this theorem to the representation T of the group $G \times G$ in the space $L^2(G, dg)$ defined by the formula

$$[T(g_1, g_2)f](g) = f(g_1^{-1} g g_2). \tag{1}$$

To do this, we need to use the following assertions.

Problem 1. Let G be a tame group. Prove that every irreducible representation of the group $G \times G$ has the form

$$T(g_1, g_2) = T_1(g_1) \otimes T_2(g_2),$$

where T_1 and T_2 are irreducible representations of the group G.

Hint. Use the properties of primary representations of tame groups (see problem 3 in 8.4).

Problem 2. Prove that the representation T defined in formula (1) has simple spectrum.

Hint. Prove that if A (B) is the von Neumann algebra generated by left (right) translations in $L^2(G, dg)$, then $A = B'$.

Thus we have a decomposition of the form

$$T = \int_{\hat{G}} \int_{\hat{G}} (T_\lambda \times T_\mu) \, d\nu(\lambda, \mu). \tag{2}$$

Problem 3. Show that the measure ν in formula (2) is concentrated on the subset Δ of the space $\hat{G} \times \hat{G}$ consisting of representations of the form $T_\lambda \times T_\lambda^*$.

Hint. Let C be the von Neumann algebra generated by left and right translations in $L^2(G, dg)$. Prove that the center Z of the algebra C coincides with $A \cap B$ (notation as in problem 2).

We consider the conjugate-linear automorphism ε of the space $L^2(G, dg)$ that takes the function $f(g)$ into the function $f^*(g) = \overline{f(g^{-1})}$. The assertion of the problem follows from the following fact: for every operator $z \in Z$, the equality $\varepsilon \cdot z \cdot \varepsilon = z^*$ obtains.

One can prove this fact by approximating the operator z by operators $L(\phi_n) = \int_G \phi_n(g) L(g) \, dg$, $\phi_n \in L^1(G, dg)$, and applying both parts of the proven equality to a vector of the form $f = f_1 * f_2$, $f_i \in L^1(G, dg) \cap L^2(G, dg)$.

Another derivation of Plancherel's formula is based on the notion of trace on a C^*-algebra. This proof can be found in Dixmier's book [15].

The problem of describing the C^*-algebra of a group G that is neither compact nor commutative is at once very interesting and very difficult. For certain classes of groups, the algebra $C^*(G)$ can be realized as a subalgebra of the algebra $C^*(\hat{G})$. The algebra $C^*(\hat{G})$ is defined as the algebra of all operator functions on $\hat{G}$ whose values at each point $\lambda \in \hat{G}$ are completely continuous operators in the corresponding space H_λ.

The case $G = SL(2, \mathbb{C})$ has been completely worked out by Fell in [76]. This article also contains many useful facts about the structure of C^*-algebras of operator functions.

The principle of duality based on the concept of a ring group has been generalized in the works of G. I. Kac [96]. However, the final formulation is much more unwieldy and much less natural. A modernized treatment of this circle of questions appears in [135] and [75].

A series of interesting and difficult questions arises in the study of the topology of the space $\hat{G}$. We have already remarked *supra* that for locally compact groups, the space $\hat{G}$ has T_0 separation if and only if G is a tame group (for a proof, see the book of Dixmier cited above [15]).

The so-called R-property means that the identity representation T_0 lies in the support of the Plancherel measure μ on $\hat{G}$. As was remarked in 9.1, this property is equivalent to amenability for locally compact groups G.

D. A. Každan has recently discovered an interesting connection between algebra properties of the group G and the topological structure of a neighborhood of the point $T_0 \in \hat{G}$.

Theorem 2 (D. A. Každan). *If G is a countable discrete group and the point $T_0 \in \hat{G}$ is isolated, then G is finitely generated and the factor group by its commutant is finite.*

The proof is based on the following observation.

Problem 4. If T_0 is an isolated point in $\hat{G}$ and a family $\mathscr{S}$ of representations of G weakly contains T_0, then T_0 is a subrepresentation of one of the representations $S \in \mathscr{S}$.

Každan's theorem has made it possible to obtain new and essential information about the structure of discrete subgroups of semisimple Lie groups. A detailed exposition of these results is given in the Séminaire Bourbaki [69].

§ 13. Induced Representations

The construction of an induced representation plays a very great rôle in the theory of representations. Suffice it to say that representations of groups of transformations in spaces of functions, vector and tensor fields, differential operators, and many other objects are examples of induced representations.

For many groups, all or almost all irreducible unitary representations are *monomial*, that is, are induced by a one-dimensional representation of a certain subgroup.

Finally, the operation of induction itself, which is a functor from the category of representations of a subgroup into the category of representations of the group, has in recent years been the object of a detailed and fruitful study.

13.1. Induced Representations of Finite Groups

We begin with an explanation of the algebraic foundation of the concept of an induced representation. In this paragraph we shall restrict ourselves to finite-dimensional representations of finite groups.

Let X be a homogeneous right G-space. In the space $L(X)$ of all functions on G, we define a representation T of the group G:

$$[T(g)f](x) = f(xg). \tag{1}$$

We fix a point $x_0 \in X$ and denote by H the stationary subgroup of this point. For every function $f \in L(X)$, we define a function F on the group G:

$$F(g) = f(x_0 g). \tag{2}$$

Problem 1. Prove that formula (2) establishes an isomorphism between the space $L(X)$ and the space $L(G, H)$ of functions on G that are constant on right cosets of H. Under this isomorphism, the representation T goes into the representation defined by the formula

$$[T(g)F](g_1) = F(g_1 g). \tag{3}$$

Note that we have thus obtained a subrepresentation of the right regular representation of G corresponding to the invariant subspace $L(G, H)$.

Representations of the type just described are sometimes called *quasi-regular*. The concept of an induced representation is obtained by generalizing this construction in the following way.

Let U be a representation of the subgroup H in a finite-dimensional space V. Let $L(G, H, U)$ denote the space of vector-functions F on G with values in V that satisfy the condition

$$F(hg) = U(h) F(g), \quad h \in H. \tag{4}$$

Problem 2. Prove that the space $L(G, H, U)$ is invariant under right translations on G and that formula (3) gives a representation of G in $L(G, H, U)$.

The representation obtained in this way is called *induced* (more precisely, *the representation of the group G induced by the representation U of the subgroup H*). It is denoted by $\text{Ind}(G, H, U)$ or simply by $\text{Ind}\, U$ if it is clear what the groups G and H are. When the explicit form of U is unknown or unimportant, we speak of a *representation induced from the subgroup H*.

§ 13. Induced Representations

Suppose that the inducing representation U is the trivial one-dimensional representation U_0 of the subgroup H. Then $\operatorname{Ind} U$ is the quasi-regular representation in the space $L(X)$, where $X = H\backslash G$. In particular, the regular representation of the group G can be written as $\operatorname{Ind}(G, \{e\}, U_0)$.

Like a quasi-regular representation, an induced representation can be realized in a space of (vector-) functions on the homogeneous space $X = H\backslash G$. For this we remark that the function $F \in L(G, H, U)$ is completely defined if we know its values at at least one point in each coset H_g. Suppose that we have chosen a representative in each such class. In other words, we have a mapping $s: X \to G$ such that $p \circ s = 1$, where p is the natural projection of G onto $H\backslash G = X$.

Problem 3. Every element $g \in G$ can be written uniquely in the form $g = h \cdot s(x)$, where $h \in H$, $x \in X$.

Hint. Set $x = x_0 g$.

To each function $F \in L(G, H, U)$ we assign the vector-function f on X with values in V defined by

$$f(x) = F(s(x)). \tag{5}$$

Problem 4. Prove that the correspondence (5) is an isomorphism between $L(G, H, U)$ and the space $L(X, V)$ of all vector-functions on X with values in V. Under this isomorphism, the representation T goes into the representation defined by the formula

$$[T(g)f](x) = A(g, x) f(xg), \tag{6}$$

where the function $A(g, x)$ is defined by the equalities

$$A(g, x) = U(h) \tag{7}$$

and

$$s(x)g = h s(xg). \tag{8}$$

Hint. Use the decomposition of problem 3, applied to the element $s(x)g \in G$.

Thus the operators of an induced representation are the superposition of translation and multiplication by an operator function.

One can show that this property is characteristic for induced representations.

Problem 5. Prove that every representation of the form (6) is equivalent to a representation induced from the subgroup H.

Hint. Prove that the function $A(g, x)$ satisfies the functional equation

$$A(g_1, x) A(g_2, x g_1) = A(g_1 g_2, x), \tag{9}$$

the general solution of which has the form

$$A(g, x) = B(x)^{-1} U(h) B(xg).$$

Here B is a certain operator function on X with values in $\operatorname{Aut} V$ and $h \in H$ is defined from the equality (8).

We note that a function A satisfying condition (9) can be interpreted as a one-dimensional cocycle of the group G with coefficients in $\operatorname{Aut} V$, and that problem 5 can be interpreted as stating that every cocycle is cohomologous to a cocycle of the form (7).

We shall now formulate some simple but important properties of induced representations in the form of problems.

Problem 6. Prove that if $U_1 \sim U_2$, then also $\operatorname{Ind} U_1 \sim \operatorname{Ind} U_2$. (Thus the operation of induction is also defined for equivalence classes of representations.)

Hint. If $C \in \mathscr{C}(U_1, U_2)$ then the mapping $F \mapsto CF$ carries $L(G, H, U_1)$ into $L(G, H, U_2)$ and belongs to $\mathscr{C}(\operatorname{Ind} U_1, \operatorname{Ind} U_2)$.

Problem 7. Prove the equivalence

$$\operatorname{Ind}(U_1 + U_2) \sim \operatorname{Ind} U_1 + \operatorname{Ind} U_2.$$

Corollary. *If $\operatorname{Ind} U$ is irreducible, then U is also irreducible.*

Problem 8 (the theorem on induction in stages). Let H be a subgroup of G and K a subgroup of H. Then for every representation U of the subgroup K we have the equivalence

$$\operatorname{Ind}(G, H, \operatorname{Ind}(H, K, U)) \sim \operatorname{Ind}(G, K, U)$$

or, symbolically

$$\operatorname{Ind}_H^G \operatorname{Ind}_K^H \sim \operatorname{Ind}_K^G.$$

Here Ind_H^G denotes the operation of going from U to $\operatorname{Ind}(G, H, U)$. (We remark that this operation is a covariant functor from the category of representations of H to the category of representations of G.)

Hint. Let V be the space of the representation U. Consider the mapping of $L(G, K, V)$ into $L(G, H, L(H, K, V))$ defined by the formula

$$f \mapsto F, \quad \text{where} \quad F(g, h) = f(hg).$$

Problem 9 (Frobenius's formula). Prove that the character of the induced representation can be computed by the formula

$$\chi_{\operatorname{Ind} U}(g) = \sum_{x \in X} \chi_U(s(x)^{-1} g s(x)), \tag{10}$$

where the character χ_U of the representation U on the right side of the equality is extended from H to the entire group G by defining it as zero off of H.

Hint. Use formulas (6)–(8).

Corollary. *If H is a normal subgroup of G, then the character of an arbitrary representation induced from H vanishes off of H.*

§ 13. Induced Representations

The following relation is fundamental for a majority of the calculations connected with induced representations of finite groups.

Theorem 1 (the Frobenius reciprocity theorem). *Let T be a representation of the group G and U a representation of the subgroup H. Then we have*

$$c(T, \operatorname{Ind} U) = c(T|_H, U). \tag{11}$$

(The notation $T|_H$ denotes the restriction of the representation T to the subgroup H.)

The proof of this theorem can easily be deduced from Frobenius's formula (9) and equality 3 of 11.1.

Another method consists in establishing an isomorphism between the spaces $\mathscr{C}(T, \operatorname{Ind} U)$ and $\mathscr{C}(T|_H, U)$. Suppose that the representation T acts in the space W and U in the space V. Then for every operator $A \in \operatorname{Hom}(W, L(G, H, V))$ we can construct an operator $B \in \operatorname{Hom}(W, V)$ by the formula

$$B(w) = [A(w)](x_0), \quad w \in W. \tag{12}$$

Problem 10. Prove that the correspondence defined by the equality (12) establishes an isomorphism between $\mathscr{C}(T, \operatorname{Ind} U) \subset \operatorname{Hom}(W, L(G, H, V))$ and $\mathscr{C}(T|_H, U) \subset \operatorname{Hom}(W, V)$.

We note special cases of Frobenius reciprocity, which are often used in applications.

Corollary 1. *If T is an irreducible representation of G and U is an irreducible representation of H, then T appears in the decomposition of* Ind U *with the same multiplicity that U appears in the decomposition of* $T|_H$.

Corollary 2. *Let T act in the space W, and let H be a subgroup of G. The following numbers are equal: the multiplicity of an irreducible component of T in the decomposition of the quasi-regular representation corresponding to the subgroup H; and the number of linearly independent invariants of the subgroup H in the space W.*

Example. We consider the representation T of the group $G \times G$ in the space $L(G)$ defined by the formula

$$[T(g_1, g_2)f](g) = f(g_1^{-1} g g_2).$$

Plainly this is the quasi-regular representation corresponding to the diagonal subgroup $\Delta \subset G \times G$ consisting of all pairs of the form (g, g). Every irreducible representation of $G \times G$ has the form $T_1 \times T_2: (g_1, g_2) \to T_1(g_1) \otimes T_2(g_2)$, where T_1 and T_2 are irreducible representations of G. By corollary 2 of theorem 1, the representation $T_1 \times T_2$ appears in the decomposition of T just as many times as there are Δ-invariant vectors for the representation $T_1 \times T_2$. The restriction of $T_1 \times T_2$ to Δ, as one easily sees, is the tensor product $T_1 \otimes T_2$ of the representations T_1 and T_2 of the group G.

By problem 5 of 7.1, the desired number is equal to $c(T_1, T_2^*)$, that is, it is one if $T_1 \sim T_2^*$ and is zero otherwise. Thus the representation T has a simple spectrum and its irreducible components have the form $T_k \times T_k^*$, where T_k runs through the set $\hat{G}$.

Frobenius reciprocity is a characteristic property of the induced representation and enables us to define this representation in categorical terms.

Let $\Pi(G,K)$ denote the category of finite-dimensional linear representations over the field K of the finite group G.

Suppose that we have a subgroup $H \subset G$ and a K-representation U of H. We consider the contravariant functor F from $\Pi(G,K)$ to the category of linear spaces over the field K defined by

$$F(T) = \mathscr{C}(T|_H, U).$$

By theorem 1, this functor can be written in the form

$$F(T) = \mathscr{C}(T, \operatorname{Ind} U).$$

Thus the functor F is representable and the induced representation $\operatorname{Ind} U$ is the object that represents this functor.

We remark that in the infinite-dimensional case the functor F is as a rule not representable. Hence attempts to extend the Frobenius reciprocity theorem to this case lead to an extension of the category of representations that was originally chosen. We shall consider this question in more detail in the sequel.

One can give yet another definition of induced representations in categorical terms. Specifically, the induced representation $\operatorname{Ind} U$ is a universal object in the following category.

For a given representation U of a subgroup H in a space V, we consider all possible mappings $\phi: V \to L$, where L is the space of a representation of the group G and ϕ commutes with the action of the subgroup H. Such mappings form a category, if we define a morphism of ϕ to ϕ' as a commutative diagram

$$\begin{array}{ccc} V & \xrightarrow{\phi} & L \\ & \phi' \searrow & \downarrow \varepsilon \\ & & L' \end{array}$$

where ε is a mapping that commutes with the action of the group G.

Problem 11. Prove that in the category described above, there is a universal repulsive object ϕ_0. This object is the mapping of the space V into the space $L(G,H,U)$ of the induced representation $\operatorname{Ind} U$ defined by the formula

$$(\phi_0(v))(g) = \begin{cases} U(g)v & \text{if } g \in H, \\ 0 & \text{if } g \notin H. \end{cases}$$

§ 13. Induced Representations

Hint. Suppose that $\phi: V \to L$ is an arbitrary object of our category. Then the required mapping

$$\varepsilon: L(G, H, U) \to L$$

is defined by the formula

$$\varepsilon(F) \to \sum_{H \backslash G} T(g^{-1}) \phi(F(g))$$

(the expression on the right side is constant on right cosets of H in G).

Finally we present an explicit construction of the space of an induced representation as a module over the group algebra $K[G]$. Let U be a representation of the subgroup H in the space V. Then V is a left $K[H]$-module. We consider the tensor product $K[G] \underset{K[H]}{\otimes} V$, where $K[G]$ is considered as a left $K[G]$-module and a right $K[H]$-module. This product will then be a left $K[G]$-module.

Problem 12. Prove that the $K[G]$-module $K[G] \underset{K[H]}{\otimes} V$ corresponds to the induced representation $\mathrm{Ind}(G, H, U)$.

Hint. Verify that the mapping $\phi_1: V \to K[G] \underset{K[H]}{\otimes} V$, defined by the equality $\phi_1(v) = 1 \otimes v$, is a universal object, and then use the preceding problem.

Another method consists in constructing an explicit isomorphism between $K[G] \underset{K[H]}{\otimes} V$ and $L(G, H, U)$. This isomorphism maps the function $F \in L(G, H, U)$ into the element

$$\sum_{H \backslash G} \delta_{g^{-1}} \otimes F(g) \in K[G] \underset{K[H]}{\otimes} V.$$

The inverse mapping carries the element $\delta_g \otimes v$ from $K[G] \underset{K[H]}{\otimes} V$ to the function

$$F(g_1) = \begin{cases} U(g_1 g) v & \text{if } g_1 g \in H, \\ 0 & \text{if } g_1 g \notin H. \end{cases}$$

The last two definitions of induced representation with the help of "reversal of arrows" lead to the definition of a dual concept. This is what is called the *coinduced* representation (another term is *produced module*). One can show, however, in the case where the algebra $K[G]$ is semisimple that the operation of coinduction coincides with the operation of induction.

13.2. Unitary Induced Representations of Locally Compact Groups

The construction of an induced representation, given in 13.1 for finite groups, can be carried over to locally compact groups with preservation of many (although not all) of the basic properties. A detailed study of this construction has been given in the works of G. Mackey, and we shall call the representations obtained in this way *induced in the sense of Mackey*.

As in the finite case, there are two possible realizations of an induced representation: 1) in a space of vector-functions on the entire group G that transform according to the given representation under left translations by elements of the subgroup H; 2) in a space of vector-functions on the right homogeneous space $X = H \backslash G$. The transition from one realization to the other for infinite-dimensional representations is far from as evident as in the finite case. Nevertheless, the criterion for inducibility which we shall give at the end of this paragraph implies the equivalence of both constructions.

To begin with, we consider the first realization. Let G be a locally compact group, let H be a closed subgroup of G, and let U_0 be a certain representation (not necessarily unitary) of the subgroup H in a Hilbert space V. By analogy with what we did in 13.1, we introduce a space $L(G, H, U_0)$ of measurable vector-functions F on G with values in V, that satisfy the condition

$$F(hg) = U_0(h) F(g), \quad h \in H, \quad g \in G. \tag{1}$$

It is clear that this space is invariant under right translations on G.

We ask when does $L(G, H, U_0)$ admit a G-invariant scalar product of the form

$$(F_1, F_2) = \int_G (G_1(g), F_2(g))_V \, d\mu(g), \tag{2}$$

where μ is a certain measure on the group G.

Problem 1. Show that for a measure μ of the form $\rho(g) d_r g$, the condition of G-invariance of the scalar product (2) reduces to the following. The quantity $\int_H \|U_0(h)\xi\|_V^2 \dfrac{\Delta_G(h)}{\Delta_H(h)} \rho(hg) d_r h$ does not depend upon g, for each $\xi \in V$.

We shall consider representations U_0 of a special form:

$$U_0(h) = \left[\frac{\Delta_H(h)}{\Delta_G(h)} \right]^{1/2} U(h), \tag{3}$$

where U is a unitary representation of H. Then the condition of problem 1 turns into the following: the quantity $\int_H \rho(hg) d_r h$ does not depend upon g.

Problem 2. Prove that there is a continuous nonnegative function ρ on G such that

$$\int_H \rho(hg) d_r h \equiv 1 \tag{4}$$

and $\operatorname{supp} \rho$ has compact intersection with every coset Hg.

Hint. A detailed solution is presented in a monograph of Bourbaki [8, Ch. VII, § 2, para. 4].

Let s be a measurable mapping of X into G having the property that $s(Hg) \in Hg$, and let v_s be the corresponding measure on X (see 9.1).

§ 13. Induced Representations

Problem 3. Let U_0 be a representation of the form (3) and let ρ be a function on G satisfying the conditions of problem 2. Then for an arbitrary function $F \in L(G, H, U_0)$ the equality

$$\int_G \|F(g)\|_V^2 \rho(g) d_r g = \int_X \|F(s(x))\|_V^2 dv_s(x) \tag{5}$$

holds.

Hint. Use formula (6) from 9.1.

Corollary. *The left side of the equality (5) does not depend upon the choice of the function ρ, but the right side depends upon the choice of the mapping s.*

We now give the definition of an induced representation. Let U be a unitary representation of H in the space V. We define the representation U_0, which is connected with U by equality (3). In $L(G, H, U_0)$ we define a norm, the square of which is given by formula (5). The Hilbert space obtained in this way will be denoted by $L^2(G, H, U)$. Consider the unitary representation T acting in $L^2(G, H, U)$ by the formula

$$[T(g)F](g_1) = F(g_1 g). \tag{6}$$

We call T *the representation induced in the sense of Mackey by the representation U.* As in 13.1, we denote it by $\mathrm{Ind}(G, H, U)$ or simply $\mathrm{Ind}\, U$.

We recall that the word "representation" for topological groups always means "continuous representation". We leave it to the reader to verify that $\mathrm{Ind}\, U$ is continuous if U is continuous. For this it is simplest to use problem 1 from 7.2 along with the following fact.

Problem 4. Prove that functions of the form

$$F(g) = \int_H U_0(h) \phi(h^{-1} g) d_l h,$$

where ϕ is a continuous vector-function on G with compact support, form a dense subset of $L^2(G, H, U)$.

We now present another realization of this representation. Fix a measurable mapping $s: X \to G$ having the property that $s(Hg) \in Hg$. With every function $F \in L^2(G, H, U)$, we associate the function $f(x) = F(s(x))$. It is evident that F is uniquely determined by f.

It follows from equality (5) that the mapping $F \to f$ is an isomorphism of $L^2(G, H, U)$ onto the space $L^2(X, v_s, V)$ of vector-functions on X with values in V, having summable square norm with respect to the measure v_s.

Problem 5. Prove that, under the isomorphism indicated above, the representation (6) goes into the representation

$$[T(g)f](x) = A(g, x) f(xg), \tag{7}$$

where the operator function $A(g,x)$ is defined by the equality

$$A(g,x) = U_0(h), \tag{8}$$

in which the element $h \in H$ is defined from the relation

$$s(x)g = hs(xg). \tag{9}$$

Hint. Use the explicit form for F in terms of f:

$$F(hs(x)) = U_0(h)f(x), \quad h \in H, \quad u \in X.$$

The realization just given suggests the following generalization of the construction of an induced representation. Let X be a right G-space, on which there exists a quasi-invariant measure μ. We consider the family of operators $T(g)$, $g \in G$, defined by formula (7) in the space $L^2(X, \mu, V)$, where V is a certain Hilbert space and $A(g,x)$ is a function on $G \times X$ with values in End V.

The equality $T(g_1 g_2) = T(g_1)T(g_2)$ holds if and only if the equality

$$A(g_1, x) A(g_2, xg_1) = A(g_1 g_2, x) \tag{10}$$

holds (for arbitrary $g_1, g_2 \in G$) for μ-almost $x \in X$. The operators $T(g)$ will be unitary if for every $g \in G$, the operators

$$B(g,x) = \left[\frac{d\mu(xg)}{d\mu(x)}\right]^{-1/2} A(g,x)$$

are unitary in the space V for almost all $x \in X$.

Problem 6. Prove that the operator function $B(g, x)$, like the function $A(g, x)$, satisfies condition (10).

Hint. Use the identity

$$\frac{d\mu(xg_1 g_2)}{d\mu(xg_1)} \cdot \frac{d\mu(xg_1)}{d\mu(x)} = \frac{d\mu(xg_1 g_2)}{d\mu(x)}.$$

We see that the function $B(g, x)$ is a one-dimensional cocycle of the group G with coefficients in the group Γ of measurable unitary operator functions on X (more precisely, classes of functions that are equal μ-almost everywhere).

Problem 7. Suppose that the cocycles B_1 and B_2 are cohomologous, that is,

$$B_1(g, x) = C(x)^{-1} B_2(g, x) C(xg)$$

for a certain $C \in \Gamma$. Prove that the corresponding representations T_1 and T_2 are unitarily equivalent.

Hint. The required equivalence is produced by the unitary operator C in $L^2(X, \mu, V)$ which acts by the formula

$$(Cf)(x) = C(x)f(x).$$

§ 13. Induced Representations

Thus we come to the problem of the structure of the cohomology group $H^1(G, \Gamma)$.

Suppose that the action of the group G on X satisfies the conditions of Glimm's theorem (see 9.1). Then the study of the cohomology group $H^1(G, \Gamma)$ can be reduced to the analogous problem for the case in which $X = H\backslash G$ is a homogeneous space.

It is found that on the homogeneous space $X = H\backslash G$ every cocycle is cohomologous to a cocycle of the special form given by formulas (8) and (9), and thus is given by a unitary representation of the subgroup H. This assertion is essentially equivalent to the following: every representation of the form (7) on the homogeneous space $X = H\backslash G$ is induced in the sense of Mackey from the subgroup H. The validity of this last follows from the criterion of inducibility introduced *infra* (see theorem 1 of this paragraph).

Suppose that the measure μ on X is ergodic with respect to G (see 9.1) but is concentrated on no orbit of this group in X. Then the cohomology classes $H^1(G, \Gamma)$ can be interpreted as classes of unitary representation of a so-called "*virtual subgroup*" of the group G.

The study of virtual subgroups was begun of G. Mackey (see his survey [115]). It is interesting that virtual subgroups of a commutative group can admit irreducible representations of dimension >1. In this connection, see the author's article "Dynamical systems, factors, and representations of groups", Uspehi mat. Nauk vol. 22, vyp. 5 (1967), 67–80. Further results have been obtained in the work [94] of R. S. Ismagilov.

We shall list a number of properties of the operation of induction in the sense of Mackey. Just as in the finite case one verifies that this operation can be carried over to classes of equivalent representations, preserves decomposition into a sum (also continuous sums), and satisfies the principle of "induction by stages" (see problems 6–8 of 13.1).

Frobenius reciprocity as formulated in corollary 1 to theorem 1 from 13.1 is preserved for compact groups, and in general does not hold for locally compact groups. We postpone a more detailed discussion of this question to 13.5.

In many cases it is important to know when a given representation T of a group G is induced in the sense of Mackey from a closed subgroup H. To formulate a criterion of inducibility, we introduce the concept of "representation of a G-space". By definition, a *unitary representation of a right G-space* X in a Hilbert space L is a pair (T, P), where T is a unitary representation of the group G in L and P is a $*$-representation of the algebra $C_0(X)$ of continuous functions on X with compact supports (and with the usual multiplication and complex conjugation as involution). Here T and P are connected by the relation

$$T(g)P(f)T(g^{-1}) = P(R(g)f), \qquad (11)$$

where

$$[R(g)f](x) = f(xg). \qquad (12)$$

Theorem 1 (criterion of inducibility). *A unitary representation T of a group G is induced in the sense of Mackey from a closed subgroup $H \subset G$ if and only if it can be extended to a unitary representation of the homogeneous space $X = H\backslash G$.*

Before proving the theorem, we shall present two equivalent formulations.

1. The representation P of the algebra $C_0(X)$ can be replaced by what is called a projection measure Δ, which associates with every Borel subset $E \subset X$ an orthogonal projection $\Delta(E)$ and has the following properties:

$$\Delta(E_1 \cap E_2) = \Delta(E_1)\Delta(E_2);$$

$$\Delta\left(\bigcup_{k=1}^{\infty} E_k\right) = \sum_{k=1}^{\infty} \Delta(E_k), \quad \text{if} \quad E_i \cap E_j = \emptyset \quad \text{for} \quad i \neq j;$$

$$P(f) = \int_X f(x)\Delta(dx) \quad \text{for} \quad f \in C_0(X).$$

In terms of the measure Δ, condition (11) means that

$$T(g)\Delta(E)T(g)^{-1} = \Delta(Eg). \tag{13}$$

Such a projection measure is called a *system of imprimitivity* with respect to T. Thus *the criterion of inducibility of the representation T from the subgroup H is the existence of a system of imprimitivity on $X = H \backslash G$ with respect to T.* It was just in this form that this criterion was first obtained by G. Mackey.

2. Instead of a unitary representation (T, P) of the G-space X, we can consider a $*$-representation Φ of a certain ancillary algebra $\mathfrak{A}$ (which is an analogue of the group algebra). This algebra is constructed in the following way.

Let $C_0(X \times G)$ be the linear topological space of all continuous functions $\alpha(x, g)$, $x \in X$, $g \in G$, having compact supports. (Convergence in this space means uniform convergence with the additional proviso that the supports of all of the functions under consideration are contained in a fixed compact set $K \subset X \times G$.)

Problem 8. There exists a unique continuous linear mapping $\Phi: C_0(X \times G) \to \text{End } V$ which for all functions of the form $\alpha(x, g) = u(x)v(g)$ is given by the formula

$$\Phi(\alpha) = P(u)T(v),$$

where $T(v) = \int_G v(g)T(g)d_r g$.

Hint. We can write the desired mapping in the form

$$\Phi(\alpha) = \int_G P(\alpha(\cdot, g)) T(g) d_r g, \tag{14}$$

where $\alpha(\cdot, g)$ denotes the element of $C_0(X)$ that is obtained from $\alpha \in C_0(X \times G)$ by fixing the second argument.

Problem 9. Define multiplication and involution in $C_0(X \times G)$ so that the correspondence $\alpha \mapsto \Phi(\alpha)$ is a $*$-representation of the algebra with involution that has just been defined.

§ 13. Induced Representations

Answer.

$$(\alpha_1 * \alpha_2)(x, g) = \int_G \alpha_1(x, g_1^{-1}) \alpha_2(xg_1^{-1}, g_1 g) d_r g_1,$$

$$\alpha^*(x, g) = \overline{\alpha(xg, g^{-1})} \Delta_G(g_1)^{-1}. \tag{15}$$

We denote the algebra defined in Problem 9 by the symbol $\mathfrak{A}$.

We shall say that a $*$-representation Φ of the algebra $\mathfrak{A}$ in a space L is *consistent* with the representation T of the group G in the same space if the relation

$$T(g_1)\Phi(f)T(g_2) = \Phi(\tilde{f}) \tag{16}$$

holds, where we define $\tilde{f}$ by the relation $\tilde{f}(x, g) = f(xg_1, g_1^{-1} g g_2^{-1}) \Delta_G(g_1)^{-1}$.

It is clear that if the representation T is induced in the sense of Mackey from a subgroup H, then the representation Φ of the algebra $\mathfrak{A}$ given by formula (14) is consistent with T.

Problem 10. Prove that every $*$-representation Φ of the algebra $\mathfrak{A}$ that is consistent with T has the form (14). Here P is a certain $*$-representation of the algebra $C_0(X)$ which forms together with T a unitary representation of the G-space X.

Hint. Use the method described in 10.2 for reconstructing a representation of the group from a representation of the group algebra.

Thus *the criterion for inducibility of a representation T of the group G is the existence of a $*$-representation Φ of the algebra $\mathfrak{A}$ that is consistent with T.*

We turn to a proof of theorem 1.

The necessity of the condition of the theorem is completely obvious. In fact, if the representation $T = \text{Ind } U$ is realized in the form (7), then we can associate with every function $f \in C_0(X)$ the operator of multiplying by this function in the space $L^2(X, v_s, V)$. Doing this, we obtain the required representation P of the algebra $C_0(X)$.

The proof sufficiency is more complicated. We first indicate a method of reconstructing the representation U from the induced representation $T = \text{Ind } U$. We will also see that this method is applicable to an arbitrary representation T that satisfies the condition of theorem 1, and leads to a certain representation U of the subgroup H. Finally we will verify that U is unitary and that T is equivalent to $\text{Ind } U$, by which we complete the proof.

Thus, suppose that T satisfies the condition of theorem 1. As described above, we construct a representation Φ of the ancillary algebra $\mathfrak{A} \approx C_0(X \times G)$. This representation is the sum of cyclic subrepresentations. Since the functor Ind commutes with the operation of sums, it suffices to prove the theorem for each summand. Hence we may suppose that Φ is a cyclic representation acting in the space L and that ξ is its source. We suppose that $T = \text{Ind } U$. Then we have $L = L^2(G, H, U)$ and the vector $\xi \in L$ is a vector-function on G with values in the space V of the representation U. We consider an arbitrary function β on $G \times G$ having compact support. The integral $\int_H \beta(hg_1, g_2) d_r h$ depends only on g_2 and on the coset Hg_1. We set

$$\alpha(Hg_1, g_2) = \int_H \beta(hg_1, g_2) d_r h.$$

We then have $\alpha \in C_0(X \times G)$. Let us compute the quantity $(\Phi(\alpha)\xi, \xi)$. Using formulas (5), (6), and (14), we obtain

$$(\Phi(\alpha)\xi, \xi) = \int_G ([\Phi(\alpha)\xi](g_1), \xi(g_1))_V \rho(g_1) d_r g_1$$

$$= \int_G \int_G \alpha(Hg_1, g_2)(\xi(g_1 g_2), \xi(g_1))_V \rho(g_1) d_r g_1 d_r g_2$$

$$= \int_H \int_G \int_G \beta(hg_1, g_2)(\xi(g_1 g_2), \xi(g_1))_V \rho(g_1) d_r h d_r g_1 d_r g_2.$$

Making the substitution $g_1 \mapsto h^{-1} g_1$ and using the relations (1), (3), (4), and formula (4) from 9.1, we can write the last integral in the form

$$\int \int \beta(g_1, g_2)(\xi(g_1 g_2), \xi(g_1))_V d_r g_1 d_r g_2.$$

Now for the function β we take a function of the form

$$\beta(g_1, g_2) = \theta_1(g_1 g_2) \overline{\theta_2(g_1)} \Delta_G(g_1).$$

Then we can write our integral as $(\hat\theta_1, \hat\theta_2)_V$, where $\hat\theta$ is the element of V given by the formula

$$\hat\theta = \int \theta(g) \xi(g) d_r g.$$

Problem 11. Prove that vectors of the form $\hat\theta$, where $\theta \in C_0(G)$, are everywhere dense in V.

Hint. Use the fact that ξ is a source in L.

Problem 12. Prove the relation

$$(L(h)\theta)\hat{} = [\Delta_H(h) \Delta_G(h)]^{1/2} U(h)\hat\theta,$$

where $[L(h)\theta](g) = \theta(h^{-1}g)$.

Hint. Use formulas (1) and (3).

The results of problems 11 and 12 permit us to reconstruct the representation U (up to equivalence) from the representation Φ. In fact, the space V is the completion of $C_0(G)$ in the norm generated by the scalar product

$$(\theta_1, \theta_2) = (\Phi(\alpha)\xi, \xi) \tag{17}$$

where

$$\alpha(Hg_1, g_2) = \int_H \overline{\theta_2(hg_1)} \theta_1(hg_1 g_2) \Delta_G(hg_1) d_r h.$$

The representation U acts in this space according to the formula

$$[U(h)\theta](g) = \Delta_H(h)^{-1/2} \Delta_G(h)^{-1/2} \theta(h^{-1}g). \tag{18}$$

§ 13. Induced Representations

Now let T be a representation of the group G that satisfies the condition of the theorem and let Φ be the corresponding $*$-representation of the algebra $\mathfrak{A}$ in the space L. In $C_0(G)$, we define a scalar product by formula (17) and a representation U of the subgroup H by the formula (18).

Problem 13. Prove that the scalar product (17) is invariant under the operators $U(h), h \in H$, given by formula (18).

Hint. Carry out a change of variables in the appropriate integral.

It is more complicated to prove that the scalar product (17) is positive-definite. To establish this, we will show that the original representation T is equivalent to the representation $\text{Ind } U$. (We admit a certain laxness here, in writing $\text{Ind } U$ for the representation obtained by inducing in the sense of Mackey from the representation U, about which we do not yet know that it is unitary.)

The space M of the representation $\text{Ind } U$ consists of vector-functions F on the group G with values in $C_0(G)$, that is, M consists of functions on $G \times G$. These functions satisfy the condition

$$F(hg_1, g_2) = \Delta_G^{-1}(h) F(g_1, h^{-1} g_2). \tag{19}$$

In M we define a scalar product by the equality

$$(F_1, F_2) = \int (F_1(g, \cdot), F_2(g, \cdot))_{C_0(G)} \rho(g) d_r g. \tag{20}$$

Formulas (19) and (20) are obviously specializations of the general formulas (1) and (5), which define the space and the operators of an induced representation.

Let L_0 be the subset of L consisting of vectors of the form $\Phi(\alpha) \xi$, $\alpha \in C_0(X \times G)$. Plainly L_0 is dense in L. We define a mapping τ of the space L_0 into M, carrying the vector $\Phi(\alpha) \xi$ into the function

$$F(g_1, g_2) = \alpha(H g_1, g_1^{-1} g_2) \Delta_G(g_1)^{-1}. \tag{21}$$

Problem 14. Prove that the mapping τ defined by equality (21) is an isometry that commutes with the action of the group G.

Hint. Use formulas (16) and (17).

Problem 15. Prove that the scalar product (17) is positive-definite.

Hint. Use the results of problems 11 and 14.

The proof of the theorem is completed.

13.3. Representations of Group Extensions

In this paragraph, we shall show that one can obtain essential information about irreducible unitary representations of group extensions with the aid of the criterion of inducibility of 13.2.

Let G be a locally compact group containing a closed normal subgroup N.

The group G acts on N and consequently also on the dual space $\hat{N}$. For the sake of definiteness, we shall regard $\hat{N}$ as a right G-space on which G acts by the following rule.

If $n \mapsto U(n)$ is a representation of the class $\lambda \in \hat{N}$, then a representation of the class λg has the form $n \mapsto U(gng^{-1})$.

We shall suppose that N is a tame group. Then every unitary representation T of the group G can be extended to a representation of the G-space $\hat{N}$.

In fact, the restriction of the representation T to N can be written as a continuous sum of the form

$$T|_N = \int_{\hat{N}} W_\lambda d\mu(\lambda), \qquad (1)$$

where W_λ is a primary representation of the form $U_\lambda \otimes S_\lambda$, U_λ belongs to the class λ, and S_λ is the trivial representation of dimension $n(\lambda)$ (see theorem 3 of 8.4). Let

$$L = \int_{\hat{N}} L_\lambda d\mu(\lambda) \qquad (2)$$

be the corresponding decomposition of the space L of the representation T. For every measurable bounded function f on $\hat{N}$, we define the diagonal operator $P(f)$:

$$(P(f)\xi)(\lambda) = f(\lambda)\xi(\lambda). \qquad (3)$$

From theorem 3 of 8.4, it follows that this definition is consistent, i.e., it does not depend upon the choice of the decompositions (1) and (2).

Problem 1. Prove that (T, P) is a representation of the G-space $\hat{N}$, that is, condition (11) of 13.2 holds.

Hint. First prove the required relation for the case in which f is the characteristic function of a certain measurable set $S \subset \hat{N}$.

Problem 2. Prove that the measure μ is quasi-invariant with respect to G.

Hint. Use the uniqueness assertion in theorem 3 of 8.4 and the fact that the representations $T|_N$ and $T(g)T|_N T(g)^{-1}$ are unitarily equivalent.

Problem 3. Prove that if the representation T is irreducible, then the measure μ is G-ergodic.

Hint. If S is a measurable G-invariant subset of $\hat{N}$ then to its characteristic function χ_S there corresponds a projection operator $P(\chi_S)$, commuting with the representation T. Hence $P(\chi_S)$ is equal to 0 or 1, and consequently $\chi_S = 0$ or $\chi_S = 1$ almost everywhere.

If in addition the action of G on $\hat{N}$ satisfies the conditions of Glimm's theorem in 9.1, then the measure μ is concentrated on a single G-orbit in $\hat{N}$.

Let X be this orbit.

The operators $P(f)$ in (3) depends only upon the restriction of f to X. Thus we obtain a representation of the homogeneous G-space X. In accordance with the criterion of inducibility in 13.2, a representation T induced from a subgroup H is the stabilizer of a certain point $x \in X$.

We shall now consider what one can say about the induced representation U. It must be irreducible, since otherwise T would also be reducible.

§ 13. Induced Representations

Furthermore, the subgroup H contains the normal subgroup N: the representations $n \mapsto U_x(n)$ and $n \mapsto U_x(n_1 n n_1^{-1})$, are obviously equivalent if $n_1 \in N$.

Problem 4. Prove that the restriction of the inducing representation U to N is a multiple of U_x.

Hint. Use the method of reconstructing U from $\operatorname{Ind} U$ described in the proof of theorem 1 of 13.2 and of the realization (1) of the restriction of T to N.

The technique described here reduces the problem of classification of unitary irreducible representations of the group G to the description of irreducible representations of a smaller group H. The restriction of these representations to N is a multiple of the given representation. The last problem, as we shall see *infra* (§ 14), is equivalent to the classification of projective representations of the factor group H/N. These considerations assume an especially simple form when the normal subgroup N is commutative. In this case, the space $\hat{N}$ is the Pontrjagin dual of the group N.

We come to the following result.

Theorem 1 (G. Mackey). *Suppose that N is a closed commutative normal subgroup in a locally compact group G and that the action of the group G on $\hat{N}$ satisfies the conditions of Glimm's theorem from 9.1. Then every irreducible representation T of the group G has the form $\operatorname{Ind}(G, H, U)$, where H is the stationary subgroup of a certain point $\chi \in \hat{N}$, and the restriction of U to N is scalar and is a multiple of the character χ.*

To illustrate the applications of theorem 1, we shall prove the following fact.

Theorem 2. *Let G be a finite nilpotent group or a connected nilpotent Lie group. Then every unitary continuous irreducible representation T of the group G is monomial, that is, is induced by a one-dimensional representation of a certain subgroup.*

The proof follows from the following lemma.

Lemma. *If a representation T is not one-dimensional, then it is induced from a certain proper subgroup of the same type.*

The assertion of the theorem follows from the lemma, since a strictly descending chain of finite groups or connected nilpotent Lie groups breaks off after a finite number of steps. (In the second case, this follows from the analogous assertion for Lie algebras.)

Let us prove the lemma. First of all we note that we may suppose the representation T to be faithful in the case of a finite group and locally faithful (that is, with discrete kernel) in the case of a Lie group. In the opposite case, the representation T would generate a representation of the factor group G/N, where N is the kernel of T. This factor group has smaller order if G is finite and smaller dimension if G is a Lie group. Hence we could use induction on the order of the group or on its dimension.

Thus let T be a faithful (resp. locally faithful) representation of a finite group (resp. Lie group).

Problem 5. Prove that the center Z of the group G is a cyclic group (resp. has dimension 1).

Hint. Under an irreducible representation T, the center of G goes into a subgroup of scalar operators.

Let $\tilde{N}$ be the center of the factor group G/Z and let N be the inverse image of $\tilde{N}$ under the projection $G \to G/Z$.

Then N is a commutative normal subgroup of G.

Problem 6. Prove that if G is a connected linear nilpotent group acting in a space V, then the stationary subgroup of an arbitrary vector $\xi \in V$ is connected.

Hint. Use the fact that the group G is exponential (see 6.4). Deduce from Engel's theorem (see 6.2) that for every X in the Lie algebra of the group G, the equation $\exp tX \xi = \xi$ either is satisfied identically in t, or has only a finite number of roots.

The fact that the action of G on N satisfies the conditions of Glimm's theorem is trivial in the finite case, and is easily verified in the case of a Lie group. We shall not linger over this, since the action of G on $\hat{N}$ will be more closely examined in § 15.

To finish the proof of the lemma, it remains to apply theorem 1 to the group G and the normal subgroup N and to note that the corresponding G-orbit in $\hat{N}$ cannot collapse to a single point. In fact, in the opposite case, the restriction of T to N would be scalar, which for a faithful (resp. locally faithful) representation T would imply the inclusion $N \subset Z$. The lemma is proved.

Groups for which all irreducible representations are monomial are called *monomial* or *M-groups*. We have thus shown that *all finite nilpotent groups and all connected nilpotent Lie groups are monomial*.

Problem 7. All finite M-groups are solvable.

Hint. Let G be a nonsolvable M-group of least order. Prove that all of its factor groups are solvable. Let S be the intersection of the kernels of all homomorphisms of G into solvable groups. Prove that S is contained in an arbitrary normal subgroup of the group G and that $[S, S] = S$. Let T be an irreducible unitary representation of G of minimal dimension that is nontrivial on S. We suppose that $T = \text{Ind}(G, H, \chi)$, where χ is a certain character of the subgroup H. Let χ_0 be the identity character. Prove that $T_0 = \text{Ind}(G, H, \chi_0)$ is trivial on S. Infer from this that $S \subset H$. Further the relation $[S, S] = S$ implies that $\chi|_S \equiv 1$, and hence T is trivial on S. The contradiction just obtained shows that there are no nonsolvable M-groups.

An analogous result holds also for Lie groups: *every connected monomial Lie group is solvable*.

The following problem will be useful to the reader who wishes practice in applying theorem 1.

Problem 8. Describe all irreducible unitary representations of the group G of affine mappings of a line (*i.e.*, of a one-dimensional linear space) over a field K, under the hypothesis that the additive group of this field is dual to itself in the sense of Pontrjagin.

Answer. The series of one-dimensional representations corresponding to characters of the multiplicative group K^* of the field K, and one infinite-dimen-

§ 13. Induced Representations 199

sional representation, acting in the space $L^2(K^*, d^*x)$ by the formula

$$[T(a,b)f](x) = \chi(bx)f(ax).$$

Here d^*x is Haar measure on K^* and a and b are parameters in the group G corresponding to the description of an affine transformation in the form $x \mapsto ax + b$, and where χ is a nontrivial additive character of the field K.

13.4. Induced Representations of Lie Groups and their Generalizations

Every Lie group is locally compact, and hence the results of 13.2 and 13.3 hold for Lie groups. On the other hand, the construction of an induced representation for Lie groups admits interesting and important generalizations that are connected with "complexification".

Before speaking of these generalizations, we shall show that the space of a representation T of a Lie group G induced from a closed subgroup H can be considered as the space of sections of a certain vector bundle over the smooth manifold $H \backslash G$.

(The connection between induced representations and bundles is not a special privilege of Lie groups. With the appropriate generalization of the concept of a bundle, it can be established for arbitrary groups.)

Problem 1. Let G be a Lie group, H a closed subgroup of G, and x a point of the homogeneous space $X = H \backslash G$. Prove that in a certain neighborhood W of the point x, one can define a smooth mapping $s: W \to G$, have the property that $xs(y) = y$ for all $y \in W$.

Hint. Let $\mathfrak{g}$ be the Lie algebra of the group G, $\mathfrak{h}$ the subalgebra corresponding to the subgroup H, and $\mathfrak{p}$ the subspace of $\mathfrak{g}$, complementary to $\mathfrak{h}$. Prove that the mapping $\tau: \mathfrak{p} \to X$ given by the formula

$$\tau(P) = x \cdot \exp P,$$

has a nonsingular derived mapping at the point $P = 0$. Hence there exists in a certain neighborhood of the point x a smooth inverse mapping τ^{-1}, which can be taken for the s that we seek.

We consider the covering of the manifold X by neighborhoods W_i in which there exist mappings $s_i: W_i \to G$ that satisfy the condition of the problem. On the intersection $W_i \cap W_j$ we define the mapping $y \to s_i(y)s_j(y)^{-1}$, assuming values in the group H. We denote this mapping by s_{ij}. If U is a finite-dimensional representation of H in the space V, then the covering $\{W_j\}$ together with the set of functions $y \to U(s_{ij}(y))$ defines a vector bundle over X, which we denote by E_U.

Problem 2. Prove that the equivalence class of the bundle E_U is defined by the equivalence class of the representation U, and in particular does not depend on the choice of the original mappings $s_i: W_i \to G$.

Hint. Use problem 2 of 5.4.

The bundle E_U can also be obtained in a different way. Namely, we apply the construction of "products over a group" from 2.1 and set $E'_U = G \underset{H}{\times} V$. (The group H acts on G by left translations and on V in agreement with the representation U.)

Problem 3. Prove that E'_U is a vector bundle over $X = H \backslash G$, equivalent to the bundle E_U.

Hint. Consider the mapping of $W_i \times V$ into $G \times V$ defined by the formula $(y, v) \to (s_i(y), v)$ and the corresponding factor-mapping of $W_i \times V$ into $G \underset{H}{\times} V$.

This method of constructing the bundle E_U allows us to define action of the group G on it. Indeed, the group G is a right G-space, where the action of G on the right commutes with the action of the group H on the left. Hence the set $E_U = G \underset{H}{\times} V$ is a right G-space.

Problem 4. Prove that the action of G on E_U commutes with projection onto the base and is linear on the fibers.

A bundle with such an action of the group G will be called a *G-bundle*. It is clear that in the space $\Gamma(E)$ of sections of an arbitrary G-bundle over X there arises a linear representation T of the group G, defined as follows:

$$[T(g)s](x) = s(xg), \quad x \in X, \quad g \in G, \quad s \in \Gamma(E). \tag{1}$$

Problem 5. Prove that every G-bundle over a homogeneous manifold $X = H \backslash G$ is equivalent to a bundle of the form E_U.

Hint. Let x_0 be a point of X for which H is the stabilizer. As U we can take the representation of H in the fiber over x_0 that arises from (1).

Let $C^\infty(G, H, U)$ be the space of all smooth functions on G with values in V that satisfy the condition

$$F(hg) = U(h)F(g), \quad h \in H, \quad g \in G. \tag{2}$$

It is evident that this space is invariant under right translations on G and hence this space yields a representation T of the group G defined by the formula

$$(T(g)F)(g_1) = F(g_1 g). \tag{3}$$

Problem 6. Prove that the representation (3) in $C^\infty(G, H, U)$ is equivalent to the representation (1) in the space $\Gamma(E_U)$ of smooth sections of the bundle E_U.

Hint. Let s be a section of the bundle E_U. With this section we associate a function f_s on G which assumes such a value $v \in V$ at a point $g \in G$ that the equivalence class of the pair (g, v) coincides with the value of the section s at the point $Hg \in H \backslash G$. Prove that the mapping $s \to f_s$ constructed in this way yields the desired isomorphism of $\Gamma(E_U)$ onto $C^\infty(G, H, U)$.

Thus the representation space of the representation induced from the subgroup H is the Hilbert space obtained from the space of smooth sections of a

§ 13. Induced Representations

certain vector bundle over the manifold $X = H \backslash G$ by introducing a G-invariant scalar product in it.

Problem 7. If the representation U is finite-dimensional and X is compact, then the Gårding space of the representation $T = \text{Ind}(G, H, U)$ coincides with the space $\Gamma(E_{U_0})$ [for the definition of U_0, see 13.2, formula (3).]

Hint. The fact that $\Gamma(E_{U_0})$ is contained in the Gårding space of the representation T is easily deduced from problem 2 in 10.5. To prove the reversed inclusion, use the fact that the covering algebra $U(\mathfrak{g})$ acts in the Gårding space of the representation T and also the fact that a generalized function on the manifold X, all derivatives of which belong to the space $L^2(X)$, is an ordinary infinitely differentiable function. We remark that this part of the assertion of the problem is also true without the hypothesis that $H \backslash G$ be compact.

If the subgroup H is connected, the space $\Gamma(E_U)$, which is isomorphic to $C^\infty(G, H, U)$, can also be defined as the space of solutions of a certain system of differential equations on the group G.

In fact, let S be the representation of H in the space $C^\infty(G, V)$ defined by the equality

$$[S(h)f](g) = U(h)f(h^{-1}g). \tag{4}$$

It is clear that $C^\infty(G, H, U)$ coincides with the set of invariant elements for the representation S. Let S_* be the corresponding representation of the Lie algebra $\mathfrak{h}$ of the group H.

Problem 8. Prove that the space $C^\infty(G, H, U)$ coincides with the space of solutions of the system of equations

$$S_*(X)f = 0, \quad X \in \mathfrak{h} \tag{5}$$

in $C^\infty(G, V)$.

Hint. The connected group H is generated by an arbitrary neighborhood of the identity, including the one that is covered by the exponential mapping (see 6.4).

Using the explicit form (4) of the operators S, one can write the equations (5) in the form

$$L_x f + U_*(X)f = 0, \quad X \in \mathfrak{h}, \tag{6}$$

where L_x is the Lie operator of left translation on the group G, corresponding to the element $X \in \mathfrak{g}$ and U_* is the representation of the algebra $\mathfrak{h}$ in the space V that corresponds to the representation U of the group H.

The conditions written in formula (6) permit "complexification" in the following sense. The representation $X \mapsto L_x$ of the Lie algebra $\mathfrak{g}$ in the form of differential operators on G can be extended by linearity to a representation of the complex hull $\mathfrak{g}_\mathbb{C} = \mathfrak{g} \underset{\mathbb{R}}{\otimes} \mathbb{C}$. Let ρ be a representation of the complex subalgebra $\mathfrak{n} \subset \mathfrak{g}_\mathbb{C}$ in a finite-dimensional space V. Then the system of equations

$$L_x f + \rho(X)f = 0, \quad X \in \mathfrak{n}, \tag{7}$$

defines a certain subspace of $C^\infty(G, V)$, which we denote by $L(G, \mathfrak{n}, \rho)$.

Consider the case in which $\mathfrak{n}$ is the complex hull of the real subalgebra $\mathfrak{h} \subset \mathfrak{g}$ that corresponds to a closed connected subgroup $H \subset G$, and in which the representation ρ corresponds to the representation U of the subgroup H. Then $L(G, \mathfrak{n}, \rho)$ coincides with $L(G, H, U)$.

To illustrate the circumstances in the case where $\mathfrak{n} \neq \bar{\mathfrak{n}}$ we consider the simplest example. Suppose that $G = \mathbf{R}^2$, $\mathfrak{g} = \mathbf{R}^2$, $\mathfrak{g}_c = \mathbf{C}^2$, and for $\mathfrak{n}$, we take the one-dimensional complex subalgebra generated by the vector $X + iY$, where X and Y are basis vectors in $\mathfrak{g}$.

For ρ we take the one-dimensional representation $\mathfrak{n}$ which takes the value $\lambda \in \mathbf{C}$ at the vector $X + iY$. Then condition (7) has the form

$$\frac{\partial f}{\partial x} + i \frac{\partial f}{\partial y} + \lambda f = 0$$

in an appropriate system of coordinates.

Setting $z = x + iy$ and introducing the formal derivatives $\frac{\partial}{\partial z} = \frac{1}{2}\left(\frac{\partial}{\partial x} - i \frac{\partial}{\partial y}\right)$, $\frac{\partial}{\partial \bar{z}} = \frac{1}{2}\left(\frac{\partial}{\partial x} + i \frac{\partial}{\partial y}\right)$, we can rewrite this condition as

$$\frac{\partial f}{\partial \bar{z}} = -\frac{\lambda}{2} f.$$

From this it follows that $f = e^{-\lambda \bar{z}/2} \phi$, where ϕ is an arbitrary analytic function of z. (We recall that the equation $\partial \phi / \partial \bar{z} = 0$ is equivalent to the Cauchy-Riemann equations for analyticity of a function ϕ.) Thus the space $C^\infty(G, \mathfrak{n}, \rho)$ in this case is isomorphic to the space of all analytic functions of z.

Under certain additional hypotheses, the general case turns out to be the superposition (in a well-defined sense) of two cases that have already been considered: the space $C^\infty(G, \mathfrak{n}, \rho)$ is isomorphic to a space of partially holomorphic sections of a certain G-bundle over a mixed manifold (see 5.4 for the definition of these concepts).

The additional hypotheses are the following.

The intersection $\mathfrak{n} \cap \bar{\mathfrak{n}}$ is a complex subalgebra in $\mathfrak{g}_c$, coinciding, as is easy to see, with the complex hull $\mathfrak{h}_c$ of the real algebra $\mathfrak{b} = \mathfrak{n} \cap \mathfrak{g}$.

We suppose that:

1) there is a closed subgroup H of the group G (not necessarily connected) whose Lie algebra is $\mathfrak{b}$;

2) the operators of the adjoint representation $\mathrm{Ad}\, h$, $h \in H$, extended by linearity over $\mathfrak{g}_c$, preserve the subalgebra $\mathfrak{n}$;

3) there exists a closed subgroup M in G such that $H \subset M$ and $\mathfrak{n} + \bar{\mathfrak{n}} = \mathfrak{m}_c$ (here $\mathfrak{m}$ is the Lie algebra of the group M and $\mathfrak{m}_c$ is the complex hull of $\mathfrak{m}$).

Suppose now that we have a representation U of the group H in a finite-dimensional space V. We suppose that the corresponding representation U_* of the algebra $\mathfrak{h}$ can be extended to a holomorphic representation ρ of the subalgebra $\mathfrak{n}$. Consider the space of solutions of the system (7) that have the property

$$f(hg) = U(h) f(g), \quad h \in H. \tag{8}$$

§ 13. Induced Representations

We denote this system of solutions by $C^\infty(G, \mathfrak{n}, H, \rho, U)$. Note that if the group H is connected, then condition (8) follows from (7).

Theorem 1. *If the conditions listed above hold, then the space $X = H\backslash G$ is a mixed manifold of type (k, l), $k = \dim G - \dim M$, $l = \frac{1}{2}(\dim M - \dim H)$.*

One can thus define a partially holomorphic G-bundle $\mathscr{E}_{U,\rho}$ (equivalent to $\mathscr{E}_U = G \underset{U}{\times} V$ in the category of smooth G-bundles) such that the representation of the group G arising in the space of partially holomorphic sections of $\mathscr{E}_{U,\rho}$ is equivalent to the representation of this group by right translations in the space $L(G, \mathfrak{n}, H, \rho, U)$.

Proof. We first define the structure of a mixed manifold on X. For this we set $Y = M\backslash G$ and denote by p the natural projection of X onto Y, which maps the class $Hg \in X$ onto the class $Mg \in Y$ that contains it. Every fiber of the projection p (that is, the subset $p^{-1}(y)$ in X) is identified with the homogeneous space $Z = H\backslash M$. Namely, if we fix an element g that carries the initial point x_0 to the point $x \in p^{-1}(y)$, then we can define the action of the group M on $p^{-1}(y)$ by the formula

$$x * m = x_0 mg = xg^{-1}mg.$$

The stationary subgroup of the point x will obviously be H.

The identification obtained in this way between $p^{-1}(y)$ and $Z = H\backslash M$ depends upon the choice of the element $g \in G$. A different choice of this element is equivalent to an automorphism of the space Z with the aid of a certain translation from H.

We now define a complex structure on the space Z. For this we use the fact that the space Z coincides locally with the complex homogeneous space $N\backslash M_c$, where N and M_c are the complex Lie groups corresponding to the Lie algebras $\mathfrak{n}$ and $\mathfrak{m}_c$. In fact, let $\tilde{N}$ and $\tilde{M}_c$ be the local Lie groups. Then the factor space $\tilde{N}\backslash\tilde{M}_c$ is a complex homogeneous manifold. If $\mathfrak{a}$ is the complementary subspace to $\mathfrak{n}$ in $\mathfrak{m}_c$, then every element m in $\tilde{M}_c$ can be written in exactly one way in the form

$$m = n \exp A, \quad n \in \tilde{N}, \quad A \in \mathfrak{a}. \tag{9}$$

The correspondence $\tilde{N}m \to A$ gives by definition a chart on $\tilde{N}\backslash\tilde{M}_c$, which maps $\tilde{N}\backslash\tilde{M}_c$ into a neighborhood of zero in the complex linear space $\mathfrak{a}$. We now remark that the space $\mathfrak{m}_c$ is generated by the subspace $\mathfrak{n}$ and the real subspace $\mathfrak{m}$. From this it follows that the $\tilde{M}$-orbit of the initial point in $\tilde{M}\backslash\tilde{M}_c$ is open. (Here $\tilde{M}$ is the local Lie group corresponding to $\mathfrak{m}$.) Furthermore, the intersection $\tilde{M} \cap \tilde{N}$ is obviously the local group corresponding to the algebra $\mathfrak{m} \cap \mathfrak{n} = \mathfrak{h}$. We thus obtain a one-to-one correspondence between neighborhoods of initial points in $\tilde{N}\backslash\tilde{M}_c$ and in $Z = H\backslash M$. This enables us to define a complex chart covering the point $z_0 \in Z$: the image of the point $z = Hm \in Z$ is the element $A \in \mathfrak{a}$ defined by the equality (9).

Problem 9. Prove that the charts obtained from translations by elements of M by the above construction form an atlas. Prove that this atlas defines on Z the structure of a complex manifold on which M acts by holomorphic transformations.

Hint. Use the fact that the exponential mapping is holomorphic.

We can now construct an atlas on the entire manifold X. For this we fix the subspace $\mathfrak{b} \subset \mathfrak{g}$ complementary to $\mathfrak{m}$. Then in a certain neighborhood of the identity, every element $g \in G$ can be uniquely written in the form

$$g = m \exp B, \quad m \in M, \quad B \in \mathfrak{b}. \tag{10}$$

Now decomposing m by formula (9), we associate with every element g from a certain neighborhood of the identity a pair (A, B), where $A \in \mathfrak{a}$, $B \in \mathfrak{b}$.

Problem 10. Prove that the elements A and B depend only on the coset $Hg \in X$, that the correspondence $Hg \to (A, B)$ defines a chart covering the initial point of the manifold X, and that the charts obtained from this one by translations from G form an atlas that defines the structure of a mixed manifold on X.

We shall prove the remaining assertions of the theorem. Let $\tilde{\rho}$ be the representation of the local group $\tilde{N}$ that corresponds to the representation ρ of the algebra $\mathfrak{n}$. Then $\mathscr{E}_\rho = \hat{M}_c \underset{N}{\times} V$ is a holomorphic $\tilde{M}_c$-bundle over $\tilde{N} \backslash \tilde{M}_c$.

One verifies without difficulty that $\mathscr{E}_\rho$ is equivalent in the category of smooth bundles to the restriction of the bundle $\mathscr{E}_U = G \underset{M}{\times} V$ to the corresponding neighborhood W of the point $z_0 \in Z$.

To define the structure of a partially holomorphic bundle on $\mathscr{E}_U$, we use the identification of its restrictions to subsets of the form Wg with holomorphic bundles which are obtained from $\mathscr{E}_\rho$ by translating by g. Then the last assertion of the theorem will be an obvious consequence of this definition.

Representations of the group G in spaces of (partially) holomorphic sections of (partially) holomorphic G-bundles have been given the name *holomorphically induced*.

It is natural to try to introduce a G-invariant scalar product in the space of such representations and so obtain a unitary representation of the group G. In a series of cases, this is indeed possible. For example, if the "complex part" Z of the base X is compact or equivalent to a bounded region in $\mathscr{C}^n$, the Hilbert space generated by $L(G, \mathfrak{n}, H, \rho, U)$ is a closed subspace in $L^2(G, H, U)$, the space of the representation induced in the sense of Mackey.

In the case where the conditions of theorem 1 are not satisfied (for example, when $\mathfrak{n} + \bar{\mathfrak{n}}$ is not a subalgebra or when the groups M or N are not closed), there is a very interesting and almost uninvestigated question about the interpretation of the space $L(G, \mathfrak{n}, \rho)$ and the representations that we get in it.

We shall now point out a further generalization of the concept of a holomorphically induced representation. We recall (see 5.4) that a space of partially holomorphic sections of a G-bundle $\mathscr{E}$ over X can be considered as the zero-dimensional cohomology space of a certain sheaf on X. This leads to the thought of studying the higher cohomology groups of such a sheaf and the representations that arise in them. Many essential results have been obtained in this direction in recent years. Here one of the most important, and up to now unsolved, problems is the proof of Langlands' conjecture that an arbitrary irreducible representation of a real semisimple Lie group G appearing in the discrete part of the decomposition of the regular representation is realized in the space of L^2-cohomo-

§ 13. Induced Representations

logy of a suitable sheaf on the space $X = G/H$, where H is a compact Cartan subgroup of G.

More details about this direction of the theory of representations can be found in the works [66], [121], [104], [112], [126], [129].

13.5. Intertwining Operators and Duality

The problem of computing the intertwining number of two induced representations plays a great rôle in the theory of representations. The problems of decomposing the restriction of an induced representation to a subgroup and of decomposing the tensor product of induced representations reduce to this problem.

For finite groups, this problem can be reduced to the solution of simple functional equations on double cosets. Let G be a finite group, H_1 and H_2 subgroups of G, and U_1 and U_2 representations of these subgroups in spaces V_1 and V_2 respectively. We shall suppose that the characteristic of the base field does not divide the order of the group G.

We consider operator functions $K(g)$ on G with values in $\mathrm{Hom}(V_2, V_1)$ that have the property

$$K(h_1 g_1 h_2) = U_1(h_1) K(g) U_2(h_2). \tag{1}$$

With every such function we associate an operator C_K, which acts by the formula

$$[C_K f](g_1) = \sum_{g_2 \in G} K(g_1 g_2^{-1}) f(g_2). \tag{2}$$

Problem 1. Prove that if $f \in L(G, H_2, V_2)$, then $C_K f \in L(G, H_1, V_1)$, and that the correspondence $K \to C_K$ is an isomorphism between the space of all operator functions with the property (1) and the space of intertwining operators for the representations $\mathrm{Ind}\, U_2$ and $\mathrm{Ind}\, U_1$.

Hint. Let P be the operator in the space V_2^G of all V_2-valued functions on G that acts according to the formula

$$[Pf](g) = \sum_{h \in H_2} U_2(h) f(h^{-1} g).$$

Prove that P maps V_2^G onto $L(G, H_2, V_2)$ and that every operator from V_2^G to V_1^G that commutes with right translates on G has the form (2).

Plainly the condition (1) can be checked separately on every double coset $H_1 g H_2$. When U_1 and U_2 are the trivial one-dimensional representations, condition (1) assumes an especially simple form. In this case, it gives the following result.

The intertwining number of two quasi-regular representations of the group G that correspond to subgroups H_1 and H_2 is equal to the number of points in the space $H_1 \backslash G / H_2$. We recall (see 2.1) that the last number coincides with the number of H_2-orbits in $X_1 = G/H_1$, of H_1-orbits in $X_2 = G/H_2$ and of G-orbits in $X_1 \times X_2$.

For operators in infinite-dimensional spaces, a representation in the form of an integral operator is not always possible. Hence the computation of the intertwining number of two infinite-dimensional induced representations presents an essentially more difficult problem. However, for the case of Lie groups, one can present a certain analogue of the method described above. Namely, if the subgroups H_1 and H_2 are such that $X_1 = G/H_1$ and $X_2 = G/H_2$ are compact manifolds, and the inducing representations U_1 and U_2 are finite-dimensional, the Gårding spaces of the representations $\operatorname{Ind} U_1$ and $\operatorname{Ind} U_2$ are nuclear spaces (see problem 7 in 13.4). For such spaces, the kernel theorem holds (see 4.1).

Therefore every intertwining operator can be written in the form

$$C_K f(g_1) = \int_G K(g_1 g_2^{-1}) f(g_2) \, d_r g_2, \qquad (2')$$

where $K(g)$ is a generalized function on the group G, with values in $\operatorname{Hom}(V_2, V_1)$, that satisfies condition (1).

Thus the problem of finding intertwining numbers is reduced to the search for generalized functions on the smooth manifold X that transform in the given way under the action of a certain Lie group G. This is a very interesting and difficult problem of analysis, which up to this time is not completely solved.

Suppose that the manifold X is compact, that the group G has only a finite number of orbits $\Omega_1, \Omega_2, \ldots, \Omega_n$ in X, and that the union $X_k = \bigcup_{i=1}^{k} \Omega_i$ is a closed submanifold in X for $k = 1, 2, \ldots, n$. (Just this case arises, for example, in the problem of classifying the irreducible representations of complex semisimple Lie groups.)

In the space $C^\infty(X)$, one can then introduce a decreasing filtration, by setting $C_k^\infty(X) = \{f, \operatorname{supp} f \in X \setminus X_k\}$.

We then obtain an increasing filtration in the space of generalized functions, invariant under the action of the group.

Going to the corresponding graded space, we reduce our problem to the following: to describe the generalized vector-functions on the manifold X that transform in the given way under the action of the Lie group G and are concentrated on the G-orbit $\Omega \subset X$. One can offer the following construction of such generalized functions. Let H be the stationary subgroup of a point $x_0 \in \Omega$. We shall suppose that there is a generalized vector-function F_0 with support at the point x_0 such that

$$F_0(xh) = \frac{\Delta_G(h)}{\Delta_H(h)} F(x) \quad \text{for} \quad h \in H. \qquad (3)$$

Then for an arbitrary function $\phi \in C_0^\infty(X)$, we define a vector-function Φ on G by the equality

$$\Phi(g) = \langle F_0, T(g)\phi \rangle,$$

§ 13. Induced Representations

where $[T(g)\phi](x) = \phi(xg)$. This vector-function Φ has the property that

$$\Phi(hg) = \frac{\Delta_H(h)}{\Delta_G(h)} \Phi(g).$$

We can then define the generalized function F by the relation

$$\langle F, \phi \rangle = \int_G \rho(g) U(g^{-1}) \Phi(g) d_r g, \tag{4}$$

where ρ is an arbitrary smooth function on G such that $\int_H \rho(hg) d_l h \equiv 1$. Just as in 13.2 (see problem 3 and its corollary) we can show that the generalized function F does not depend upon the choice of the function ρ and has the property

$$F(xg) = U(G)^{-1} F(x). \tag{5}$$

Problem 2. Prove that every generalized vector-function on X with support on Ω that satisfies (5) is obtained by the above construction from a certain generalized function F_0 with support at the point x_0. If we further demand that the function F_0 depend only on the values of the basic function and its derivatives in directions transversal to $\Omega \subset X$, then the correspondence between F_0 and F is one-to-one.

Hint. Use the fact that every generalized function can be approximated by a generalized function with finite support.

The problem can also be solved by using the results of F. Bruhat in [67].

We remark that the identification of generalized functions that transform in the given way is necessary but in general not sufficient for solving the problem of the intertwining number. The point is that the operators corresponding to the given functions by formula (2') act from the Gårding space of one representation to the conjugate space of the Gårding space of the other representation. It is rather complicated to determine when such an extension is possible and also to determine when the extension will be a positive operator. The problem in [139] of singling out the unitary irreducible representations of complex semisimple Lie groups from the (known) absolutely irreducible representations comes up against exactly this problem.

We shall now consider possible generalizations of Frobenius reciprocity. As we have already pointed out in 13.2, this reciprocity does not hold in the infinite-dimensional case.

Interesting variants of analogues of Frobenius reciprocity have been proposed by G. Mackey [115], C. Moore [118], and I. M. Gel'fand and I. I. Pjateckiĭ-Šapiro (see [18]). G. I. Ol'šanskiĭ [124] has recently found a simple and general formulation of a generalized Frobenius reciprocity.

Theorem 1 (G. I. Ol'šanskiĭ). *Let G be a locally compact group and H a closed subgroup of G such that the factor space $X = H\backslash G$ is compact and admits a G-invariant measure μ. Let T be a unitary irreducible representation of the group G and*

U a finite-dimensional unitary representation of the subgroup H. Then we have

$$c(T, \text{Ind } U) = c(T^\infty|_H, U), \tag{6}$$

where $T^\infty|_H$ denotes the representation of H in the Gårding space of the representation T.

We remark that this theorem uses the concept of Gårding space for an arbitrary locally compact group (compare the remark at the end of 12.1).

We shall give here a proof of the theorem for the case in which G is a Lie group. Suppose that $A \in \mathscr{C}(T, \text{Ind } U)$. Then in view of problem 5 of 10.5, the operator A carries the Gårding space L^∞ of the representation T into the Gårding space of the representation $\text{Ind } U$. In view of problem 7 of 13.4, the latter space coincides with $C^\infty(G, H, U)$. Hence, given a vector ξ in L^∞, we can define a vector $A\xi(e)$ in the space of the representation U. We have

$$AT(h)\xi(e) = (\text{Ind } U)(h)A\xi(e) = A\xi(h) = U(h)A\xi(e)$$

for $h \in H$. Hence the correspondence $\xi \mapsto A\xi(e)$ is an operator in $\mathscr{C}(T^\infty|_H, U)$. It is clear that the mapping of $\mathscr{C}(T, \text{Ind } U)$ into $\mathscr{C}(T^\infty|_H, U)$ constructed in this way is linear and has kernel zero.

We shall now show that every operator $B \in \mathscr{C}(T^\infty|_H, U)$ is obtained in this way.

For a vector $\xi \in L^\infty$ we set

$$\tilde{A}\xi(g) = B(T(g)\xi).$$

The operator defined in this way carries ξ into the function $\tilde{A}\xi$ on the group G. This function satisfies the condition

$$\tilde{A}\xi(hg) = U(h)\tilde{A}\xi(g).$$

Since the correspondence $g \to T(g)\xi$ is smooth (see problem 2 in 10.5), the function $\tilde{A}\xi$ belongs to $C^\infty(G, H, U)$. Since $X = H \backslash G$ is compact, the last space is continuously embedded in the space $L^2(G, H, U)$ of the representation $\text{Ind } U$.

We have thus constructed an operator $\tilde{A} \in \mathscr{C}(T^\infty, \text{Ind } U)$. We shall show that it can be extended to an operator $A \in \mathscr{C}(T, \text{Ind } U)$.

Problem 3. Prove that the operator $\tilde{A}$ admits a closure.

Hint. One must verify that the conditions $\xi_n \to 0$ in L, $\xi_n \in L^\infty$, $\tilde{A}\xi_i \to f$ in $L^2(G, H, U)$ imply that $f = 0$. Use the method of smoothing (see 10.5) and the relation

$$\tilde{A}T(\phi) = (\text{Ind } U)(\phi)\tilde{A} \quad \text{for} \quad \phi \in C_0^\infty(G).$$

Problem 4. Let $\bar{A}$ be the closure of $\tilde{A}$. Prove that $\bar{A}$ is a bounded operator.

Hint. Use the method of proof of theorem 2 in 7.3.

§ 13. Induced Representations

Thus we have constructed an operator $\bar{A} \in \mathscr{C}(T, \text{Ind } U)$. It is easy to see that the operator corresponding to it in $\mathscr{C}(T^\infty|_H, U)$ coincides with the original operator B.

The theorem is proved.

13.6. Characters of Induced Representations

The principal result of this paragraph is formula (5), which expresses the generalized character of an induced representation $T = \text{Ind } U$ in terms of the generalized character of the representation U.

This formula is an analogue of Frobenius's formula (see problem 9 in 13.1) and reduces to it when G is a finite group.

We recall that Frobenius's formula has the form

$$\chi_T(g) = \sum_{x \in H \backslash G} \tilde{\chi}_U(s(x) g s(x)^{-1}),$$

where $\tilde{\chi}_U$ is the function on G coinciding with χ_U on H and equal to zero off of H.

Now let G be a locally compact group, H a closed subgroup of G, and $X = H \backslash G$ a right homogeneous G-space. We fix a Borel mapping $s: X \to G$ having the property that $s(x) \in x$ for all $x \in X$. Then every element $g \in G$ can be written uniquely in the form

$$g = h s(x), \quad h \in H, \quad x \in X, \tag{1}$$

and there exists a Borel measure v_s on X such that

$$d_l g = d_l h \, dv_s(x) \tag{2}$$

(see formulas 5' and 6' of 9.1).

If χ is a generalized function on the subgroup H, we denote by χ^x the generalized function on the group G defined by the equality

$$\langle \chi^x, \phi \rangle = \langle \chi, \phi_x \rangle \tag{3}$$

where $\phi_x(h) = \phi(s(x)^{-1} h s(x))$.

Theorem 1. *Let U be a unitary representation of the subgroup H in a separable space V and let U_0 be the representation connected with U by the equality*

$$U_0(h) = \left[\frac{\Delta_H(h)}{\Delta_G(h)} \right]^{1/2} U(h). \tag{4}$$

Then:

1) *if the character χ_T of the representation $T = \text{Ind}(G, H, U)$ is defined on a function $\phi \in C_0(G)$, then for almost all $x \in X$, the generalized functions $\chi_{U_0}^x$ are defined on ϕ and the equality*

$$\langle \chi_T, \phi \rangle = \int_X \langle \chi_{U_0}^x, \phi \rangle \, dv_s(x) \tag{5}$$

holds;

2) *if the function $\phi \in C_0(G)$ is positive-definite, lies in the domain of definition of the characters $\chi_{U_0}^x$ for almost all $x \in X$, and if the integral on the right side of the equality* (5) *converges, then ϕ lies in the domain of definition of the character χ_T.*

Proof. We first explore the effect of changing the choice of the mapping s on the quantities appearing in the statement of the theorem. Upon replacing $s(x)$ by $s'(x) = h(x)s(x)$ the function ϕ_x undergoes the transformation generated in $C_0(H)$ by the inner automorphism $h \to h(x)^{-1}hh(x)$. This automorphism multiplies the measure $d_l h$ by $\Delta_H(h(x))^{-1}$. Hence the character $\chi_{U_0}^x$ is multiplied by $\Delta_H(h(x)^{-1})$ when we go from s to s'. The measure $v_{s'}$, as is easy to see, is connected with the measure v_s by the equality $dv_{s'} = dv_s(x) \Delta_H(h(x)^{-1})$. This means that the integrand in the right side of formula (5) does not depend upon the choice of s.

We use the explicit formula for the representation T given in 13.2:

$$[T(g)f](x) = U_0(h)f(y), \tag{6}$$

where $h \in H$ and $y \in X$ are defined from the equality

$$s(x)g = hs(y). \tag{7}$$

From this it is easy to find an expression for the operator $T(\phi) = \int_G T(g)\phi(g)d_l g$ as an integral operator in $L^2(X, V)$ with operator kernel.

Problem 1. Prove that the operator $T(\phi)$ has the form

$$T(\phi)f(x) = \int_X K_\phi(x, y)f(y)dv_s(y), \tag{8}$$

where

$$K_\phi(x, y) = \int_H \phi(s(x)^{-1}hs(y))U_0(h)d_l h. \tag{9}$$

Hint. Use formulas (2), (6), and (7).

Up to this point, we have assumed about the mapping s only that it is Borel. We now require that s be continuous on a certain open subset $X_0 \subset X$ of complete measure [that is, $v_s(X \setminus X_0) = 0$].

We shall not linger over the proof that such a choice is possible, since the proof is rather formidable, and the assertion itself in all examples needed in the sequel is sufficiently evident.

The assertion of theorem 1 follows from the equalities (8) and (9) and a property of integral operators with continuous kernel, which we will state as a separate theorem.

Let X be a locally compact topological space with a Borel measure μ, let $K(x, y)$ be a measurable operator function on $X \times X$, the values of which are linear operators in a certain separable Hilbert space V. We consider the following integral operator $\mathscr{K}$ in the space $L^2(X, \mu, V)$ of square integrable V-valued functions on X:

$$[\mathscr{K}f](x) = \int_X K(x, y)f(y)d\mu(y). \tag{10}$$

§ 13. Induced Representations

Theorem 2. *Suppose that the operator function $K(x, y)$ is continuous in both variables in the weak operator topology. In order for the operator $\mathcal{K}$ defined by formula (10) to have a trace it is necessary, and for positive-definite operators it is also sufficient, that the trace $\operatorname{tr} K(x, x)$ exist for almost all $x \in X$ and that the integral*

$$\int_X \operatorname{tr} K(x, x) d\mu(x). \tag{11}$$

converge.

If the trace of $\mathcal{K}$ exists, then it coincides with the value of the integral (11).

We shall give the proof of theorem 2 in several steps. First we consider the case where $\dim V = 1$, that is, $K(x, y)$ is a continuous numerical function. In this case, the statement of the theorem turns into the well-known formula for the trace of a kernel integral operator with continuous kernel. We recall the scheme of deriving this formula. By the definition of a kernel operator, the function $K(x, y)$ has the form

$$K(x, y) = \sum_i \lambda_i \phi_i(x) \overline{\psi_i(y)} \tag{12}$$

as an element of $L^2(X \times X)$. Here ϕ_i and ψ_i are functions of norm one in $L^2(X)$, and $\sum_i |\lambda_i| < \infty$. The trace of the operator $\mathcal{K}$ with kernel $K(x, y)$ is easily seen to be equal to the expression

$$\sum_i \lambda_i (\phi_i, \psi_i)_{L^2(X)}. \tag{13}$$

If the series (12) were to converge on the diagonal $\Delta_X \subset X \times X$ in the norm of $L^1(\Delta_X, \mu)$, then the expression (13) would coincide with the required quantity (11). However we know only that the series (12) converges in the mean on $X \times X$. To circumvent this difficulty, one ordinarily applies the technique of smoothing. Namely, one constructs a directed family of operators $S_\alpha, \alpha \in A$, having the following properties:

1) the norms of the operators S_α admit a common (finite) upper bound;
2) $\lim\limits_{\alpha \in A} S_\alpha = 1$ in the strong operator topology of the space $L^2(X, \mu)$;
3) every operator S_α maps $L^2(X, \mu)$ continuously into $C_0(X)$.

We shall not give here the explicit construction of such a family. We note only that if X is a manifold and the measure μ in a local chart is ordinary Lebesgue measure, then we can take S_α to be, within this chart, the operator of convolution with an appropriate positive continuous function ϕ_α.

Problem 2. Prove that if $\mathcal{K}$ is a kernel operator, then the operators $\mathcal{K}_\alpha = S_\alpha \mathcal{K} S_\alpha^*$ are also kernel operators and that $\lim\limits_{\alpha \in A} \operatorname{tr} \mathcal{K}_\alpha = \operatorname{tr} \mathcal{K}$.

Hint. Use the decomposition (12) of the operator $\mathcal{K}$ as a sum of operators of rank 1.

Problem 3. Prove that the operator $\mathcal{K}_\alpha$ is an integral operator with continuous kernel $K_\alpha(x, y)$ and that $\operatorname{tr} \mathcal{K}_\alpha = \int_X K_\alpha(x, x) d\mu(x)$.

Hint. Use the decomposition (12) and property 3) of the operators S_α.

Problem 4. Prove that if $\mathscr{K}$ is a kernel integral operator with continuous kernel $K(x, y)$ vanishing outside of a certain compact set, then

$$\operatorname{tr} \mathscr{K} = \int_X K(x, x) d\mu(x).$$

Hint. Use the preceding problem and the fact that $K_a(x, y)$ converges uniformly to $K(x, y)$.

Problem 5. Let $\mathscr{K}$ be a kernel integral operator with continuous kernel $K(x, y)$. For an arbitrary function $\phi \in C_0(X)$, prove that

$$\left| \int_X \phi(x) K(x, x) d\mu(x) \right| \leq \|\mathscr{K}\|_1 \cdot \|\phi\|_{C(X)},$$

where $\|\mathscr{K}\|_1$ denotes the kernel norm of the operator $\mathscr{K}$.

Hint. Use the result of problem 4 and the inequality $\|P_\phi K\|_1 \leq \|P_\phi\| \|\mathscr{K}\|_1$, where P_ϕ is the operator of multiplying by ϕ in $L^2(X, \mu)$.

The first assertion of the theorem, in the case $\dim V = 1$, follows in an evident way from problem 5. The second assertion is easily deduced from the following fact.

Problem 6. If $\mathscr{K}$ is a positive operator with continuous kernel $K(x, y)$ vanishing outside of a certain compact set, then $\mathscr{K}$ is a kernel operator and

$$\|\mathscr{K}\|_1 = \operatorname{tr} \mathscr{K} = \int_X K(x, x) d\mu(x).$$

Hint. The hypotheses of the problem imply that $K(x, x) \geq 0$ and $|K(x, y)|^2 \leq K(x, x) K(y, y)$. Furthermore, if λ_i are the eigenvalues of the operator K and ϕ_i are the corresponding normalized eigenfunctions, then the operator with kernel

$$K(x, y) - \sum_{i=1}^{N} \lambda_i \phi_i(x) \overline{\phi_i(y)}$$

is positive for every N. Infer from this that the series $\sum \lambda_i \phi_i(x) \overline{\phi_i(y)}$ converges uniformly and that $\sum_i \lambda_i < \infty$.

We turn to the case $\dim V > 1$.

Let $\{e_n\}$ be an orthonormal basis in the space V and P_n the projection operator on e_n. Then, if $\mathscr{K}$ is a kernel operator, the operators $\mathscr{K}_n = P_n \mathscr{K} P_n$ are also, and furthermore we have $\operatorname{tr} \mathscr{K} = \sum_n \operatorname{tr} \mathscr{K}_n$.

The operator $\mathscr{K}_n$ can be considered as an integral operator in $L^2(X, \mu)$ with kernel

$$K_n(x, y) = (K(x, y) e_n, e_n)_V.$$

By hypothesis, this kernel is continuous. Hence

$$\operatorname{tr} \mathscr{K}_n = \int_X K_n(x, x) d\mu(x)$$

§ 13. Induced Representations

and
$$\operatorname{tr} \mathscr{K} = \sum_n \int_X K_n(x, x) d\mu(x).$$

We will show that the sum and the integral in the last expression can be interchanged. For this it suffices to show that the series $\sum_n K_n(x, x)$ converges absolutely in $L^1(X, \mu)$. This in turn is equivalent to the inequality

$$\left| \sum_n \int_X K_n(x, x) \phi_n(x) d\mu(x) \right| < \infty$$

for an arbitrary family of functions $\phi_n \in L^\infty(X, \mu)$, with a common bound on their moduli. The last is equivalent to the assertion that $\Phi \mathscr{K}$ is a kernel operator, where Φ is the operator in $L^2(X, \mu, V)$ defined by the formula

$$\Phi f(x) = \sum_n \phi_n(x) P_n f(x).$$

Since
$$\|\Phi\| = \sup \|\phi_n\|_{L^\infty(X, \mu)} < \infty,$$

the operator $\Phi \mathscr{K}$ is indeed a kernel operator and our assertion is proved. Thus we have proved the equality

$$\operatorname{tr} \mathscr{K} = \int_X \sum_n (K(x, x) e_n, e_n)_V d\mu(x). \tag{14}$$

We now show that the operators $K(x, x)$ are kernel operators for almost all x, and hence the equality (14) turns into the required equality (5). We consider the space
$$L^1(X, \mu, V \hat{\otimes} V')$$

of μ-measurable operator functions K on X with values in the space $V \hat{\otimes} V'$ of kernel operators in the space V and with the norm

$$\|K\| = \int_X \|K(x)\|_1 d\mu(x).$$

The bilinear mapping of
$$L^2(X, \mu, V) \times L^2(X, \mu, V')$$
into
$$L^1(X, \mu, V \hat{\otimes} V'),$$

defined by the formula $(\phi, \psi) \to \phi(x) \otimes \psi(x)$ is continuous and consequently can be extended to a mapping of the space

$$L^2(X, \mu, V) \hat{\otimes} L^2(X, \mu, V').$$

This last space is obviously isomorphic to the space of kernel operators in $L^2(X, \mu, V)$. Hence we can associate with every kernel operator $\mathcal{K}$ the function

$$K_0(x) \in L^1(X, \mu, V \hat{\otimes} V'),$$

so that $\operatorname{tr} \mathcal{K} = \int_X \operatorname{tr} K_0(x) d\mu(x).$

Problem 7. Let $A(x)$ be a continuous bounded operator function on X, the values of which are operators in the space V. Denote by $\mathcal{A}$ the operator in $L^2(X, \mu, V)$ acting by the formula

$$(\mathcal{A} f)(x) = A(x) f(x).$$

Prove that

$$\operatorname{tr} \mathcal{A} \mathcal{K} = \int_X \operatorname{tr} A(x) K_0(x) d\mu(x).$$

Hint. Consider first the case in which $\mathcal{K}$ has finite rank.

Problem 8. Prove that $K_0(x)$ coincides with $K(x, x)$ almost everywhere.
Hint. In the opposite case, the nonequality

$$(K_0(x) e_n, e_m) \neq (K(x, x) e_n, e_m)$$

would hold for all x in a set U of positive measure and for certain e_m and e_n. This contradicts the assertion of problem 8.

Thus the first statement of the theorem is proved.

It remains to prove the last statement of the theorem for the case $\dim V > 1$. This follows from the fact that the positive operator $\mathcal{K}$ has a trace if and only if all of the operators $\mathcal{K}_n = P_n \mathcal{K} P_n$ have traces, and $\sum_n \operatorname{tr} \mathcal{K}_n < \infty$.

Corollary. *The character χ_T of the representation $T = \operatorname{Ind}(G, H, U)$ is concentrated on the closure of the set $\bigcup_{g \in G} gHg^{-1}$. In particular, if H is a normal subgroup, χ_T is concentrated on H.*

In fact, the support of the generalized function $\chi_{U_0}^*$ is contained in $s(x)^{-1} H s(x)$, and the support of χ_T is contained in the closure of the union of the supports of the generalized functions $\chi_{U_0}^*$ for $x \in X$.

The set $\bigcup_{g \in G} gHg^{-1}$ has a simple geometric description: it is the set of elements $g \in G$ having a fixed point in X.

This leads us to the following thought. Suppose that the generalized function χ_T is a measure given by a continuous density with respect to Haar measure, so that it makes sense to speak of the value of the character χ_T at a point g. Then this value may be connected with the structure of fixed points of the transformation g. We give here one result of this sort.

Problem 9. Let G be a Lie group, H a closed subgroup of G, and U a finite-dimensional unitary representation of the group H. Let us suppose that there is an open subset G_0 of G such that all of the elements $g \in G_0$ have only a finite number of nondegenerate fixed points in X. (Nondegeneracy means that unity is not an eigenvalue of the derived mapping g_* in the tangent space to a fixed point.) Prove that then the restriction of the character χ_T of the representation $T = \mathrm{Ind}(G, H, U)$ to G_0 is a measure, the density of which with respect to $d_l g$ is given by the formula

$$\rho_T(g) = \sum \frac{\chi_U(s(x_i) g s(x_i)^{-1})}{|\det(g_*(x_i)^{1/2} - g_*(x_i)^{-1/2})|}, \tag{15}$$

where the summation is carried out over all fixed points x_i of the transformation g.

Hint. Let $g_0 \in G_0$ and let x_0 be a fixed point for the transformation g_0. Prove that the mapping of $X \times H$ into G given by the formula $(x, h) \mapsto s(x)^{-1} h s(x)$ is a diffeomorphism in a neighborhood of the point (x_0, h_0), and that the measure $dv_s(x) d_l h$ goes under this diffeomorphism into a measure $d\lambda(g)$, whose derivative with respect to $d_l g$ is equal to the quantity $|\det(1 - (g_0)_*)|$ at the point g_0. Use also the formulas (4) and (5) and the relation $\dfrac{\Delta_G(h_0)}{\Delta_K(h_0)} = |\det(g_0)_*|$.

We point out the connection between formula (15) and the formula given by M. Atiyah and R. Bott in [58] for the traces of operators in spaces of sections of smooth bundles. Atiyah and Bott consider operators $T = D \cdot S$, where S is a translation operator in a smooth bundle, and D is a pseudo-differential operator. When the translation has only a finite number of nondegenerate fixed points on the base, it is shown that the linear functional $f(D) = \mathrm{tr}(DS)$ can be extended uniquely (by continuity in an appropriate topology) from the subspace of kernel operators with smooth kernel to the space of all pseudo-differential operators. In the case where D is a differential operator (in particular, an operator of multiplication by a function), the authors give an explicit formula expressing the trace of the operator T in terms of the coefficients of the operator D at the fixed points of S. One can show that the Atiyah-Bott formula, like formula (15), coincides with the expression that we obtain by writing T as an integral operator with a generalized kernel and then computing the trace quite formally by formula (11).

§ 14. Projective Representations

14.1. Projective Groups and Projective Representations

We consider an n-dimensional linear space V over a field K. The group of all linear invertible mappings of V onto itself is isomorphic to the group $GL(n, K)$ of nonsingular matrices of order n with elements in K.

Let $P(V)$ be the corresponding projective space, that is, the set of all one-dimensional subspaces in V. The group $GL(n, K)$ does not act effectively[1] on $P(V)$. The scalar matrices (and only these) give the identity transformation of $P(V)$.

Hence the group of all projective transformations (i.e., automorphisms of $P(V)$) is isomorphic to the group $PGL(n, K) = GL(n, K)/C$, where C is the set of scalar matrices.

The group $PGL(n, K)$ admits also another important realization.

Problem 1. Let $\text{Mat}_n(K)$ be the full matrix algebra of order n over the field K. Prove that every automorphism of $\text{Mat}_n(K)$ has the form

$$X \mapsto AXA^{-1}. \tag{1}$$

Hint. Let P_i, $i=1,\ldots,n$, be the diagonal matrix in which the i-th element is equal to 1 and the remaining elements are equal to zero.
Use the relations $P_i P_j = \begin{cases} 0 & \text{for } i \neq j, \\ P_i & \text{for } i = j, \end{cases}$ and prove that the images of P_i under an arbitrary automorphism are projection operators onto one-dimensional subspaces.

Corollary. *The group of automorphisms of the algebra $\text{Mat}_n(K)$ is isomorphic to $PGL(n, K)$.*

In fact, the transformation (1) is the identity if and only if A is a scalar matrix.

We now consider a complex Hilbert space H and the corresponding projective space $P(H)$ (the set of one-dimensional subspaces of H).

We can introduce a distance in $P(H)$:

$$\rho(h_1, h_2) = \|P_1 - P_2\|, \tag{2}$$

where P_i is the orthogonal projection onto the one-dimensional subspace h_i, $i=1, 2$.

Problem 2. Let ξ_i be the unit vector in the subspace h_i, $i=1, 2$. Prove that

$$\rho(h_1, h_2)^2 = 1 - |(\xi_1, \xi_2)|^2.$$

Corollary. *The condition $\rho(h_1, h_2) = 1$ is equivalent to orthogonality of h_1 and h_2.*

Problem 3. Prove that every isometry of $P(H)$ has the form

$$h \mapsto Uh, \tag{3}$$

where $h \in P(H)$ and U is a unitary or a conjugate-unitary operator in H.

Hint. Let $\{\xi_\alpha\}_{\alpha \in A}$ be an orthonormal basis in H and h_α the one-dimensional subspace generated by the vector ξ_α. For an arbitrary isometry S of the space

---

[1] An action of a group G on a set X is called *effective* if every $g \neq e$ maps into a transformation of X different from the identity.

§ 14. Projective Representations

$P(H)$, there exists a transformation S_0 of the form (3) such that SS_0^{-1} keeps all of the points h_α, $\alpha \in A$, fixed. Infer from this and from problem 2 that for an arbitrary subset $A_0 \subset A$, the transformation SS_0^{-1} carries into itself the set $P_{A_0}(H)$ of all $h \in P(A)$ that are generated by vectors of the form $\xi = \sum_{\alpha \in A_0} c_\alpha \xi_\alpha$. Show that the transformation SS_0^{-1} either has the form (3), where U is a diagonal unitary operator in the basis $\{\xi_\alpha\}$, or is the product of such a transformation and complex conjugation (in the same basis).

Thus the group $P\tilde{U}(H)$ of all isometries of $P(H)$ is isomorphic to the factor group $\tilde{U}(H)/C$, where $\tilde{U}(H)$ is the group of all unitary and conjugate-unitary operators in H and C is its center, which consists of the scalar unitary operators.

The group $\tilde{U}(H)$ is a topological group under the strong operator topology [which coincides with the weak topology on $\tilde{U}(H)$]. It is evident that C is a closed subgroup of $\tilde{U}(H)$. Hence the factor group $P\tilde{U}(H) = \tilde{U}(H)/C$ is a Hausdorff topological group under the factor topology.

Problem 4. Prove that the connected component of the identity in $P\tilde{U}(H)$ is the group $PU(H) = U(H)/C$, where $U(H)$ is the group of unitary operators in H.

Hint. Use the fact that $U(H)$ is connected in the strong topology, a fact easily provable from the spectral theorem for unitary operators.

Just as in the finite-dimensional case, we can realize $P\tilde{U}(H)$ as the automorphism group of an operator algebra.

Problem 5. Let $\mathfrak{B}(H)$ denote the C^*-algebra of all bounded operators on H. Prove that every automorphism of $\mathfrak{B}(H)$ has the form

$$X \mapsto UXU^{-1}, \tag{4}$$

where $U \in \tilde{U}(H)$.

Hint. Use the method used to solve problem 1, taking also into account the self-adjointness of the operators P_i and of their images under an automorphism.

Corollary. *The group of all automorphisms of $\mathfrak{B}(H)$ is isomorphic to $P\tilde{U}(H)$.*

As was already stated in 7.1, a projective representation of a group G is a homomorphism of G into a group $PGL(n, K)$. For every element of $PGL(n, K)$ we choose an element of $GL(n, K)$ that projects into it. We thus see that every projective representation can be described as a mapping $T: G \to GL(n, K)$, which has the property that

$$T(g_1)T(g_2) = c(g_1, g_2)T(g_1 g_2). \tag{5}$$

Here $c(g_1, g_2)$ is a function on $G \times G$ with values in $K^* = K \setminus \{0\}$.

If we choose a different representative of $T(g)$, we multiply the function $c(g_1, g_2)$ by an arbitrary function of the form

$$c_0(g_1, g_2) = \frac{b(g_1)b(g_2)}{b(g_1 g_2)}. \tag{6}$$

Furthermore, the function $c(g_1, g_2)$ is not arbitrary. It must satisfy the identity

$$c(g_1, g_2)c(g_1g_2, g_3) = c(g_1, g_2g_3)c(g_2, g_3), \tag{7}$$

which follows from equating the operators $T(g_1g_2)T(g_3)$ and $T(g_1)T(g_2g_3)$.

Recalling the definition of 2.5, we see that a projective representation T defines uniquely a certain element h_T of the two-dimensional cohomology group $H^2(G, K^*)$ of G with values in K^*.

An arbitrary cocycle $c(g_1, g_2)$ belonging to the class h_T is called a *multiplier* of the projective representation T. Multipliers that belong to the zero class, i.e., that have the form (6), are called trivial. It is clear that a projective representation with trivial multiplier is equivalent to a linear representation (that is, to a representation with multiplier identically equal to 1).

A *projective unitary representation* of a group G is a homomorphism of this group into the group $P\tilde{U}(H)$.

If G is a topological group, then we shall suppose that this homomorphism is continuous.

It follows from the result of problem 4 that for connected groups G, a projective unitary representation is a homomorphism of G into $PU(H)$.

The properties of projective groups set forth in problems 1 and 5 are the basic reason why projective and projective unitary representations of a group G arise in diverse problems connected with this group. We shall explain here in more detail how projective (resp. projective unitary) representations arise in the study of linear (resp. unitary) representations of group extensions. We shall look at other situations leading to projective representations, in 17.2 and 18.2.

Recall that we have reduced the problem of classifying the representations of group extensions to the following problem (see 13.3). Suppose that we have a group H, a normal subgroup N of H, and a representation U of the subgroup N having the following property: for every $h \in H$, the representations $U(n)$ and $U_h(n) = U(hnh^{-1})$ are equivalent. It is required to describe all irreducible representations T of the group H whose restrictions to N are multiples of U. We shall now show that the latter problem is equivalent to the classification of the projective representations of the factor group $K = H/N$.

For each class $k \in K$, we choose a representative $\sigma(k) \in H$. Then every element $h \in H$ can be written uniquely in the form

$$h = \sigma(k) \cdot n, \quad k \in K, \quad n \in N. \tag{7'}$$

The space V in which the desired representation T acts can conveniently be written in the form of a Hilbert product $V_1 \otimes V_2$ in such a way that the operators $T(n)$, $n \in N$, assume the form

$$T(n) = 1_{V_1} \otimes U(n). \tag{8}$$

We remark that the operators $T(n)$, $n \in N$, can be written in the form (8) if and only if $T|_N$ is a multiple of U.

§ 14. Projective Representations

Note next that the representations

$$U(n) \quad \text{and} \quad U_k(n) = U(\sigma(k)n\sigma(k)^{-1})$$

are equivalent for all $k \in K$. Hence there exist operators $W(k)$ in the space V_2 such that

$$U(\sigma(k)n\sigma(k)^{-1}) = W(k)U(n)W(k)^{-1}. \tag{9}$$

We compare the operators $T(\sigma(k))$ and $1_{V_1} \otimes W(k)$ according to their action on the operators $T(n)$, $n \in N$. It is not hard to deduce from the relations (8) and (9) that the operator $T(\sigma(k))(1_{V_1} \otimes W(k)^{-1})$ commutes with all of the operators $T(n)$, $n \in N$. Since U is irreducible, it follows that this operator has the form $S(k) \otimes 1_{V_2}$ (for the unitary case, see problem 1 from 4.5 and for the finite-dimensional linear case, see theorem 2 of 8.2). Thus the desired representation T must have the form

$$T(h) = S(k) \otimes W(k)U(n), \tag{10}$$

where $k \in K$ and $n \in N$ are defined by the equality (7').

We now consider what can be said about the operators $S(k)$ and $W(k)$. Condition (9) defines the operator $W(k)$ up to a scalar multiple. Hence the operators $S(k)$ are also defined only up to a scalar multiple. We thus obtain a mapping of K into a group of projective transformations. We shall show that this mapping is a homomorphism. This follows from the following simple fact.

Problem 6. The quality $A \otimes B = A_1 \otimes B_1$ is possible only if $A = \lambda A_1$, $B = \lambda^{-1}B_1$ for a certain scalar λ.

Hint. Compare the matrix elements of both operators.

In fact, since $\sigma(k_1)\sigma(k_2)$ is comparable with $\sigma(k_1 k_2)$ modulo N, the operator $T(\sigma(k_1)\sigma(k_2))$ has the form

$$S(k_1 k_2) \otimes W(k_1 k_2)U(n)$$

for a certain $n \in N$. On the other hand, this operator is equal to the product of $T(\sigma(k_1))$ and $T(\sigma(k_2))$, that is, to

$$S(k_1)S(k_2) \otimes W(k_1)W(k_2).$$

Hence, in view of problem 6, we have

$$S(k_1)S(k_2) = \lambda(k_1, k_2)S(k_1, k_2). \tag{11}$$

This means that S is a projective representation with multiplier $\lambda(k_1, k_2)$ defined by the relation

$$W(k_1)W(k_2) = \lambda(k_1, k_2)^{-1}W(k_1 k_2)U(n). \tag{12}$$

Here $n \in N$ is defined by the equality

$$\sigma(k_1)\sigma(k_2) = \sigma(k_1 k_2)n. \tag{13}$$

We recapitulate the arguments just given as follows.

Theorem 1. *Formula (10) establishes a one-to-one correspondence between representations T of the group H whose restrictions to N are multiples of an irreducible representation U and projective representations S of the group $K = H/N$ with multiplier (12). The representation T is irreducible if and only if the corresponding projective representation S is also irreducible.*

14.2. Schur's Theory

The projective representations of the group G can be constructed in the following way. Let

$$1 \to G_0 \to \tilde{G} \to G \to 1$$

be a central extension of the group G with the aid of a commutative subgroup G_0. We consider a linear representation $\tilde{T}$ of the group $\tilde{G}$ over a field K and suppose that all of the operators $\tilde{T}(g_0)$, $g_0 \in G_0$, are scalar.

For every $g \in G$, we choose an arbitrary inverse image $\tilde{g} \in \tilde{G}$. Clearly we obtain a projective representation of the group G by mapping the element $g \in G$ into the operator $\tilde{T}(\tilde{g})$.

We shall say that the projective representation $\tilde{T}$ obtained by this construction is *linearized by the group* $\tilde{G}$ or *is obtained from a linear representation of* $\tilde{G}$.

Problem 1. Prove that every projective representation T of the group G is linearized by a certain group $\tilde{G}$.

Hint. For $\tilde{G}$, take a central extension of G with the aid of K^*, which corresponds to the class h_T (see 2.5).

It turns out that for a finite group G, there exists a *universal* central extension $\tilde{G}$ which linearizes all projective representations of the group G. Namely, we have

Theorem 1 (I. Schur). *Let G be a finite group and K an algebraically closed field (of arbitrary characteristic). There exists a central extension $\tilde{G}$ of the group G with the aid of a finite Abelian group G_0 such that every projective representation of the group G over the field K is obtained from a linear representation of the group $\tilde{G}$ that is scalar on G_0.*

Proof. If we do not demand finiteness of the group G_0, then the theorem can be proved with no restrictions on the group G and the field K by the following simple method. We consider the group $Z^2(G, K^*)$ of all two-dimensional cocycles on G with values in K^*. To every pair of elements g_1, g_2 in G, there corresponds the homomorphism

$$\alpha_{(g_1, g_2)} : Z^2(G, K^*) \to K^*, \tag{1}$$

which maps the cocycle c into the number $c(g_1, g_2)$.

§ 14. Projective Representations

Problem 2. Prove that the correspondence $(g_1, g_2) \to \alpha_{(g_1, g_2)}$ is a two-dimensional cocycle on G with values in the group M of all homomorphisms of $Z^2(G, K^*)$ into K^*.

We now construct an extension G_1 of the group G with the aid of the subgroup of M that corresponds to the cocycle α of problem 2. We recall that this extension consists of pairs (g, m), $g \in G$, $m \in M$, which are multiplied by the rule

$$(g_1, m_1)(g_2, m_2) = (g_1 g_2, m_1 m_2 \alpha(g_1, g_2)). \tag{2}$$

If T_1 is a linear representation of the group G_1 whose restriction to the subgroup of M is scalar and has the form

$$T_1(1, m) = m(c), \quad c \in Z^2(G, K^*), \tag{3}$$

then the operators $T(g) = T_1(g, 1)$ satisfy condition (5) of 14.1. The converse also holds: if the operators $T(g)$ satisfy condition (5) of 14.1, then the formula

$$T_1(g, m) = T(g) \cdot m(c) \tag{4}$$

defines a linear representation of the group G_1.

Thus all projective representations of the group G are obtained from linear representations of the group G_1.

The group M that appears in this construction is needlessly large. In particular it can be infinite even if the group G is finite. Let us show that one can replace M by a smaller group without damage to the properties of this group that we need. In fact let $Z_1^2(G, K^*)$ be a subgroup of $Z^2(G, K^*)$ that has nonvoid intersection with every cohomology class of cocycles. In other words, the natural projection of $Z_1^2(G, K^*)$ in $H^2(G, K^*)$ is an epimorphism. To obtain all projective representations of the group G, it suffices to consider only cocycles in $Z_1^2(G, K^*)$.

Let M_1 be the subgroup of M that is the annihilator of $Z_1^2(G, K)$. If a projective representation T corresponds to a cocycle $c \in Z_1^2(G, K^*)$, then the linear representation T_1 defined by formula (4) is trivial on $M_1 \subset G_1$. This means that it generates a representation of the group $\tilde{G} = G_1 / M_1$.

In accordance with the hypothesis already made, every projective representation of the group G is equivalent to a representation with a multiplier in $Z_1^2(G, K^*)$. Hence every projective representation of G is obtained from a linear representation of the group $\tilde{G}$. The last-named group is evidently a central extension of G with the aid of the group $G_0 = M/M_1$.

The problem naturally arises of finding the smallest possible subgroup $Z_1^2(G, K)$ having the property that we need. This group cannot be too small, since $H^2(G, K^*)$ is a factor group of it.

It turns out that under the conditions of the theorem (that is, for a finite group G), this natural bound is attained. Namely, there exists a subgroup $Z_1^2(G, K^*)$ of $Z^2(G, K^*)$ which projects isomorphically onto $H^2(G, K^*)$.

This assertion is easily inferred (see for example, [13], § 53) from the theory of extensions set down in 2.4 and the following fact.

Problem 3. Let G be a finite group. Prove that $H^2(G, K^*)$ is a finite group, the order of which does not divide the characteristic of the field K and is a divisor of the order of G.

Hint. Let n be the order of G and $c \in Z^2(G, K^*)$. Prove that $c(g_1, g_2)^n = \dfrac{b(g_1)b(g_2)}{b(g_1 g_2)}$, where $b(g) = \prod\limits_{g_1 \in G} c(g, g_1)$. Verify also that if an element $h \in H^2(G, K^*)$ has order k, then one can choose a cocycle that represents h in such a way that $c(g_1, g_2)^k \equiv 1$. Finally, if p is the characteristic of the field K, then p-th roots of p-th powers exist and are unique in K^*. Therefore the group $H^2(G, K^*)$ enjoys the same property.

The connection described here between projective representations of the group G over the field K and the group $H^2(G, K^*)$ may prove useful for the computation of the last-named group.

Problem 4. Prove that if G is a free group, then $H^2(G, K^*) = 0$.

Hint. Every projective representation of a free group is equivalent to a linear representation.

Problem 5. Prove that if the group $H^2(G, K^*)$ is trivial, then the group $H^2(G \times G, K^*)$ is isomorphic to the group of all mappings $c: G \times G \to K^*$ that are multiplicative in each argument and also have the property that

$$c(g_2, g_1) = c(g_1, g_2)^{-1}.$$

Hint. Let T be a projective representation of $G \times G$ over K. Set

$$C_T(g_1, g_2) = T(g_1, 1)T(1, g_2)T(g_1, 1)^{-1}T(1, g_2)^{-1}.$$

Use the fact that the restrictions of T to the subgroups $G \times 1$ and $1 \times G$ are equivalent to linear representations.

14.3. Projective Representations of Lie Groups

Schur's theory of projective representations of finite groups can be carried over to topological groups. The concept of continuous or Borel cohomology of topological groups, which arises here, has been studied in a recent work of C. Moore [119].

We shall analyze here only the simplest (but at the same time the most important for applications) case, where the group G under study is a connected Lie group.

Let T be a complex projective representation of G. We consider the group G_1 whose elements are pairs of the form (g, A), where $g \in G$ and A is one of the linear operators which generate the projective transformation $T(g)$. The rule of multiplication in G_1 is componentwise:

$$(g_1, A_1)(g_2, A_2) = (g_1 g_2, A_1 A_2).$$

§ 14. Projective Representations

The mapping $(g, A) \to g$ is obviously an epimorphism of G_1 onto G, the kernel of which is the set of pairs of the form $(e, \lambda \cdot 1)$, where $\lambda \in \mathbb{C}^*$. We thus obtain the extension

$$1 \to \mathbb{C}^* \to G_1 \to G \to 1. \tag{1}$$

The representation T is obtained from the linear representation of $\tilde{G}$ which maps the pair (g, A) into the operator A.

We can produce an analogous construction for a unitary projective representation T [that is, a homomorphism into the group $PU(H)$].

In this case, the rôle of G_1 is assumed by the set of pairs (g, U), where U is a unitary operator giving the transformation $T(g)$ of the projective space $P(H)$. We arrive at the extension

$$1 \to \mathbb{T} \to G_1 \to G \to 1. \tag{2}$$

The representation T is obtained from the linear unitary representation of G_1 that carries (g, U) into U.

We shall now show that the group G_1 constructed above is equipped with the structure of a Lie group, so that (1) and (2) are exact sequences in the category of Lie groups. For definiteness, we shall take up the case of a unitary representation. Let $(g, U) \in G_1$. We choose unit vectors ξ and η in the representation space for which $(U\xi, \eta) \neq 0$.

Problem 1. Prove that there exists a neighborhood W of the point g such that $(V\xi, \eta) \neq 0$ for all unitary operators V that define the projective transformation $T(g)$, $g \in W$.

Hint. Use the continuity of the mapping

$$T: G \to PU(H).$$

If h_1 and h_2 are the subspaces of H generated by ξ and η, then for W we can take the set of all $g \in G$ for which

$$\rho(T(g)h_1, h_2) < 1$$

(see problem 2 of 14.1).

Let W_1 be the inverse image of the neighborhood W in G_1 under the natural projection of G_1 onto G.

The mapping $(g, U) \to \left(g, \dfrac{(U\xi, \eta)}{|(U\xi, \eta)|}\right)$ is a one-to-one mapping of W_1 onto $W \times \mathbb{T}$. We carry over to W_1 the manifold structure that it possesses on $W \times \mathbb{T}$.

Problem 2. Prove that the atlas on G_1 constructed as above consists of continuously connected charts and defines the structure of a Lie group on G_1.

Hint. Use the continuity of the function

$$g \to |(U(g)\xi, \eta)| = \rho(T(g)h_1, h_2)$$

and the theorem of Gleason-Montgomery-Zippin from 6.1.

Thus the group G_1 is a Lie group which is an extension of the group G by the aid of **T**.

Let $\tilde{G}$ be the simply connected covering group of the group G and let $\tilde{G}_1$ be the simply connected covering group of the group G_1.

We then have the following commutative diagram with exact rows:

$$\begin{array}{ccccccc} 1 \leftarrow & \tilde{G} & \leftarrow \tilde{G}_1 & \leftarrow \mathbf{R} & \leftarrow 1 \\ & \| & \downarrow \downarrow \downarrow & & \| \\ 1 \leftarrow & G & \leftarrow G_1 & \leftarrow \mathbf{T} & \leftarrow 1, \end{array}$$

where the vertical arrows denote the natural projections.

As is known, the group $\tilde{G}_1$ is completely determined from its Lie algebra $\mathfrak{g}_1$, which is an extension of the Lie algebra $\mathfrak{g}$ with the aid of the one-dimensional Lie algebra of **R**. (See 6.3.)

One can construct a theory of extensions of Lie algebras analogous to the theory of group extensions described in 2.4. We present here the final result in the form that we need. Let $Z^2(\mathfrak{g}, \mathbf{R})$ be the set of bilinear skew-symmetric real functions c on $\mathfrak{g} \times \mathfrak{g}$ that have the property

$$c([X_1, X_2], X_3) + c([X_2, X_3], X_1) + c([X_3, X_1], X_2) = 0. \tag{3}$$

Let $B^2(\mathfrak{g}, \mathbf{R})$ be the subspace of $Z^2(\mathfrak{g}, \mathbf{R})$ consisting of the functions of the form

$$c(X_1, X_2) = \langle F, [X_1, X_2] \rangle, \quad F \in \mathfrak{g}^*. \tag{4}$$

The factor space $H^2(\mathfrak{g}, \mathbf{R}) = Z^2(\mathfrak{g}, \mathbf{R})/B^2(\mathfrak{g}, \mathbf{R})$ is called the space of two-dimensional cohomology of the algebra $\mathfrak{g}$ with coefficients in **R**.

Theorem 1. *There exists a one-to-one correspondence between the elements of $H^2(\mathfrak{g}, \mathbf{R})$ and equivalence classes of extensions of the algebra $\mathfrak{g}$ with the aid of* **R**.

More precisely, given the class $h \in H^2(\mathfrak{g}, \mathbf{R})$, containing an element $c \in Z^2(\mathfrak{g}, \mathbf{R})$, we associate with it the class of equivalent extensions that contains the extension defined by the formula

$$[(X_1, t_1), (X_2, t_2)] = ([X_1, X_2], c(X_1, X_2)),$$
$$X_i \in \mathfrak{g}, \quad t_i \in R, \quad i = 1, 2.$$

If the Lie algebra $\mathfrak{g}$ is the sum of a semisimple subalgebra $\mathfrak{g}_1$ and a solvable ideal $\mathfrak{g}_2$, then $H^2(\mathfrak{g}, \mathbf{R})$ is isomorphic to the subspace of $H^2(\mathfrak{g}_2, \mathbf{R})$, generated by the cocycles $c \in Z^2(\mathfrak{g}_2, \mathbf{R})$ which enjoy the property

$$c([X_1, Y], X_2) = c(X_1, [Y, X_2]), \quad X_i \in \mathfrak{g}_2, \quad Y \in \mathfrak{g}_1. \tag{5}$$

Corollary. *Every projective representation of a connected and simply connected semisimple Lie group is obtained from a linear representation of this group.*

§ 14. Projective Representations

In fact, it follows from the last statement of theorem 1 that $H^2(\mathfrak{g}, \mathbf{R})=0$ for a semisimple Lie algebra $\mathfrak{g}$. Hence the algebra $\mathfrak{g}_1$ in this case is the direct sum of $\mathfrak{g}$ and $\mathbf{R}$, while the group G_1 is the direct product of G and $\mathbf{R}$.

The following is a useful exercise in mastering the material of the present paragraph.

Problem 3. Let P be the group of all affine transformations of a two-dimensional complex space that preserve a nondegenerate bilinear skew form on this space, and let P_0 be the connected component of the identity in P.

Construct a universal extension G of the group P_0 that linearizes all projective representations of this group (compare theorem 1 of 14.2).

Hint. The Lie algebra $\mathfrak{p}$ of the group P can be conveniently realized as the set of all complex matrices of the third order having the form

$$X = \begin{pmatrix} a & c & u \\ c & -a & v \\ 0 & 0 & 0 \end{pmatrix}$$

with the commutation operation

$$[X_1, X_2] = X_1 X_2 - X_2 X_1.$$

The subalgebra $\mathfrak{p}_1$ defined by the condition $u=v=0$ is semisimple and corresponds to the subgroup of homogeneous transformations (Lorentz transformations). The ideal $\mathfrak{p}_2$ defined by the condition $a=0$ corresponds to the subgroup of translations. Every bilinear skew-symmetric form c on $\mathfrak{p}_2$ that satisfies condition (5) has the form

$$c(X_1, X_2) = \operatorname{Re} \lambda(u_1 v_2 - v_1 u_2), \quad \lambda \in \mathbf{C}.$$

The corresponding universal extension can be realized by complex matrices of the form

$$\begin{pmatrix} 0 & -v & u & w \\ 0 & a & b & u \\ 0 & c & -a & v \\ 0 & 0 & 0 & 0 \end{pmatrix}.$$

The group G that corresponds to this Lie algebra has a simple geometric meaning: it is the subgroup of $Sp(4, \mathbf{C})$ that leaves one nonzero vector fixed.

Problem 4. Find the universal extension of the group of motions of the plane.

Hint. Prove an analogue of theorem 1 for the case in which $\mathfrak{g}_1$ is the Lie algebra of the subgroup of rotations and $\mathfrak{g}_2$ is the Lie algebra of the subgroup of parallel translations.

§ 15. The Method of Orbits

At the basis of the method of orbits lies the following "experimental fact": the theory of infinite-dimensional representations of every Lie group is closely connected with a certain special finite-dimensional representation of this group. This representation acts in the dual space $\mathfrak{g}^*$ of the Lie algebra $\mathfrak{g}$ of the group under study. We will call it a *co-adjoint* or briefly a *K-representation*.[1]

Orbits of a Lie group in the space of a K-representation are symplectic manifolds. They can be interpreted as phase spaces of a Hamiltonian mechanical system for which the given Lie group is the group of symmetries. In 15.2, we shall give a classification of all homogeneous symplectic manifolds with a given group of symmetries.

It turns out that unitary irreducible representations of the group G are connected with orbits of this group in the K-representation. The construction of the representation in an orbit is given in 15.3.

This is a generalization of the procedure of quantization that is used in quantum mechanics. This point of view is explained in more detail in 15.4.

The author sees the significance of the method of orbits not only in the specific theorems obtained by this method, but also in the great collection of simple and intuitive heuristic rules that give the solution of the basic questions of the theory of representations. With the passage of time, these rules will be elevated to the level of strict theorems, but already now their value is indisputable.

We shall show in 15.5 how the operations of restriction to a subgroup and induction from this subgroup can be described with the aid of the natural projection $p: \mathfrak{g}^* \to \mathfrak{h}^*$, where $\mathfrak{h}$ is the Lie algebra of the subgroup H.

As we shall see in 15.6, generalized characters of irreducible unitary representations admit a simple expression in the form of an integral over the corresponding orbit. In many cases, this allows us to write an explicit expression for the Plancherel measure.

Finally, in 15.7 we show that infinitesimal characters of irreducible unitary representations of a group G can be computed as values of G-invariant polynomials on the corresponding orbits.

15.1. The Co-Adjoint Representation of a Lie Group

Let G be a Lie group, $\mathfrak{g}$ its Lie algebra, and $\mathfrak{g}^*$ the dual space to $\mathfrak{g}$. The group G acts in $\mathfrak{g}$ with the aid of the adjoint representation Ad (see 6.3) and in $\mathfrak{g}^*$ with the aid of the co-adjoint representation, or, briefly, the K-representation. If the Lie algebra $\mathfrak{g}$ is realized in the form of the algebra of left-invariant vector-fields on G, then it is natural to realize $\mathfrak{g}^*$ in the form of the space of left-invariant differential forms of the first order on G. The K-representation of the group G acts in the space of 1-forms by right translations.

---

[1] The Russian word for co-adjoint is "koprisoedinennoe" (Author's footnote for the translation).

§ 15. The Method of Orbits

We analyze in more detail an example which will also be useful in the consideration of the general case. Let $G = GL(n, \mathbf{C})$ be the group of all nonsingular complex matrices of the n-th order. Since G is an open subset in the linear space $\mathrm{Mat}_n(\mathbf{C})$ of all matrices, we can identify the tangent space to G at an arbitrary point $g \in G$ with $\mathrm{Mat}_n(\mathbf{C})$. Then a vector field on G will be simply a matrix function.

Problem 1. Prove that every left-invariant vector field on G has the form

$$v_A(X) = XA, \quad A \in \mathrm{Mat}_n(\mathbf{C}), \tag{1}$$

and that the field v_A goes into the field $v_{Y^{-1}AY}$ under the action of right translation by $Y \in G$.

Hint. Verify that the field $v(X)$ on G goes into the field $v'(X) = Y_1 v(Y_1^{-1} X Y_2^{-1}) Y_2$, under the action of Y_1 on the left and Y_2 on the right.

The dual space of $\mathrm{Mat}_n(\mathbf{C})$ can conveniently be identified with $\mathrm{Mat}_n(\mathbf{C})$ through the bilinear (over $\mathbf{R}$) form

$$\langle X, Y \rangle = \mathrm{Re}\,\mathrm{tr}\, XY. \tag{2}$$

Then 1-forms on G will also be written in the form of matrix functions.

Problem 2. Prove that every left-invariant 1-form on G has the form

$$\omega_B(X) = BX^{-1}, \quad B \in \mathrm{Mat}_n(\mathbf{C}) \tag{3}$$

and that the form ω_B under translation by $Y \in G$ goes into the form $\omega_{YBY^{-1}}$.

Thus in the example under consideration the K-representation is equivalent to the adjoint representation and the orbits of the group G in the K-representation are classes of similar matrices.

Now let G be an arbitrary Lie group. Replacing G if necessary by a group locally isomorphic to G, we may suppose that G is a subgroup of $GL(n, \mathbf{C})$ (see 6.2).

We remark that under the K-representation, elements of the center of G go into the identity operator. Hence locally isomorphic groups have isomorphic linear groups as images under the K-representation.

The Lie algebra $\mathfrak{g}$ of the group G will then be a subalgebra of $\mathrm{Mat}_n(\mathbf{C})$.

Let $\mathfrak{g}^\perp$ be the orthogonal complement to $\mathfrak{g}$ under the bilinear form (2), let V be any subspace of $\mathrm{Mat}_n(\mathbf{C})$ complementary to $\mathfrak{g}^\perp$, and let P be the projection on V parallel to $\mathfrak{g}^\perp$. Then one can identify $\mathfrak{g}^*$ with V in such a way that the K-representation assumes the form

$$K(g)X = P(gXg^{-1}), \quad X \in V, \quad g \in G. \tag{4}$$

Example. Let G be the subgroup of $GL(n, \mathbf{C})$ consisting of all real upper triangular matrices with entries of 1 on the main diagonal. Then for V we can take the subspace of real lower triangular matrices with zeros on the main diagonal. The action of the operator P in this case is then the replacement of all elements of the matrix on or above the main diagonal by zeros.

Let $\mathcal{O}(G)$ be the set of orbits of the Lie group G in the K-representation, with the factor topology of the natural topology in $\mathfrak{g}^*$. As a rule, the topological space $\mathcal{O}(G)$ does not enjoy Hausdorff's separation property. In the examples considered above, it has T_0 separation. As in 19.2, one can produce examples of groups G for which $\mathcal{O}(G)$ is not even T_0.

We turn now to the study of orbits of the Lie group G in the K-representation. We shall first show that there is a closed nonsingular G-invariant 2-form on every such orbit.

We need certain general observations about differential forms on homogeneous manifolds.

Let G be a Lie group, H a closed subgroup of G, and $M = H\backslash G$ the homogeneous manifold of right cosets of H in G. Let p denote the natural projection of G onto M, which maps the point $g \in G$ onto the coset Hg. The derived mapping $p_*(e)$ is a linear operator from the Lie algebra $\mathfrak{g}$ of the group G onto the tangent space to M at the initial point H. It is clear that the kernel of this operator coincides with the Lie algebra $\mathfrak{h}$ of the subgroup H. Thus the tangent space $T_H M$ is identified with the factor space $\mathfrak{g}/\mathfrak{h}$. The translation $g \in G$ carries the point H into Hg. The derived mapping $g_*(H)$ is an isomorphism of $T_H M$ onto $T_{Hg} M$, and the mapping $g^*(H)$, which is dual to it, will be an isomorphism of $T^*_{Hg} M$ onto $T^*_H M$. Using these isomorphisms, we can give tensor fields on M in the form of functions on G with values in a tensor algebra over $\mathfrak{g}/\mathfrak{h}$. Namely, if Φ is a tensor field of type (k, l) over M, then the value of the function ϕ at a point $g \in G$ is a tensor $\phi(g)$ of type (k, l) over $T_H M$ which at the vectors $\xi_1, \ldots, \xi_l$ and covectors $\eta_1, \ldots, \eta_k$ has the value

$$\Phi(Hg)(g_*(H)\xi_1, \ldots, g_*(H)\xi_l, g^*(H)^{-1}\eta_1, \ldots, g^*(H)^{-1}\eta_k). \tag{5}$$

Problem 3. Prove that a function ϕ on G with values in $T^{k,l}(\mathfrak{g}/\mathfrak{h})$ corresponds to a tensor field of type (k, l) on M if and only if it satisfies the condition

$$\phi(hg) = \rho_{k,l}(h)\phi(g), \tag{6}$$

where $\rho_{k,l}$ is the natural representation of H in $T^{k,l}(\mathfrak{g}/\mathfrak{h})$, generated by the adjoint representation of H in $\mathfrak{g}/\mathfrak{h}$.

Hint. Use formula (5) to reconstruct Φ from ϕ and show that the consistency of this reconstruction is equivalent to condition (6).

In particular, differential forms of order l on M are given by functions on G with values in

$$\wedge^l((\mathfrak{g}/\mathfrak{h})^*) \approx \wedge^l(\mathfrak{h}^\perp),$$

satisfying the condition

$$\phi(hg) = \rho_l(h)\phi(g), \tag{6'}$$

where ρ_l is the l-th exterior power of the natural representation of H in the space $\mathfrak{h}^\perp$.

The form of writing used above is convenient since actions of the group G have a very simple appearance.

§ 15. The Method of Orbits

Problem 4. Prove that if the tensor field Φ corresponds to the function ϕ, then the field obtained from Φ by translating by $g \in G$ corresponds to the function obtained from ϕ by translation on the right by g.

We note that the connection just described between tensor fields and functions is a special case of writing sections of a G-bundle on $H \backslash G$ in the form of vector-functions on G, which transform in the specified way under left translations on H (see 13.4).

Corollary. *G-invariant differential forms on $M = H \backslash G$ correspond uniquely to H-invariant elements in $\wedge(\mathfrak{h}^\perp)$.*

In fact, it follows from problem 4 that the function on G corresponding to a G-invariant form is constant with values in $\wedge(\mathfrak{h}^\perp)$, and from condition (6) it follows that the value of this function is an H-invariant element.

Problem 5. Let Φ be an invariant k-form on $M = H \backslash G$, and suppose that to Φ there corresponds the exterior form $\phi \in \wedge^k(\mathfrak{h}^\perp)$. Then to the differential $d\Phi$ there corresponds the form $d\phi \in \wedge^{k+1}(\mathfrak{h}^\perp)$, given by the formula

$$d\phi(X_1, \ldots, X_{k+1}) = \frac{1}{k+1} \sum_{i<j} (-1)^{i+j+1} \phi([X_i, X_j], \ldots, \hat{X}_i, \ldots, \hat{X}_j, \ldots). \tag{7}$$

Hint. Use formula (3) from 5.3, the relation

$$L_\xi \iota(\eta) - \iota(\eta) L_\xi = \iota([\xi, \eta]),$$

and the fact that if the form Φ is G-invariant, and the vector field ξ corresponds to the element X of the Lie algebra of the group G, then $L_\xi \Phi = 0$. Compare also formula (4) from 5.3.

We return to the consideration of orbits of the Lie group G in the K-representation.

Let Ω be one of these orbits, let F be an arbitrary point in Ω, and let G_F be the stabilizer of this point.

Problem 6. Prove that the Lie algebra $\mathfrak{g}_F$ of the group G_F coincides with the kernel of the bilinear skew-symmetric form B_F on $\mathfrak{g}$ defined by the formula

$$B_F(X, Y) = \langle F, [X, Y] \rangle. \tag{8}$$

Hint. By definition, the kernel of the form B_F consists of the elements $X \in \mathfrak{g}$ such that $B_F(X, Y) = 0$ for all Y in $\mathfrak{g}$.

Prove the relation

$$B_F(X, Y) = \left\langle \frac{d}{dt} K(\exp tX) F \bigg|_{t=0}, Y \right\rangle \tag{9}$$

and deduce from it the assertion of the problem.

Since the quantity $B_F(X, Y)$ depends only on the images of X and Y in the factor space $\mathfrak{g}/\mathfrak{g}_F$, we obtain a skew-symmetric bilinear form on $\mathfrak{g}/\mathfrak{g}_F$, which we denote by $\tilde{B}_F$. The kernel of the form $\tilde{B}_F$ is the image of the kernel of the form B_F in $\mathfrak{g}/\mathfrak{g}_F$, and in view of problem 6 consists only of zero. This means that the form $\tilde{B}_F$ is nondegenerate.

Problem 7. Prove that $\tilde{B}_F$ is a G_F-invariant element in $\wedge^2(\mathfrak{g}/\mathfrak{g}_F)^*$.
Hint. Use the definition of G_F and the relations

$$\rho_2(g)\tilde{B}_F(X, Y) = \tilde{B}_F(Ad(g^{-1})X, Ad(g^{-1})Y) = \langle K(g)F, [X, Y] \rangle. \tag{10}$$

From the assertions of problem 6 and the corollary to problem 4, it follows that there is a nondegenerate G-invariant 2-form B_Φ on $\Omega = G_F \backslash G$, which corresponds to a G_F-invariant element $\tilde{B}_F$ in $\wedge^2(\mathfrak{g}/\mathfrak{g}_F)^*$.

Our construction of the form B_Ω depended upon the choice of the point F. We shall now show that the form B_Ω itself does not depend upon this choice. This fact follows from the following explicit expression for B_Ω.

Problem 8. Let ξ_X denote the vector field on Ω corresponding to the element $X \in \mathfrak{g}$. Then for an arbitrary point $F' \in \Omega$ we have the equality

$$B_\Omega(F')(\xi_X(F'), \xi_Y(F')) = \langle F', [X, Y] \rangle. \tag{11}$$

Hint. Use formula (5) and the fact that the vector $\xi_F(X)$ goes into $X \bmod \mathfrak{g}_F$ under the identification of $T_F\Omega$ with $\mathfrak{g}/\mathfrak{g}_F$.

We shall now show that the form B_Ω is closed. Computing the differential of this form in accordance with formula (7) from problem 5, we come to the expression

$$\tfrac{1}{3}[B_F([X, Y], Z) + B_F([Y, Z], X) - B_F([X, Z], Y)]$$
$$= \tfrac{1}{3}\langle F, [[X, Y], Z] + [[Y, Z], X] + [[Z, X], Y]\rangle,$$

which is equal to zero in view of Jacobi's identity. This means that $dB_\Omega = 0$ and the form B_Ω is closed.

We summarize our investigations.

Theorem 1. *On every orbit Ω of the Lie group G in its K-representation, there exists a nondegenerate closed G-invariant 2-form B_Ω, defined by formula (11).*

We point out a geometrically intuitive corollary of theorem 1.
All G-orbits in the K-representation have even dimension.
As a matter of fact, a nondegenerate skew-symmetric form can exist only in an even-dimensional space.

For another way to "explain" the appearance of symplectic structures on G-orbits, see the author's recent paper "Lie algebras enjoying the property of locality", in "Funkcionalnyĭ Analiza i ego Priloženija", Vol. 9, No. 2. There is also a detailed exposition in Preprint No. 64 (1975) of Institute of Applied Mathematics of the Academy of Sciences of the USSR, submitted for publication to "Uspehi mat. Nauk".

15.2. Homogeneous Symplectic Manifolds

A *symplectic manifold* is a smooth real manifold M of even dimension on which there exists a closed nondegenerate differential 2-form B.

As an example of a symplectic manifold, we cite the space T^*N of the cotangent bundle over an arbitrary manifold N. The form B is defined on T^*N in the following way.

Let $\tilde{U}$ be an arbitrary chart on N and $q_1, \ldots, q_k$ the corresponding local coordinates. In each of the spaces T_m^*N, $m \in \tilde{U}$, we choose coordinates $p_1, \ldots, p_k$, corresponding to the basis $\dfrac{\partial}{\partial q_1}, \ldots, \dfrac{\partial}{\partial q_k}$ in $T_m N$.

Then the collection $(q_1, \ldots, q_k; p_1, \ldots, p_k)$ will be a local system of coordinates in the region $U \subset T^*N$ which is the inverse image of $\tilde{U}$ under the natural projection $p: T^*N \to N$. We define a differential 1-form σ_U in U, setting

$$\sigma_U = \sum_{i=1}^{k} p_i dq_i. \qquad (1)$$

Problem 1. Let $\tilde{U}$ and $\tilde{V}$ be two arbitrary charts on N, and U and V the corresponding charts on T^*N. Prove that the forms σ_U and σ_V coincide on the intersection of the regions U and V.

Hint. Let $(q_1, \ldots, q_k; p_1, \ldots, p_k)$ be coordinates in the region U, and $(\tilde{q}_1, \ldots, \tilde{q}_k; \tilde{p}_1, \ldots, \tilde{p}_k)$ coordinates in the region V. It is clear that the $\tilde{q}_i$ are functions only of $q_1, \ldots, q_k$, and that the $\tilde{p}_i$ depend linearly on $p_1, \ldots, p_k$. Prove that

$$\tilde{p}_i(q_1, \ldots, q_k; p_1, \ldots, p_k) = \sum_{j=1}^{k} \frac{\partial q_j}{\partial \tilde{q}_i} p_j.$$

Thus there exists a unique 1-form σ on the entire manifold T^*N whose restriction to an arbitrary chart U coincides with σ_U.

We set $B = d\sigma$. Then B is a closed form, since $dB = d^2\sigma = 0$. Furthermore, the form B is nondegenerate, since in a local system of coordinates U it has the form

$$B_U = \sum_{i=1}^{k} dp_i \wedge dq_i. \qquad (2)$$

The presence of a nondegenerate 2-form permits us to establish a one-to-one correspondence between vector and covector fields on a symplectic manifold. Namely, to each vector field ξ, there exists a covector field (i.e., a 1-form) $\iota(\xi)B$ [for the definition of the operation $\iota(\xi)$, see 5.3].

A vector field ξ on a symplectic manifold M with a form B is called *Hamiltonian* if

$$L_\xi B = 0. \qquad (3)$$

In other words, a field ξ is Hamiltonian if the family of transformations induced by this field preserves the form B.

Problem 2. Prove that a vector field is Hamiltonian if and only if the corresponding 1-form $\iota(\xi)B$ is closed.

Hint. Use formula (3) from 5.3.

We call the field ξ *strictly Hamiltonian* if the corresponding 1-form $\iota(\xi)B$ is exact, that is,

$$\iota(\xi)B + dF = 0 \tag{4}$$

for a certain function F on M. The function F is called the *generating function of the field* ξ and is defined by the equality (4) up to an additive constant. Every real function F on M is the generating function of a strictly Hamiltonian field on M, which we denote by ξ_F.

Problem 3. Prove that the commutator of two Hamiltonian vector fields is a strictly Hamiltonian vector field.

Hint. As generating function of the field $[\xi, \eta]$, one can take $\iota(\xi)\iota(\eta)B = 2B(\xi, \eta)$. Use the hint for problem 6 of 15.1.

Let F and G be two smooth real functions on M. Then we have the equalities

$$\xi_F G = 2B(\xi_F, \xi_G) = -\xi_G F. \tag{5}$$

In fact, we have $\xi_F G = -\langle dG, \xi_F \rangle = -(\iota(\xi_F)B)(\xi_F) = -2B(\xi_G, \xi_F) = 2B(\xi_F, \xi_G)$. The second equality is proved similarly.

The common value of the three expressions entering in (5) is called the *Poisson bracket* of the functions F and G and is denoted by $\{F, G\}$.

Problem 4. Prove that if the form B has the form (2) in a certain local system of coordinates, then the Poisson bracket is defined by the equality

$$\{F, G\} = \sum_{i=1}^{k} \left(\frac{\partial F}{\partial p_i} \frac{\partial G}{\partial q_i} - \frac{\partial F}{\partial q_i} \frac{\partial G}{\partial p_i} \right). \tag{6}$$

As a matter of fact, the hypothesis of the problem is no restriction. By a well-known theorem of Darboux, every closed nondegenerate 2-form has the form (2) in an appropriate local system of coordinates.

We remark that if the one-dimensional cohomology group $H^1(M, \mathbf{R})$ of the manifold M is trivial, then every closed 1-form is exact, and consequently every Hamiltonian vector field is strictly Hamiltonian. In the general case, strictly Hamiltonian fields form a subspace $H_0(M)$ in the space $H(M)$ of all Hamiltonian fields. The codimension of $H_0(M)$ in $H(M)$ is equal to the first Betti number $b_1(M)$ of the manifold M.

The space $C^\infty(M, \mathbf{R})$ of smooth real functions on M forms an infinite-dimensional Lie algebra under the Poisson bracket. In fact, Jacobi's identity

$$\{F, \{G, H\}\} + \{G, \{H, F\}\} + \{H, \{F, G\}\} = 0 \tag{7}$$

follows from the following argument.

Since the correspondence $F \to \xi_F$ is defined by the form B, it commutes with all transformations that preserve this form. Such transformations are called

§ 15. The Method of Orbits

canonical. But every Hamiltonian field η is locally the derivative of a certain family of canonical transformations. Hence the equality

$$\xi_{L_\eta F} = L_\eta \xi_F$$

holds, or equivalently,

$$\xi_{\eta F} = [\eta, \xi_F].$$

Substituting ξ_G for the field η and applying both sides of the equality so obtained to the function H, we find the desired identity (7).

Problem 5. Prove that the mapping $F \mapsto \xi_F$ is a homomorphism of the Lie algebra $C^\infty(M, \mathbf{R})$ onto the Lie algebra $H_0(M)$.

Hint. The equality

$$\xi_{\{F,G\}} = [\xi_F, \xi_G], \tag{8}$$

which one must establish, turns into the identity (7) upon applying both sides to an arbitrary function $H \in C^\infty(M, R)$.

We thus have an exact sequence of Lie algebras:

$$0 \to \mathbf{R} \xrightarrow{i} C^\infty(M, \mathbf{R}) \xrightarrow{j} H_0(M) \to 0, \tag{9}$$

where i is the natural embedding of $\mathbf{R}$ into $C^\infty(M, \mathbf{R})$ as the subspace of constants, and j carries F into ξ_F.

Suppose now that a Lie group G acts transitively on the manifold M, in such a way that all of the transformations of this group are canonical (preserve the form B). We shall then say that M is a *homogeneous* symplectic manifold. To every element X of the Lie algebra $\mathfrak{g}$ of the group G, there then corresponds a Hamiltonian field ξ_X on M.

If all of the fields ξ_X, $X \in \mathfrak{g}$, are strictly Hamiltonian and if we are able to choose the generating functions F_X of these fields in such a way that the equality

$$F_{[X,Y]} = \{F_X, F_Y\} \tag{10}$$

holds, we say that M is a *strictly homogeneous* symplectic manifold.

In other words, a homogeneous symplectic manifold is strictly homogeneous if the following diagram of Lie algebras is commutative:

$$\begin{array}{ccc}
& 0 & \\
& \downarrow & \\
0 \to \mathbf{R} \to C^\infty(M, \mathbf{R}) & \to & H_0(M) \to 0 \\
\uparrow & & \downarrow \\
\mathfrak{g} & \to & H(M) \\
& & \downarrow \\
& & H^1(M, R) \\
& & \downarrow \\
& & 0
\end{array}$$

As an example of a strictly homogeneous symplectic manifold with a group of motions G, consider an arbitrary orbit of the group G in its K-representation.

In fact, we saw in 15.1 that every such orbit Ω is a homogeneous symplectic manifold. It remains to verify that Ω is a strictly homogeneous symplectic manifold. This follows from the following assertion.

Problem 6. Let η_X be the vector field on Ω that corresponds to the element $X \in \mathfrak{g}$. Then for generating function ϕ_X of this field, we can take the restriction to Ω of the linear function $F \mapsto \langle F, X \rangle$ on $\mathfrak{g}^*$. We have the identity

$$\{\phi_X, \phi_Y\} = \phi_{[X,Y]}.$$

Hint. The first assertion follows from the definition of the form B_Ω and of generating functions, the second from the equality (5).

It turns out that orbits in the K-representation actually exhaust all strictly homogeneous symplectic manifolds for which the group of motions is a connected Lie group G. In fact, let M be such a manifold. As above, we denote by F_X the generating function of the vector field ξ_X on M that corresponds to the element X of the Lie algebra $\mathfrak{g}$ of the group G. Suppose that the relations (10) hold.

We consider the mapping $\phi: M \to \mathfrak{g}^*$ defined by the formula

$$\langle \phi(m), X \rangle = F_X(m). \tag{11}$$

Problem 7. Prove that the mapping ϕ commutes with the action of the group G.

Hint. Since G is connected, it suffices to prove that ϕ commutes with elements of the form $\exp Y$, $Y \in \mathfrak{g}$. For this it in turn suffices to prove that the derived mapping ϕ_* carries the vector field ξ_Y on M into the vector field η_Y on $\mathfrak{g}^*$. This follows from (10).

Problem 8. Prove that the mapping ϕ is a locally homeomorphic mapping of M onto one of the G-orbits in $\mathfrak{g}^*$.

Hint. The fact that $\phi(M)$ is a G-orbit in $\mathfrak{g}^*$ follows from problem 7. Among the fields ξ_X, $X \in \mathfrak{g}$, at an arbitrary point $m \in M$, there are $k = \dim M$ independent fields $\xi_{X_1}, \ldots, \xi_{X_k}$. Their generating functions $F_{X_1}, \ldots, F_{X_k}$ have linearly independent differentials at this point. Hence the mapping ϕ is locally a homeomorphism.

Thus our manifold M is a "covering" of a certain orbit Ω of the group G in $\mathfrak{g}^*$. If the orbit Ω is simply connected, then it admits no nontrivial connected coverings. In the opposite case, as is known, there are as many coverings as there are subgroups of the fundamental group $\pi_1(\Omega)$. Every such covering is obtained by factoring the universal (simply connected) covering $\tilde{\Omega}$ by the corresponding subgroup $\Gamma \subset \pi_1(\Omega)$ (compare 6.1).

The result just obtained enables us to describe all homogeneous symplectic manifolds in terms of orbits, provided that the group of motions of the manifold is a connected Lie group G.

If M is such a manifold, its universal covering manifold $\tilde{M}$ will be a homogeneous symplectic manifold with respect to the group $\tilde{G}$ (the simply connected covering group of G).

§ 15. The Method of Orbits

Every Hamiltonian field on $\tilde{M}$ is strictly Hamiltonian. However, the generating functions of the fields ξ_X, $X \in \mathfrak{g}$, in general satisfy the relation (10) only up to an additive constant.

We consider the Lie algebra (under the Poisson bracket) generated by all generating functions of all of the fields ξ_X, $X \in \mathfrak{g}$. Evidently this Lie algebra $\mathfrak{g}_1$ is an extension of $\mathfrak{g}$ with the aid of $\mathbf{R}$. Let G_1 be the corresponding simply connected Lie group.

Problem 9. Prove that the manifold $\tilde{M}$ is strictly Hamiltonian with respect to the group G_1.

Hint. Define the action of G_1 on M so that elements of the normal subgroup corresponding to the ideal $\mathbf{R} \subset \mathfrak{g}_1$ go into the identity transformation.

We have thus proved the following theorem.

Theorem 1. *Every homogeneous symplectic manifold whose group of motions is a connected Lie group G is locally isomorphic to an orbit in the K-representation of the group G or a central extension of G with the aid of $\mathbf{R}$.*

15.3. Construction of an Irreducible Unitary Representation by an Orbit

All of the methods known up to this time for constructing irreducible representations of groups consist in the application (perhaps repeated) of the three basic operations:

  1) restriction to a subgroup;
  2) extension from a subgroup;
  3) induction from a subgroup (and also various generalizations of this operation; see 13.4).

Beginning with a certain collection of the "simplest" representations, we obtain all representations by these operations. It seems evident that for Lie groups (as for finite groups) our "simplest" representations are the one-dimensional representations. However, this assertion has not been proved.

We consider here only the case in which the irreducible representation of the group G is obtained from a one-dimensional representation in one step with the aid of the operation of induction or its generalizations: holomorphic induction or representation in cohomologies.

As we saw in 13.4, such a representation T is described by a collection $(\mathfrak{n}, H, \rho, U)$ where: $\mathfrak{n}$ is a complex subalgebra of $\mathfrak{g}_c$ (the complex hull of the Lie algebra $\mathfrak{g}$ of the group G); H is a closed subgroup of G whose Lie algebra has the form $\mathfrak{h} = \mathfrak{n} \cap \mathfrak{g}$; ρ is a one-dimensional holomorphic representation of $\mathfrak{n}$, defined by the formula

$$\rho(X) = 2\pi i \langle F, X \rangle, \quad F \in \mathfrak{g}^*; \tag{1}$$

U is a unitary one-dimensional representation of H which has the form

$$U(\exp X) = e^{\rho(X)} \tag{2}$$

in a neighborhood of the identity.

We will say that the *orbit* Ω *of the group G in* $\mathfrak{g}^*$ *going through the point F corresponds to the representation T.*

We will study the possibility of going back from the orbit to the representation.

We shall suppose that the group G is connected and simply connected. Let Ω be one of the orbits of the group G in $\mathfrak{g}^*$. We choose an arbitrary point F on Ω. Let G_F denote the stabilizer of F in G and $\mathfrak{g}_F$ the Lie algebra of the group G_F. As we saw in 15.1, $\mathfrak{g}_F$ coincides with the kernel of the skew-symmetric bilinear form

$$B_F(X, Y) = \langle F, [X, Y] \rangle. \tag{3}$$

We shall say that the subalgebra $\mathfrak{n} \subset \mathfrak{g}$ is subordinate to the functional $F \in \mathfrak{g}^*$ if the form B_F vanishes identically on $\mathfrak{n}$. This definition makes sense also for complex subalgebras of $\mathfrak{g}_c$, if we agree that the form B_F is extended over $\mathfrak{g}_c$ by linearity.

Problem 1. Prove that the subalgebra $\mathfrak{n}$ is subordinate to F if and only if the mapping

$$X \to \langle F, X \rangle$$

is a one-dimensional representation of the Lie algebra $\mathfrak{n}$.

It follows from the assertion of the problem and from the equality (1) that the subalgebra $\mathfrak{n}$ appearing in the definition of T is subordinate to F.

Problem 2. Prove that the codimension in $\mathfrak{g}$ (resp. the complex codimension in $\mathfrak{g}_c$) of a real subalgebra $\mathfrak{n} \subset \mathfrak{g}$ (resp. of a complex subalgebra $\mathfrak{n} \subset \mathfrak{g}_c$) that is subordinate to a functional $F \in \mathfrak{g}^*$ is not less than half of the dimension of the G-orbit in $\mathfrak{g}^*$ that passes through the point F.

Hint. The codimension of a maximal isotropic subspace for B_F in $\mathfrak{g}_c$ is equal to

$$\tfrac{1}{2} \operatorname{rang} B_F = \tfrac{1}{2}(\dim \mathfrak{g} - \dim \mathfrak{g}_F) = \tfrac{1}{2} \dim \Omega.$$

It has been observed that in all known cases, when the subalgebra $\mathfrak{n}$ is connected with an irreducible representation T of the group G, the estimate in problem 2 is attained, that is,

$$\operatorname{codim} \mathfrak{n} = \tfrac{1}{2} \dim \Omega. \tag{4}$$

Furthermore, the subalgebra $\mathfrak{n}$ enjoys the following property:

$$F + \mathfrak{n}^\perp \subset \Omega. \tag{5}$$

This is known as Pukanszky's condition: see 15.5, *infra*. Here $\mathfrak{n}^\perp$ denotes the annihilator of $\mathfrak{n}$ in $\mathfrak{g}^*$ (that is, the set of all functions $F \subset \mathfrak{g}^*$ whose extensions over $\mathfrak{g}_c$ vanish identically on $\mathfrak{n}$).

Geometrically, condition (5) asserts that an orbit Ω contains along with a point F the linear submanifold $p^{-1}(pF)$, where p is the natural projection of $\mathfrak{g}^*$ onto $\mathfrak{n}^*$.

A subalgebra $\mathfrak{n}$ that is subordinate to F is called *admissible* if it enjoys properties (4) and (5).

§ 15. The Method of Orbits

Problem 3. Prove that if there exists a real (resp. complex) admissible subalgebra for a functional $F \in \Omega$, then all other points of Ω also have this property.

Hint. If $\mathfrak{n}$ is an admissible subalgebra for F, then $Ad_g \mathfrak{n}$ is an admissible subalgebra for $K(g)F$.

Thus, the presence of admissible subalgebras for $F \in \mathfrak{g}^*$ is a property of the orbit Ω that contains F. Admissible subalgebras do not always exist.

Let $Sp(2n+2, \mathbf{R})$ be a symplectic group (see 5.1) and let $G = St(n, \mathbf{R})$ be the subgroup of this group that leaves fixed a certain nonzero vector in $\mathbf{R}^{2n+2}$. The Lie algebra $\mathfrak{g} = \mathfrak{st}(n, \mathbf{R})$ of the latter group can be realized by matrices of the form

$$X(A, \xi, c) = \begin{pmatrix} 0 & \xi' & c \\ 0 & A & J\xi \\ 0 & 0 & 0 \end{pmatrix}, \tag{6}$$

where $A \in \mathfrak{sp}(2n, \mathbf{R})$, $\xi \in \mathbf{R}^{2n}$, $c \in \mathbf{R}$, and J has the form $\begin{pmatrix} 0 & 1_n \\ -1_n & 0 \end{pmatrix}$.

We set

$$F_\lambda(X(A, \xi, c)) = \lambda c, \quad \lambda \in \mathbf{R}.$$

Then for $\lambda \neq 0$, there is no admissible subalgebra in $\mathfrak{g}_c$ that is subordinate to F_λ. In fact, the form B_{F_λ} has the form

$$B_{F_\lambda}(X(A_1, \xi_1, c_1), X(A_2, \xi_2, c_2)) = 2\lambda \xi_1' J \xi_2 \tag{7}$$

and consequently the subspace $\mathfrak{g}_{F_\lambda}$ is defined by the condition $\xi = 0$.

Problem 4. Prove that every maximal isotropic subspace n for B_{F_λ} in $\mathfrak{g}_c$ consists of elements of the form

$$X(A, \xi, c), \quad A \in \mathfrak{sp}(2n, \mathbf{C}), \quad \xi \in V, \quad c \in \mathbf{C}. \tag{8}$$

Here V is a certain n-dimensional subspace of $\mathbf{C}^{2n}$ isotropic with respect to the form with matrix J.

Hint. Use formula (7).

Since the group $Sp(2n, \mathbf{R})$ and its Lie algebra $\mathfrak{sp}(2n, \mathbf{R})$ act irreducibly in $\mathbf{C}^{2n}$, none of the subspaces (8) can be a subalgebra.

The group $St(n, \mathbf{R})$ that figures in the above example is a group of general type. It admits a solvable normal subgroup $N(n)$, called the *Heisenberg group* or the *special nilpotent group*, and a complementary semisimple subgroup isomorphic to $Sp(2n, \mathbf{R})$. It is interesting to remark that for semisimple and solvable groups, admissible complex subalgebras always exist.

Michèle Vergne[1] has given a simple construction of an admissible subalgebra. It is based on the following facts.

---
[1] C. R. Acad. Sci. Paris 270, 173–175, 704–707.

Problem 5. Let $s: 0 = V_0 \subset V_1 \subset \ldots \subset V_n = V$ be a chain of linear spaces such that $\dim V_k = k$. Let us suppose that there is a bilinear skew-symmetric form B in V, and let B_k be the restriction of B to V_k. Then $W(s, B) = \sum_{k=1}^{n} \ker B_k$ is a maximal isotropic subspace in V for the form B.

Hint. Prove by induction that $|\text{rang } B_k - \text{rang } B_{k+1}| = 1$, and prove by induction that $W(s, B) \cap V_k = W(s_k, B_k)$.

Problem 6. Prove that if V is a Lie algebra, if the subspaces V_k are ideals in V, and the form B has the form (3), then $W(s, B)$ is a subalgebra in V.

We state without proof the following result of M. Vergne.

Theorem 1. *Let G be a connected real solvable Lie group and $\mathfrak{g}$ its Lie algebra. For every functional $F \in \mathfrak{g}^*$, one can find an admissible complex subalgebra $\mathfrak{n} \subset \mathfrak{g}_c$ having the following properties:*

1) $\mathfrak{n} = W(s, B_F)$ *for a certain chain s of ideals in $\mathfrak{g}_c$;*
2) $\mathfrak{n}$ *is invariant with respect to the stabilizer G_F of the point F;*
3) $\mathfrak{n} + \bar{\mathfrak{n}}$ *is a subalgebra in $\mathfrak{g}_c$;*
4) *if $\mathfrak{x}$ is the nilpotent radical of $\mathfrak{g}$, then $\mathfrak{n} \cap \mathfrak{x}_c$ is an admissible subalgebra subordinate to the functional $f = F|_{\mathfrak{x}}$;*
5) *if G is an exponential group, then $\mathfrak{n}$ is the complex hull of a certain real admissible subalgebra $\mathfrak{h} \subset \mathfrak{g}$, having the form $\mathfrak{h} = W(s_0, B_F)$, where s_0 is a certain chain of subalgebras in $\mathfrak{g}$.*

We return to the problem of constructing a representation from an orbit.

We shall suppose that for the element F of the orbit Ω, there exists a subalgebra $\mathfrak{n} \subset \mathfrak{g}_c$ that has properties 2) and 3) of theorem 1. We suppose also that there are closed subgroups H and M in G such that $G_F \subset H \subset M$, and that the following relations hold:

$$\mathfrak{n} \cap \bar{\mathfrak{n}} = \mathfrak{h}_c, \quad \mathfrak{n} + \bar{\mathfrak{n}} = \mathfrak{m}_c, \quad H = G_F H^0. \tag{9}$$

Here $\mathfrak{h}$ and $\mathfrak{m}$ are the Lie algebras of the groups H and M respectively, and H^0 is the connected component of the identity in the group H.

For the construction of the representation T as in 13.4, we must have a representation ρ of the algebra $\mathfrak{n}$ and a representation U of the group H, connected by the identity

$$U(\exp X) = e^{\rho(X)}, \quad X \in \mathfrak{h}. \tag{10}$$

For ρ we take the one-dimensional representation given by the equality (1). Then formula (10) defines U in a certain neighborhood of the identity of the group H. The question arises as to the existence and uniqueness of an extension of U over the entire group H. The answer to this question turns out to depend upon topological properties of the orbit Ω. To formulate the exact result we introduce some terminology. We consider the set S of functionals of the form

$$F_1 = K(g)F, \quad g \in H,$$

and call S the *fiber going through the initial point F*.

§ 15. The Method of Orbits

Problem 7. Prove that for arbitrary $g_1, g_2 \in G$, the sets $K(g_1)S$ and $K(g_2)S$ either coincide or are disjoint.

Hint. Use the fact that $gHg^{-1} = H$ for $g \in G_F$.

Thus we have a partition of the orbit Ω into fibers of the form $K(g)S$. Let Y denote the corresponding factor space and Γ the fundamental group $\pi_1(Y)$.

We call an orbit Ω of the group G in $\mathfrak{g}^*$ *integral*, if the form B_Ω belongs to an integer cohomology class. (This means that the integral of the form B_Ω over an arbitrary two-dimensional cycle in Ω is equal to an integer.)

Theorem 2. *In order for the local representation defined by the equality* (10) *to be extensible to a one-dimensional unitary representation of the group H, it is necessary that the orbit Ω going through the point $F \in \mathfrak{g}^*$ be integral.*

The set of possible extensions, if nonvoid, is parametrized by the characters of the group Γ introduced above.

Proof. We use the well-known connection between topological properties of the Lie group G, its closed subgroup K, and the homogeneous manifold $X = G/K$. If the group G is connected and simply connected, we have the following isomorphisms:

$$H^2(X, \mathbf{R}) \approx H^1(K^0, \mathbf{R}), \tag{11}$$

$$\pi_1(X) \approx \pi_0(K) = K/K^0. \tag{12}$$

In the language of differential forms, the isomorphism (11) can be described as follows. Let p be the natural projection of G onto X and B a differential form on X belonging to the class $h \in H^2(X, \mathbf{R})$. Then the form p^*B on G has the form $p^*B = d\sigma$ (since for a simply connected Lie group, we have $H^2(G, R) = 0$). The 1-form σ on G is defined by this condition up to a summand of the form df, $f \in C^\infty(G)$. One can verify that the restriction σ_0 of the form σ to the subgroup K^0 is a closed form. The cohomology class of this restriction is an element of the group $H^1(K^0, \mathbf{R})$ corresponding to the original class h.

Problem 8. Prove that when $K = G_F$, $X = \Omega$, $B = B_\Omega$, the form σ_0 can be taken to be the left-invariant 1-form on G_F that corresponds to the linear functional $F|_{\mathfrak{g}_F}$.

Hint. Use formula (7) of 15.1.

The first statement of the theorem now follows from the fact that the isomorphism (11) preserves integral cohomology classes. In fact, the assertion of problem 8 shows that if a local representation (10) admits an extension f over the subgroup G_F^0, then the form σ_0 has the form

$$\sigma_0 = \frac{1}{2\pi i} d \ln f. \tag{13}$$

Hence the integral of the form σ_0 over an arbitrary one-dimensional cycle in G_F^0 is equal to an integral (namely, $1/(2\pi)$ times the change in the argument of f upon tracing out this cycle). Conversely, if the form σ_0 belongs to an integral class, then the function f defined on G_F^0 by the equality (13) and the condition $f(e) = 1$ is single-valued and is an extension of the local representation (10) over the subgroup G_F^0.

We shall now prove the second assertion of the theorem. Since a representation of a connected Lie group is uniquely defined by the corresponding local representation, any two extensions agree on H^0. Consequently, all extensions are obtained from a single extension by multiplying by characters of the group H/H^0. The space Y of "fibers" introduced above is a homogeneous manifold with group of motions G and stationary subgroup H. For the case $K = H$, $X = Y$, the isomorphism (11) gives us the equality $\pi_1(Y) = \pi_0(H) = H/H^0$. The theorem is proved.

Remark 1. If the set $S = K(H)F$ introduced above is simply connected, then theorem 2 admits the following useful strengthened form.

1) The integral character of the orbit is not only necessary but also sufficient for the existence of an extension of the local representation (10) to a one-dimensional representation of H.

2) The group Γ figuring in the second part of the theorem is isomorphic to the fundamental group of the orbit Ω.

Remark 2. There is also a differential-geometric proof of theorem 2, based on the interpretation of the form B_Ω as the curvature form for a certain connection in a linear bundle over Ω. This proof can be found in the work [110] of B. Kostant (see also 15.4).

It remains for us to study the dependence of the representation constructed above on the choice of the point F on the orbit Ω and of the admissible subalgebra $\mathfrak{n}$ subordinate to F.

Problem 9. Prove that representations T_i constructed from functionals F_i, subalgebras $\mathfrak{n}_i$, subgroups H_i, and one-dimensional representations ρ_i and U_i ($i = 1, 2$) are equivalent if

$$F_1 = K(g)F_2, \quad \mathfrak{n}_1 = \operatorname{Ad} g \mathfrak{n}_2, \quad H_1 = gH_2 g^{-1},$$
$$\rho_1(X) = \rho_2(\operatorname{Ad} g X), \quad U_1(h) = U_2(g^{-1}hg)$$

for a certain $g \in G$.

Hint. Consider the inner automorphism of the group G that corresponds to the element g.

Thus the choice of the point F plays no essential rôle.

It seems evident that the choice of the admissible subalgebra also has no influence over the equivalence class of the representation constructed above. This assertion has not been proved in full generality, but it is not contradicted by the examples known at the present time. In the next paragraph, we shall give some "physical" arguments to support it[1].

It has also not been proved that the construction always leads to an irreducible representation (although here too there are no counterexamples).

More precise results are known for special classes of groups.

---

[1] The attempt to raise this to the level of a theorem has led to a deep generalization of the concept of Fourier transform. See for example L. Hörmander, Fourier integral operators, 1 (Acta Math. 127 (1971), 79–183).

§ 15. The Method of Orbits

Theorem 3 (B. Kostant—L. Auslander). *Let G be a connected and simply connected solvable Lie group. The following assertions hold.*

1) The group G belongs to type I if and only if the space $\mathcal{O}(G)$ is T_0 and all forms B_Ω are exact.

2) If G is of type I, all irreducible representations of G are obtained by the above construction from orbits of the group G in $\mathfrak{g}^*$. To every orbit Ω there corresponds a family of irreducible representations, parametrized by the characters of the group $\pi_1(\Omega)$.

3) Representations that correspond to different orbits or different characters of the fundamental group of the orbit are necessarily inequivalent.

We note that if the group G is exponential, then all G-orbits in $\mathfrak{g}^*$ are homeomorphic to euclidean space. It therefore follows in particular from theorem 3 that exponential groups are of type I and that for these groups, there is a one-to-one correspondence between the sets $\hat{G}$ and $\mathcal{O}(G)$.

If the group G is compact, connected, and simply connected, then G-orbits in $\mathfrak{g}^*$ are simply connected. The condition that the orbits be integral picks out a countable set of orbits in $\mathcal{O}(G)$. We have

Theorem 4 (Borel-Weil-Bott). *All irreducible representations of a compact, connected, and simply connected Lie group G correspond to integral G-orbits of maximal dimension in $\mathfrak{g}^*$.*[1]

We note also that representations of the principal and singular series of noncompact semisimple groups correspond to integral orbits of these groups in K-representations. The problem of whether or not these representations are irreducible is not completely solved.

The question of the connection of the topologies in the sets $\hat{G}$ and $\mathcal{O}(G)$ is very interesting. It has been only partially solved, even in the case of exponential groups (namely, it is known that the mapping of $\mathcal{O}(G)$ into $\hat{G}$ is continuous).

15.4. The Method of Orbits and Quantization of Hamiltonian Mechanical Systems

The fundamental object in classical Hamiltonian mechanics is the phase space, which is a smooth symplectic manifold M. It is usually constructed as a cotangent sheaf over the configuration space N (see 15.2), although this is not required for the Hamiltonian formalism.

Physical quantities are real functions on M, and a state of the system is a point on M. The change of the system with time is described by a strictly Hamiltonian vector field, the generating function of which is called the energy of the system and

---

[1] We give the theorem in a version convenient for our purposes. For more details, see [102], [104].

is denoted by H. Thus the equation that describes the change with time of a quantity F has the form

$$\dot{F} = \{H, F\}. \tag{1}$$

A group G is a group of symmetry of a given system if it acts on M by canonical transformations.

In quantum mechanics, the rôle of the phase space is assumed by a projective space $P(V)$, where V is a certain Hilbert space. Physical quantities are self-adjoint operators on V. The value of the quantity A at the state defined by a unit vector $\xi \in V$ is a random variable with distribution function $p(t) = (E_t \xi, \xi)$ where E_t is the spectral projection measure for the operator A. Thus, in the state defined by the vector ξ, the only quantities A that possess a determined value a are those for which ξ is an eigenvector with eigenvalue a.

Change of the system with time is given by a group of unitary operators in V, which has the form

$$U(t) = e^{\frac{ith}{2\pi} H}.$$

Here h is Planck's constant and H is a certain self-adjoint operator, called the energy operator. Change of the quantity F with time is described by the equation

$$\dot{F} = \frac{ih}{2\pi} [H, F]. \tag{2}$$

A group G is a group of symmetry of the system if it acts in V by unitary transformations.

Quantization is the process of constructing from a given classical system a quantum system that corresponds to it. Unfortunately, the word "corresponding" does not here possess a definite meaning. Classical mechanics is in a certain sense an idealization of quantum mechanics—it arises by taking the limit as $h \to 0$. Therefore there obviously cannot be a unique procedure of quantization. Nevertheless, in "sufficiently good" cases (for example, in the group situation that we need), one can expect that the final result does not depend upon the method selected for quantization.

A majority of the existing methods of quantization are subsumed under the following scheme. Consider the physical quantities associated with a system. Among these we single out a certain set of *primary quantities*, forming a Lie algebra under Poisson brackets. We suppose that when we go over to quantum mechanics, the commutation relations among primary quantities are preserved in the following sense. Let h be Planck's constant and $\hat{F}$ the quantum-mechanical operator corresponding to the primary classical quantity F. Then the following relation must be satisfied:

$$\widehat{\{F_1, F_2\}} = \frac{ih}{2\pi} [\hat{F}_1, \hat{F}_2]. \tag{3}$$

§ 15. The Method of Orbits

This means that the correspondence $F \mapsto \dfrac{2\pi}{ih} \hat{F}$ is an operator representation of the Lie algebra of primary quantities. Ordinarily, constants are included among the primary quantities, and one requires that the relation

$$\hat{1} = 1 \quad \text{(identity operator)} \tag{4}$$

hold.

For systems of the type of a cotangent sheaf, one takes as primary quantities the set of linear functions of the coordinates $p_1, \ldots, p_k$ and arbitrary functions of $q_1, \ldots, q_k$.

The remaining quantities are written in the form of functions of the primary quantities. Then their quantum analogues are functions of (noncommuting) operator variables. Sometimes a definite meaning can be attributed to these expressions. Then one says that the corresponding quantity can be quantized.

To construct a representation of the Lie algebra of primary quantities, it is natural to use the fact that the correspondence $F \mapsto \xi_F$ (see 15.3) is a representation of the Lie algebra $C^\infty(M)$ of all smooth functions on M (under the Poisson bracket). Hence, setting $\hat{F} = \dfrac{ih}{2\pi} \xi_F$, we satisfy the relations (3). To satisfy relation (4) as well, we adjust the definition of $\hat{F}$ as follows:

$$\hat{F} = \frac{ih}{2\pi} \xi_F + F + \alpha(\xi_F). \tag{5}$$

Here α is a certain 1-form. Now (4) is satisfied, but relation (3) may fail.

Problem 1. In order for the operators $\hat{F}$ defined by formula (5) to satisfy the relations (3), it is necessary and sufficient that the equality

$$d\alpha = \omega \tag{6}$$

hold.

Hint. Use the equality

$$2d\alpha(\xi_1, \xi_2) = \xi_1 \alpha(\xi_2) - \xi_2 \alpha(\xi_1) - \alpha([\xi_1, \xi_2]),$$

which follows from formula (4) of 5.3.

We note that in the case of a cotangent bundle, we can take α to be the form $\sum_{j=1}^{k} p_j dq_j$. In this case, formula (5) assumes the following form:

$$\hat{F} = \frac{ih}{2\pi} \sum_{j=1}^{k} \left(\frac{\partial F}{\partial p_j} \frac{\partial}{\partial q_j} - \frac{\partial F}{\partial q_j} \frac{\partial}{\partial p_j} \right) + F - \sum_{j=1}^{k} p_j \frac{\partial F}{\partial p_j}. \tag{5'}$$

In particular, we have

$$\hat{p}_j = \frac{ih}{2\pi} \frac{\partial}{\partial q_j}, \quad \hat{q}_j = -\frac{ih}{2\pi} \frac{\partial}{\partial p_j} + q_j. \tag{7}$$

It turns out that in the case of an inexact form ω, a generalization of the above construction leads naturally to a certain one-dimensional bundle E over M. The operators $\hat{F}$ act in the space $\Gamma(E)$ of sections of this bundle.

To construct E, we introduce a covering of M by sufficiently fine open sets U_j so that all of these sets and their pairwise intersections are connected and simply connected.

Since ω is a closed form, it has the form $d\alpha_j$ in every neighborhood U_j. Here α_j is a certain 1-form in U_j. On the intersection $U_j \cap U_k$, there are defined two 1-forms: α_j and α_k. Since we have $d\alpha_j = \omega = d\alpha_k$, the difference $\alpha_j - \alpha_k$ is closed and consequently has the form

$$\alpha_j - \alpha_k = da_{jk},$$

where a_{jk} is a certain function in $U_j \cap U_k$.

We set

$$g_{jk} = e^{2\pi i a_{jk}/h}.$$

Then the operators

$$\hat{F}_j = \frac{ih}{2\pi}\xi_F + F + \alpha_j(\xi_F)$$

on the intersection $U_j \cap U_k$ enjoy the property

$$g_{jk} \circ \hat{F}_k = \hat{F}_j \circ g_{jk}. \tag{8}$$

We shall suppose that the functions g_{jk} satisfy the following relations:

$$\left. \begin{array}{l} g_{jj} \equiv 1 \quad \text{in} \quad U_j, \quad g_{jk}g_{kj} \equiv 1 \quad \text{in} \quad U_j \cap U_k, \\ g_{jk}g_{kl}g_{lj} \equiv 1 \quad \text{in} \quad U_j \cap U_k \cap U_l. \end{array} \right\} \tag{9}$$

Then we can take these functions as transition functions for the construction of a 1-dimensional bundle E over M (see 5.4). The space of this bundle is obtained by piecing together the sets $\tilde{U}_j = U_j \times \mathbf{C}$ with respect to the following equivalence:

$$\tilde{U}_j \ni (x_j, z_j) \sim (x_k, z_k) \in \tilde{U}_k \Leftrightarrow x_j = x_k, \quad z_j = g_{jk}(x_k) z_k.$$

We recall that a section of the bundle E is given by a collection of functions ϕ_j in U_j, having the property that

$$\phi_j(x) = g_{jk}(x)\phi_k(x) \quad \text{for} \quad x \in U_j \cap U_k. \tag{10}$$

The set of smooth section of E forms a linear space $\Gamma(E)$ while the operators

$$\hat{F}_E : \{\phi_j\} \to \{\tilde{F}_j \phi_j\}, \tag{11}$$

in view of (8), give a representation of the Lie algebra $C^\infty(M)$ in $\Gamma(E)$.

§ 15. The Method of Orbits

We now consider the problem of when the conditions (9) are satisfied, and how many different representations can be constructed by the above method, starting from a given form ω.

Theorem 1. *In order for the form ω to correspond to at least one bundle E, it is necessary and sufficient that the integral of the form ω on an arbitrary two-dimensional cycle be an integer multiple of the number h.*

(This assertion is a mathematical expression of the conditions of integrality that appear in quantum mechanics.)

Proof. We consider the function $c_{jkl} = a_{jk} + a_{kl} + a_{lj}$, defined in $U_j \cap U_k \cap U_l$. It follows from the definition of the functions a_{jk} that $dc_{jkl} = 0$, that is, c_{jkl} is a constant quantity. Since all a_{jk} are defined only up to constant summands, the quantities c_{jkl} are defined up to a summand of the form $c_{jk} + c_{kl} + c_{lj}$. We have obtained what is called a two-dimensional Čech cocycle on the manifold M (with respect to the covering $\{U_j\}$), defined up to a cocycle cohomologous to zero. Thus we are given a certain element of the cohomology group $H^2(M, \mathbf{C})$. It is proved in algebraic topology that this is exactly the element that corresponds to the form ω. Hence the condition of the theorem is equivalent to the requirement that all of the quantities c_{jkl} can be made multiples of h. This in turn is equivalent to the relations (9). The theorem is proved.

We now consider how much freedom there is in constructing representations in the space of sections $\Gamma(E)$ of the bundle E, if it is already known that such a bundle exists.

Problem 2. Prove that there is a one-to-one correspondence between classes of equivalent representations and elements of the cohomology group $H^1(M, \mathbf{C}^*)$.

Hint. Our construction depended upon the choice of the forms α_j and of the functions a_{jk}. Instead of the forms α_j, we could have taken forms $\alpha_j + \beta_j$, where the β_j are arbitrary closed forms. Let $\beta_j = db_j$ and $h_j = e^{2\pi i b_j/h}$. In going from α_j to $\alpha_j + \beta_j$ the transition functions g_{jk} are changed into $g_{jk}h_jh_k^{-1}$. This leads to the replacement of E by an equivalent bundle $\tilde{E}$ and to the replacement of the representation $\hat{F}_E$ by an equivalent representation $\hat{F}_{\tilde{E}} = h \circ \hat{F}_E \circ h^{-1}$. [An isomorphism of $\Gamma(E)$ onto $\Gamma(\tilde{E})$ is given by the mapping $h: \{\phi_j\} \to \{h_j\phi_j\}$]. The freedom of choice of a_{jk} leads to replacement of the functions g_{jk} by $g_{jk} \cdot z_{jk}$, where z_{jk} are certain nonzero complex numbers. These numbers satisfy the condition $z_{jk}z_{kl}z_{lj} = 1$, which is inferred from (9). Furthermore, the collections $\{z_{jk}\}$ and $\{z'_{jk}\}$ define equivalent representations if $z'_{jk} = z_{jk}w_jw_k^{-1}$.

We remark that the group $H^1(M, \mathbf{C}^*)$ is isomorphic to the group of all multiplicative mappings of the fundamental group $\pi_1(M)$ into the group $\mathbf{C}^*$ of nonzero complex numbers.

In particular, if M is simply connected, this group is trivial.

Up to this point we have said nothing about the space in which the operators $\hat{F}$ act. This space must be a Hilbert space, and the operators $\hat{F}$ for real quantities F must be self-adjoint. Furthermore, we are interested in elementary systems, for which the family of operators $\hat{F}$ is irreducible. The space $\Gamma(E)$ of all sections of the bundle E is too large to yield an irreducible representation. Computation of dimensions shows that irreducible representations must act in a space of functions (or sections of a bundle) on a manifold of dimension half the dimension of M.

Consider the case $M = T^*N$. The standard quantization uses the space $V = L^2(N)$, in which functions on M depending only on projection on N (that is, functions of the coordinates q_j) act by ordinary multiplication, and functions on M linear in the coordinates p_j act as vector fields (that is, differential operators of the first order).

It is known that every nonsingular closed 2-form ω can be locally written in the form $\omega = \sum_j dp_j \wedge dq_j$, in appropriate coordinates. However, a "separation of variables" in the large into "p" and "q" may not exist. From the geometric point of view, such a separation of variables has the following interpretation. At each point $x \in M$, one gives a *Lagrange distribution*, that is, a subspace of half dimension $L_x \subset T_x M$, on which the form ω vanishes. From the analytic point of view, this means the selection in each neighborhood $U \subset M$ of a certain maximal commutative subalgebra $P_0(U)$ in the Lie algebra $P(U)$ of all smooth functions on U (with respect to the Poisson bracket). A theorem of Frobenius gives the connection between these two approaches. This theorem asserts that every Lagrange distribution that satisfies certain conditions of integrability has the form $x \mapsto \left\{ \dfrac{\partial}{\partial p_1}, \ldots, \dfrac{\partial}{\partial p_k} \right\}$ in appropriate local coordinates. This means that vectors in L_x are tangent to the level surfaces of an arbitrary function of the coordinates $q_1, \ldots, q_k$.

Thus, in the general case, we replace the global separation of variables into the p_j and q_j by the specification of a certain integrable Lagrange distribution L on M. If such a distribution is given, we can consider the algebra $P_0(U)$ of functions in every neighborhood $U \subset M$, defined by the condition that these functions are annihilated by the vectors in L. When the neighborhood U is sufficiently small, $P_0(U)$ will be a maximal commutative subalgebra of $P(U)$. Let $\Gamma_0(E, U)$ denote the space of all sections ϕ of the bundle E over U for which

$$\hat{F}_E \phi = F \phi \quad \text{for} \quad F \subset P_0(U). \tag{12}$$

Let $\Gamma_0(E)$ denote the space of those sections of E over M whose restrictions to an arbitrary neighborhood U belong to $\Gamma_0(E, U)$.

For the reader familiar with the concept of an affine connection in a bundle, we remark that $\Gamma_0(E)$ can be defined as the space of sections annihilated by covariant differentiation along an arbitrary vector in L. In fact, the operator $\hat{F}$ enjoys all of the properties of a covariant derivative along the field ξ_F.

We shall suppose the following about all primary quantities F. First, the operators $\hat{F}_E$ carry the space $\Gamma_0(E)$ into itself (this requirement is equivalent to the condition $\{F, P_0(U)\} \subset P_0(U)$ for an arbitrary neighborhood U). Second, for real F, the operators $\hat{F}_E$ are self-adjoint under a certain scalar product in $\Gamma_0(E)$.

Then for the space V, we can take the completion of $\Gamma_0(E)$ in this scalar product.

We obtain an important generalization of the above construction by "going over to the complex domain". This means that we define the Lagrange distribution

§ 15. The Method of Orbits

L in such a way that L_x is a complex subspace of half dimension in the complex hull of the space T_xM. Let L be an integrable complex Lagrange distribution with the added condition that $L+\bar{L}$ is also an integrable distribution. It turns out that L has locally the following form. There exists a local system of coordinates

$$q_1, \ldots, q_r, \quad p_1, \ldots, p_r, \quad z_1, \ldots, z_{k-r},$$

where p_j, q_j, are real and z_j are complex, such that the form ω has the form

$$\omega = \sum_{j=1}^{r} dp_j \wedge dq_j + \frac{i}{2} \sum_{j=1}^{k-r} dz_j \wedge d\bar{z}_j,$$

and L is generated by the vectors

$$\frac{\partial}{\partial p_1}, \ldots, \frac{\partial}{\partial p_r}, \quad \frac{\partial}{\partial \bar{z}_1}, \ldots, \frac{\partial}{\partial \bar{z}_{k-r}}.$$

In this case, the algebra $P_0(U)$ consists of functions of $q_1, \ldots, q_r, z_1, \ldots, z_{k-r}$ that are analytic in the variables z_j. The space $\Gamma_0(E)$ can be interpreted as the space of sections that are constant along certain real directions and are analytic along certain complex directions.

A further generalization of this construction is possible. Namely, one can go from $\Gamma_0(E)$ to spaces of higher cohomology of the sheaf of germs of sections of $\Gamma_0(E)$. In this connection, see [104], [121], [129].

We now recall that given a form ω on M, satisfying the integral conditions of theorem 1, we constructed a whole family of representations $\hat{F}_E$, indexed by the elements of the group $H^1(M, \mathbf{C}^k)$. If the collection $\{\phi_j\}$ gives a section of the bundle E, then the collection $\{|\phi_j|^2\}$ gives a section of a certain different bundle E' over M (with transition functions $g'_{jk} = |g_{jk}|^2$). Ordinarily we define the scalar product in $\Gamma_0(E)$ so that the scalar square of the section $\{\phi_j\}$ depends only upon the section $\{|\phi_j|^2\}$ of the bundle E'.

Therefore, if the desired scalar product exists for $\hat{F}_E$, it also exists for all representations obtained from $\hat{F}_E$ by the action of the subgroup $H^1(M, \mathbf{T})$. There is reason to believe (and this has in fact been proved in many special cases) that on other bundles, the required scalar product does not exist. Supposing that this has been proved, we come to the following result.

Theorem 2. *A classical system (M, ω) admits a quantization of the type described above if and only if the form ω has integrals that are multiples of h on an arbitrary two-dimensional cycle in M. All such quantizations are indexed by elements of the group $H^1(M, \mathbf{T})$.*

We remark that the group $H^1(M, \mathbf{T})$ coincides with the group of characters (that is, multiplicative mappings into the circle $\mathbf{T}$) of the fundamental group $\pi_1(M)$ of the manifold M.

Let G be a Lie group. We consider what representations of G can be obtained by starting with G-elementary classical systems as listed in 15.2 and then quantizing. We shall be concerned only with quantizations for which the primary quantities include the generating functions of the vector fields that correspond to elements of the Lie algebra $\mathfrak{g}$. Under this hypothesis, we can construct a representation of G by the formula

$$T(\exp X) = e^{-2\pi i \hat{X}/h},$$

where exp is the canonical mapping of the Lie algebra into the group (the exponential). The relation (3) and the self-adjointness of the operators $\hat{X}$ guarantee that the condition

$$T(g_1 g_2) = T(g_1) T(g_2)$$

holds and that the operators $T(g)$ are unitary in a certain neighborhood of the identity. This "local" representation admits a unique extension to a representation of the group G if G is simply connected. In the contrary case, there can arise a many-valued representation, which will be single-valued on the simply connected covering group $\tilde{G}$ of the group G.

From what was said in 15.2, it follows that all of the classical systems we need are integral orbits Ω in $\mathfrak{g}^*$.

Let Ω be such an orbit and E a bundle over Ω. In order to construct a space $\Gamma_0(E)$ invariant under G, we must choose a G-invariant integrable Lagrange distribution L on Ω.

Such distributions can be described simply.

Problem 3. Suppose that $\phi \in \Omega$ and $\dim \Omega = 2k$. There exists a one-to-one correspondence between real (resp. complex) integrable G-invariant Lagrange distributions on Ω and subalgebras of codimension k in $\mathfrak{g}$ (resp. $\mathfrak{g}_c$) that are subordinate to ϕ.

Hint. The desired correspondence between distributions L and subalgebras $\mathfrak{h}$ is established by the formula

$$L_\phi = p(\mathfrak{h}).$$

Here p is the natural projection of $\mathfrak{g}$ (resp. $\mathfrak{g}_c$) onto $T_\phi \Omega \approx \mathfrak{g}/\mathfrak{g}_\phi$ (resp. onto the complex hull of $T_\phi \Omega$), which arises from the action of G on Ω. The Lagrange property for L is equivalent to the assertion that $\mathfrak{h}$ be subordinate to the functional ϕ. The integrability of L is equivalent to the assertion that $\mathfrak{h}$ is a subalgebra.

The reader has no doubt already noted the analogy between the operation of quantization described here and the construction of a representation on an orbit in the preceding section. The mechanical interpretation described here allows us to give a "physical sense" to the construction of a representation on an orbit described in 15.3. Furthermore, it allows us to use other methods of quantum mechanics in the theory of representations. For example, it would be most interesting to obtain a direct proof of the universal form (Φ) for characters of irreducible representations (see 15.6 *infra*) with the help of the apparatus of Feynman integrals. In the simplest examples, such a proof has been obtained.

§ 15. The Method of Orbits

15.5. Functorial Properties of the Correspondence between Orbits and Representations

Let G as usual be a connected and simply connected Lie group, and let H be a closed subgroup of G. The following two problems play a large rôle in the theory of representations.

a) Let T be a given irreducible representation of the group G. Find the decomposition into irreducible components of its restriction to the subgroup H.

b) Let U be an irreducible representation of the subgroup H. Find the decomposition into irreducible components of the induced representation $T = \text{Ind}(G, H, U)$.

In a certain sense, these two problems are dual to each other. (For finite and compact groups, this duality is expressed by the theorem of Frobenius—see 13.5.) We note also that the decomposition into irreducible components of the tensor product of two irreducible representations of the group G is a special case of the first problem. A special case of the second problem is the decomposition into irreducible components of representations that are realized in sections of G-bundles over homogeneous spaces (in particular, in functions, vector fields, differential forms, and so on).

Let $\mathfrak{g}$ be the Lie algebra of the group G and $\mathfrak{h}$ the subalgebra corresponding to the subgroup H. Every functional $F \in \mathfrak{g}^*$ can be restricted to $\mathfrak{h}$. We obtain in this way a canonical projection $p: \mathfrak{g}^* \to \mathfrak{h}^*$.

It is natural to expect that if there is a sufficiently good correspondence between the sets $\hat{G}$ and $\mathcal{O}(G)$, then the operations of restriction to a subgroup and induction from a subgroup will be connected with the projection $p: \mathfrak{g}^* \to \mathfrak{h}^*$. Up to the present time, an exact description of this connection is known only for nilpotent groups. In this case we have

Theorem 1. *Let G be a connected and simply connected nilpotent Lie group, and let H be a connected closed subgroup of G. Then the construction described in 15.3 yields a one-to-one correspondence between the sets $\hat{G}$ and $\mathcal{O}(G)$, $\hat{H}$ and $\mathcal{O}(H)$.*

If an irreducible unitary representation T of the group G corresponds to an orbit $\Omega \in \mathcal{O}(G)$, then its restriction to H is decomposed into the direct integral of irreducible unitary representations of the subgroup H corresponding to those orbits $\omega \in \mathcal{O}(H)$ that belong to $p(\Omega)$.

If an irreducible unitary representation U of the subgroup H corresponds to the orbit $\omega \in \mathcal{O}(H)$, then the induced representation $T(G, H, U)$ is decomposed into the direct integral of irreducible representations of G that correspond to those orbits $\Omega \in \mathcal{O}(G)$ that have nonvoid intersection with $p^{-1}(\omega)$.

Problem 1. If the subgroup H is a normal subgroup, then the last condition is equivalent to the inclusion $\Omega \subset p^{-1}(\omega)$.

Hint. In this case, the set $p^{-1}(\omega)$ is G-invariant.

Suppose that the representation U is one-dimensional. Then the orbit ω corresponding to it consists of a single point $f \in \mathfrak{h}^*$. Theorem 1, applied to this special case, yields

Corollary[1]. *In order for a monomial representation to be irreducible, it is necessary and sufficient that the set $p^{-1}(f)$ be contained entirely in a single G-orbit Ω.*

Thus condition (5) of 15.3 is satisfied for the subalgebra $\mathfrak{h}$ and an arbitrary functional $F \in p^{-1}(f)$.

We point out the scheme of the proof of theorem 1. (The reader will find the details in [99].) First of all, it suffices to prove the theorem for the case where the group H has codimension 1 in G. The general case is reduced to this one by considering the chain of subgroups

$$H = H_0 \subset H_1 \subset \ldots \subset H_{n-1} \subset H_n = G,$$

the dimension of which increases by unity at each step.

Further, if $\dim G - \dim H = 1$, then H is a normal subgroup of G and the group G is a semidirect product of H and a certain subgroup S that is isomorphic to $\mathbf{R}$. Let X_0 denote a basis element of the Lie algebra $\mathfrak{s}$ of the group S, and let F_0 be a functional in $\mathfrak{h}^\perp$ that is equal to 1 at X_0.

Problem 2. Let Ω be an orbit of the group G in $\mathfrak{g}^*$ and let p be projection of $\mathfrak{g}^*$ onto $\mathfrak{h}^*$. Then two cases are possible.

1) The set $\omega = p(\Omega)$ is an H-orbit in $\mathfrak{h}^*$, and the inverse image $p^{-1}(\omega)$ is the union of the family of G-orbits $\Omega_t = \Omega + tF_0$, $t \in \mathbf{R}$.

2) The set $p(\Omega)$ is the union of the family ω_t, $t \in \mathbf{R}$ of H-orbits in $\mathfrak{h}^*$, the inverse images $p^{-1}(\omega_t)$ are wholly contained in Ω, and the group S acts in $\mathfrak{h}^*$ in such a way that $K(\exp \tau X_0)\omega_t = \omega_{t+\tau}$.

Hint. Prove that the intersection of the orbit Ω with an arbitrary line of the form $F + \mathfrak{h}^\perp$ either consists of a single point or contains all of this line. This follows from the fact that the action of G in $\mathfrak{g}^*$ is described in canonical coordinates by polynomial functions.

We suppose now that theorem 1 has been proved for all groups of dimension $< n$. Suppose that $\dim G = n$, so that $\dim H = n - 1$.

Problem 3. Let T be an irreducible representation of the group G. Then two cases are possible.

1) The restriction U of the representation T to H is an irreducible representation of H. The representation $\mathrm{Ind}(G, H, U)$ is decomposed into the direct integral of irreducible representations T_t, $t \in \mathbf{R}$, each of which coincides with U on H.

2) The restriction of T to H is decomposed into the direct integral of irreducible representations U_t, $t \in \mathbf{R}$, of the subgroup H. Each of the representations $\mathrm{Ind}(G, H, U_t)$ is equivalent to T. The group S acts on U_t by the formula

$$U_t(\exp X_0 \, h \exp(-\tau X_0)) = U_{t+\tau}(h).$$

Hint. Use the inductive hypothesis and the results of 13.3.

Finally, using the explicit form of the correspondence between orbits and representations (with the help of the construction of 15.3), one can establish that

---

[1] As L. Pukanszky proved in [127], this is true for all exponential groups.

the two cases figuring in problem 2 correspond precisely to the two cases of problem 3. This completes the proof.

A very interesting problem is the clarification of the fundamental properties of the correspondence between orbits and representations for other classes of groups. In the first place, it is natural to pose this question for groups for which the correspondence between $\hat{G}$ and $\mathcal{O}(G)$ is known (see theorems 3 and 4 of 15.3). As simple examples show, a literal translation of theorem 1 for other groups ceases to be valid. It seems likely that the character of the correspondence between the operations of restriction and induction of representations on the one hand, and projection or formation of inverse images of orbits, on the other, depends upon topological properties of the orbit Ω and its decomposition into inverse images of points under the projection p.

15.6. The Universal Formula for Characters and Plancherel Measures

One of the important problems of the theory of group representations is obtaining explicit formulas for generalized characters of irreducible representations (see 11.2).

At the present time, there are an enormous number of results, solving this problem for specific groups and representations or for whole classes of groups and representations. However, these results are outwardly very different for different types of groups. The method of orbits indicates an approach to solving this problem for all Lie groups. The idea of this approach consists in considering a generalized function I_Ω on a Lie group G which is defined on a G-orbit Ω in $\mathfrak{g}^*$ in the following way.

Let V be an open region in G, covered by a canonical system of coordinates, and let U be its inverse image in the Lie algebra $\mathfrak{g}$.

For U we can take the set of those elements $X \in \mathfrak{g}$ for which the pure imaginary eigenvalues of the operator $\operatorname{ad} X$ (if there are such) do not exceed π in absolute value.

In particular, for exponential groups, there are no nonzero pure imaginary eigenvalues, and we can set $U = \mathfrak{g}$, $V = G$.

The mapping $\exp: U \to G$ is a diffeomorphism and allows us to carry over basic and generalized functions from U to V and back. It is known that the complement of V has measure 0 in G if G is solvable, compact, or complex semisimple.

We define the generalized function I_Ω on V by the formula

$$\langle I_\Omega, \phi \rangle = \int_\Omega \left(\int_U \phi(\exp X) e^{2\pi i \langle F, X \rangle} dX \right) d\beta_\Omega(F). \tag{1}$$

Here dX is ordinary Lebesgue measure on $\mathfrak{g}^*$, and β_Ω is the measure on the orbit Ω defined by the $2k$-form

$$\frac{1}{k!} B_\Omega \wedge \ldots \wedge B_\Omega \quad (2k \text{ factors}). \tag{2}$$

It turns out that the function I_Ω is closely connected with the generalized character χ_Ω of a certain unitary representation T of the group G. In many cases, the connection between I_Ω and χ_Ω can be described in the following terms. *There exists a function $p_\Omega \in C^\infty(V)$ invariant under inner automorphisms, assuming the value 1 at the group identity, different from zero on V and having the property that*

$$\chi_\Omega = p_\Omega^{-1} I_\Omega. \tag{Φ}$$

We call this formula the *universal formula for a character* or, briefly, the formula (Φ).

More precisely, we shall say that *formula (Φ) holds for a representation T if the two following conditions are satisfied.*

1) Let ϕ be a function in $C_0^\infty(G)$ that lies in the domain of definition of the character χ_Ω. Then the integral

$$\int_\Omega \left(\int_U \phi(\exp X) p_\Omega^{-1}(\exp X) e^{2\pi i \langle F, X \rangle} dX \right) d\beta_\Omega(F) \tag{3}$$

converges and its value is equal to $\langle \chi_\Omega, \phi \rangle = \operatorname{tr} T(\phi)$.

2) *If the integral (3) converges for a function $\phi \in C_0^\infty(G)$ and the operator $T(\phi)$ is positive, then ϕ lies in the domain of definition of χ_Ω.*

When these conditions are satisfied, we shall say that the representation T corresponds to the orbit Ω, and we shall denote it by T_Ω. This correspondence between orbits and representations appears to coincide with the correspondence described in 15.3. Up to the present time, a complete proof of this assertion has not been obtained.

We remark that formula (Φ) clarifies the nature of the generalized function χ_Ω, which turns out to be closely connected with the geometry of the orbit Ω. Thus, if Ω is compact, χ_Ω is an ordinary smooth function (this case corresponds to a finite-dimensional representation). Suppose next that Ω is a cylinder. This means that if F belongs to Ω, then $F + L$ is also contained in Ω, where L is a certain subspace in $\mathfrak{g}^*$. In this case, χ_Ω contains a factor of the type of a δ-function. In particular, if $L = \mathfrak{h}^\perp$ for a certain subalgebra $\mathfrak{h} \subset \mathfrak{g}$, then χ_Ω is concentrated on $\exp \mathfrak{h}$. If the orbit Ω is a closed submanifold in $\mathfrak{g}^*$, then the integral (3) converges for an arbitrary function $\phi \in C_0^\infty(G)$. In the contrary case, the convergence of this integral is an additional necessary condition for the function ϕ to belong to the domain of definition of χ_Ω.

Often this condition can be stated as follows. The Fourier transform of the function $X \to p_\Omega(\exp X)^{-1} \phi(\exp X)$ vanishes on the set $\partial \Omega$, the boundary of Ω (coinciding with the boundary of the closure of Ω). See [102].

Formula (Φ) is also useful in studying the topology of the set $\hat{G}$ and in comparing it with the topology of $\mathcal{O}(G)$.

The validity of formula (Φ) is known at the present time for a fairly wide class of representations. In particular, this formula has been proved for all representations of compact simply connected groups, exponential groups, the group of real matrices of the second order, and representations of the principal series of noncompact semisimple groups (see [72], [73], [102], [103], [128]).

§ 15. The Method of Orbits

For representations corresponding to orbits of maximal dimension, it has turned out that a single universal function p_Ω can be selected:

$$q(\exp X) = \det \mathscr{S}(\operatorname{ad} X). \tag{4}$$

Here $\mathscr{S}$ denotes the function

$$\mathscr{S}(t) = \frac{\operatorname{sh}(t/2)}{t/2} = \sum_{k=0}^{\infty} \frac{(t/2)^{2k}}{(2k+1)!}. \tag{5}$$

The problem of deriving formula (Φ) from the "physical" considerations of 15.4 is very interesting. One may hope that the method of quantization with the help of the continuous integrals proposed by Feynman will allow us to give a unified proof of the results set forth above. Also one may hope for hints as to the necessary changes in formula (Φ) in the cases where it is invalid (for example, for representations of the degenerate series of noncompact semisimple groups).

We now proceed to the computation of the Plancherel measure for Lie groups. We recall (see 12.4) that Plancherel's formula for unimodular groups has the form

$$\int_G |\phi(g)|^2 dg = \int_{\hat{G}} \operatorname{tr}[T(\phi)T(\phi)^*] d\mu([T]), \tag{6}$$

where $[T]$ denotes the equivalence class of the representation T.

Let $A(G)$ be the subspace of $C(G)$ generated by functions of the form $\phi = \phi_1 * \phi_2$, where

$$\phi_i \in L^1(G, dg) \cap L^2(G, dg).$$

Problem 1. Prove that for an arbitrary function $\phi \in A(G)$, the inversion formula

$$\phi(g) = \int_G \operatorname{tr}[T_\lambda(\phi) T_\lambda(g)^*] d\mu(\lambda) \tag{7}$$

holds.

Hint. It suffices to verify this equality for $g = e$. In this case it assumes the form

$$\phi(e) = \int_G \operatorname{tr} T_\lambda(\phi) d\mu(\lambda), \tag{8}$$

and for functions of the form $\phi = \phi_1 * \phi_2$, it follows from (6).

The equality (8) may be thought of as a decomposition of the δ-function on the group G in terms of characters of irreducible representations.

In the case where formula (Φ) is valid, this equality takes on a simple and obvious sense: it is a decomposition of Lebesgue measure on $\mathfrak{g}^*$ in terms of canonical measures on orbits.

In fact, let us suppose that there is a one-to-one correspondence between orbits and representations and that for representations corresponding to orbits of maximal dimension, one can choose the universal function (4) for the function p_Ω. (All of these hypotheses are satisfied, for example, for exponential groups.)

If the group G is unimodular, then Lebesgue measure dF on $\mathfrak{g}^*$ is invariant under a K-representation.

We suppose further that the space of orbits $\mathcal{O}(G)$ enjoys T_0 separation. Then the measure dF can be uniquely written in the form

$$dF = \int_{\mathcal{O}(G)} \beta_\Omega d\mu(\Omega).$$

Going to the Fourier transform, we obtain the equality

$$\delta(X) = \int_{\mathcal{O}(G)} I_\Omega(\exp X) d\mu(\Omega). \tag{9}$$

Since orbits of maximal dimension fill up a set of full measure in $\mathfrak{g}^*$, the integration in (9) can be carried out only on the set $\mathcal{O}_{\max}(G)$ of orbits of maximal dimension. Finally, using formula (Φ) and the equality $q(e) = 1$, we arrive at the relation

$$\delta(g) = \int_{\mathcal{O}_{\max}(G)} \chi_\Omega(g) d\mu(\Omega). \tag{10}$$

This obviously gives the required decomposition (8).

In the case where the topological space $\mathcal{O}(G)$ is a T_0 space, the subset $\mathcal{O}_{\max}(G)$ is a smooth manifold. We can define a local chart on $\mathcal{O}_{\max}(G)$ in the following way.

Let $\lambda_1, \ldots, \lambda_k, \phi_1, \ldots, \phi_{2n}$ be a system of local coordinates on $\mathfrak{g}^*$ (not necessarily linear).

Let us suppose that the equations

$$\phi_1 = \phi_2 = \ldots = \phi_{2n} = 0 \tag{11}$$

define a submanifold S in $\mathfrak{g}^*$, transversal to the orbits of maximal dimension. Then, in a sufficiently small neighborhood, S intersects each orbit passing through this neighborhood in exactly one point. Consequently, we can identify a certain domain in $\mathcal{O}_{\max}(G)$ with a neighborhood on S. As local coordinates in this neighborhood, we may take the restrictions of the functions $\lambda_1, \ldots, \lambda_k$ to S.

Another method of introducing local coordinates on $\mathcal{O}_{\max}(G)$ is to consider G-invariant functions on $\mathfrak{g}^*$. Both methods are closely connected with each other. Indeed, the introduction of coordinates by the first method is one of the most convenient means of explicitly constructing G-invariant functions on $\mathfrak{g}^*$.

Example. Let $G = GL(n, \mathbb{C})$, $\mathfrak{g} \approx \mathfrak{g}^* = \mathrm{Mat}_n \mathbb{C}$. As we saw in 15.1, G-orbits in $\mathfrak{g}^*$ are classes of similar matrices. For the manifold S, it is convenient to take the set of matrices of the form

$$\begin{pmatrix} 0 & 1 & 0 & \ldots & 0 & 0 \\ 0 & 0 & 1 & \ldots & 0 & 0 \\ \cdot & \cdot & \cdot & & \cdot & \cdot \\ 0 & 0 & 0 & \ldots & 0 & 1 \\ p_1 & p_2 & p_3 & \ldots & p_{n-1} & p_n \end{pmatrix},$$

and for coordinates on S we take the parameters $p_1, \ldots, p_n$.

§ 15. The Method of Orbits

Problem 2. Prove that every orbit in $\mathcal{O}_{\max}(G)$ intersects S in exactly one point. The parameters $p_1, \ldots, p_n$ coincide (up to their sign) with the coefficients of the characteristic polynomial of an arbitrary matrix from the corresponding orbit.

Hint. Verify that a matrix A belongs to an orbit of maximal dimension if and only if there exists a vector ξ such that $\xi, A\xi, \ldots, A^{n-1}\xi$ are linearly independent.

We also need the concept of the Pfaffian of a skew-symmetric matrix A. Let V be a $2n$-dimensional space with a basis $\xi_1, \ldots, \xi_{2n}$. Let $\tilde{A} = \sum_{k,j} a_{kj} \xi_k \wedge \xi_j$ be a homogeneous element of the second degree in the exterior algebra over V. Then the n-th exterior power of $\tilde{A}$ has the form

$$\tilde{A} \wedge \tilde{A} \wedge \ldots \wedge \tilde{A} = c \cdot \xi_1 \wedge \xi_2 \wedge \ldots \wedge \xi_{2n}, \tag{12}$$

since the space $\wedge^{2n} V$ is one-dimensional. The coefficient c in the equality (12) is a polynomial of degree n in the coefficients a_{kj}. It is called the *Pfaffian* of the matrix A and is denoted by $\operatorname{Pf} A$.

Problem 3. Prove the identity

$$(\operatorname{Pf} A)^2 = \det A.$$

Hint. Use the fact that every element $\tilde{A} \in \wedge^2 V$ has the form

$$\tilde{A} = \sum_{k=1}^{n} \lambda_k \xi_k \wedge \xi_{n+k}$$

in an appropriate basis.

We return to the computation of the Plancherel measure. Under the hypotheses set down above, and with the coordinates $\lambda_1, \ldots, \lambda_k$ on $\mathcal{O}_{\max}(G)$, we have the following result.

Let $F(\lambda)$ be a point of the orbit with parameters $\lambda_1, \ldots, \lambda_k$ lying on the manifold S. The values of the Poisson brackets of the functions $\phi_1, \ldots, \phi_{2n}$ at the point $F(\lambda)$ form a skew-symmetric matrix Φ:

$$\Phi_{ij} = \{\phi_i, \phi_j\}(F(\lambda)). \tag{13}$$

Problem 4. The Plancherel measure for the group G is given by the formula

$$\mu = \rho(\lambda, 0) \operatorname{Pf} \Phi(\lambda) d\lambda_1 \wedge \ldots \wedge d\lambda_k. \tag{14}$$

Here $\rho(\lambda, \phi)$ is the density of Lebesgue measure dF on $\mathfrak{g}^*$ in the coordinates λ and ϕ:

$$dF = \rho(\lambda, \phi) d\lambda_1 \wedge \ldots \wedge d\lambda_k \wedge d\phi_1 \wedge \ldots \wedge d\phi_{2n}. \tag{15}$$

Hint. Compare the values of the differential forms defining the measures dF, β_Ω and μ at the point $F(\lambda)$.

Formula (14) assumes its simplest form when the coordinates $\lambda_1, \ldots, \lambda_k$ and $\phi_1, \ldots, \phi_{2n}$ are linear, that is, when they are vectors in $\mathfrak{g}$. In this case we have $\rho(\lambda, \phi) = \text{const}$ and

$$\Phi_{ij}(\lambda) = \langle F(\lambda), [\phi_i, \phi_j] \rangle = \sum_{m=1}^{k} \lambda_m c_{ij}^m,$$

where c_{ij}^m are the structure constants of the Lie algebra. The required measure μ assumes the form

$$\mu = P(\lambda) d\lambda_1 \wedge \ldots \wedge d\lambda_k, \tag{16}$$

where P is a homogeneous polynomial of degree n in $\lambda_1, \ldots, \lambda_k$, and whose coefficients are explicitly expressed in terms of the structure constants of the algebra $\mathfrak{g}$.

It is interesting to note that a formal application of the formula (16) to complex and compact semisimple groups leads to the correct expression for Plancherel's measure if we replace integration by summation for those parameters λ that have to be integers.

No less interesting is the fact that for real semisimple groups, this procedure leads to a correct answer only for the discrete series of representations.

15.7. Infinitesimal Characters and Orbits

In certain cases, the method of orbits allows us to calculate explicitly the infinitesimal character of an irreducible representation.

We recall (see 11.3) that the infinitesimal character of a representation T is a homomorphism λ_T of the center $Z(\mathfrak{g})$ of the algebra $U(\mathfrak{g})$ into the field of complex numbers, connected with the representation T by the formula

$$T(X) = \lambda_T(X) \cdot 1. \tag{1}$$

We also use theorem 2 of 10.4, asserting that the linear spaces $Z(\mathfrak{g})$ and $I(\mathfrak{g}) \subset S(\mathfrak{g})$ (the space of G-invariant elements in $S(\mathfrak{g})$) are isomorphic.

To every orbit Ω of the group G in $\mathfrak{g}^*$, we assign a linear functional λ_Ω on $Z(\mathfrak{g})$ in the following way. We can identify the algebra $S(\mathfrak{g})$ with the algebra $P(\mathfrak{g}^*)$ of polynomial functions on $\mathfrak{g}^*$. Namely, to every $X \in \mathfrak{g}$, we assign the following linear functional on $\mathfrak{g}^*$:

$$F \mapsto 2\pi i \langle F, X \rangle. \tag{1'}$$

We extend this mapping by linearity and multiplicativity to an isomorphism of $S(\mathfrak{g})$ onto $P(\mathfrak{g}^*)$. The subspace $I(\mathfrak{g})$ will then be realized by polynomials that are constant on orbits. We define λ_Ω as the functional which assigns to an invariant polynomial its value on the orbit Ω.

§ 15. The Method of Orbits

We note that this functional is not *a priori* multiplicative, since we do not know whether or not the isomorphism of $Z(\mathfrak{g})$ and $I(\mathfrak{g})$ is always an algebra isomorphism.[1]

We will say that a representation T corresponds to an orbit Ω if

$$\lambda_T = \lambda_\Omega \tag{2}$$

(which implies of course that λ_Ω is a multiplicative functional).

The correspondence between orbits and representations obtained in this fashion is cruder than the correspondences presented *supra*. In fact, it is not the orbit itself that is important for us now, but the common level set of the G-invariant polynomials in which the orbit lies.

For algebraic Lie groups (that is, for subgroups of $GL(n, \mathbf{R})$ or $GL(n, \mathbf{C})$ defined by algebraic equations) one can prove that orbits of maximal dimension are "almost distinguished" by G-invariant rational functions on $\mathfrak{g}^*$ (that is, the corresponding level sets consist of a finite number of orbits).

For complex semisimple and nilpotent groups, we have a stronger assertion. Almost all common level sets of G-invariant polynomials are G-orbits.

Thus in general an orbit is not defined by the values assumed on it by G-invariant polynomials. This is a geometric illustration of the fact that an irreducible representation is not in general defined by its infinitesimal character.

The validity of formula (2) is closely connected with the validity of formula (Φ) in 15.6. Namely, suppose that formula (Φ) holds for all orbits of maximal dimension with the universal function q for the function p_Ω. Then the isomorphism between $Z(\mathfrak{g})$ and the invariant polynomials on $\mathfrak{g}^*$ is an algebra isomorphism and admits a simple analytic interpretation. We consider the transformation that maps a smooth function ϕ with compact support on G into the function $\tilde{\phi}$ on $\mathfrak{g}^*$ by the formula

$$\tilde{\phi}(F) = \int_\mathfrak{g} q^{-1}(\exp X)\phi(\exp X)e^{2\pi i \langle F, X\rangle} dX . \tag{3}$$

It turns out *that under this transformation, Laplace operators on G go into operators of multiplication by G-invariant polynomials on $\mathfrak{g}^*$*.

If a representation T corresponds to an orbit Ω in the sense of 15.6, then its infinitesimal character coincides with λ_Ω.

Conversely, suppose that Laplace operators on the group G enjoy the property stated above. Then all of the functionals λ_Ω are multiplicative, and both sides of the formula (Φ) satisfy one and the same system of differential equations.

Thus each of the formulas (2) and (Φ) can be used to prove the other.

For compact groups, formula (2) has been known for a long time, and has served as the basis for proving formula (Φ) (see [102]).

For exponential groups, on the other hand, formula (2) was obtained in [72] as a consequence of formula (Φ).

---

[1] A proof of this fact has recently been obtained by M. Duflo [74].

We note some unsolved problems that arise naturally in studying infinitesimal characters by the method of orbits.

1. Generalize the concept of infinitesimal character and the equality (2) to the center of the Lie skew field $D(\mathfrak{g})$, constructed from the Lie algebra $\mathfrak{g}$ (see 10.4).

2. Give a purely algebraic proof of the property of Laplace operators on a Lie group that is described above.

Third Part. Various Examples

§ 16. Finite Groups

16.1. Harmonic Analysis on the Three-Dimensional Cube

In a mathematical institute, there is a model of a cube. One of the mathematicians at the institute numbers the faces of the cube with the integers 1, 2, 3, 4, 5, 6. A second mathematician comes to the institute on the following day and replaces each number by the arithmetic mean of the adjacent numbers. The first mathematician notices this on the third day and responds by also replacing each number by the arithmetic mean of the adjacent numbers. What numbers will appear on the faces of the cube after a month, under the hypothesis that the two mathematicians go to the institute on alternate days?

The solution of this rather frivolous problem is a model of the application of the theory of representations to diverse problems of mathematics, mechanics, and physics that possess symmetry of one kind or another.

The first step of the solution is to state the problem in the language of representation theory. Let G be the group of rotations of the cube, and W_F the space of functions on the set of faces of the cube. In the space W_F, there is defined a natural representation T of the group G. Let H_F be a subgroup of G, consisting of all rotations of the cube that carry one face onto itself. Then T is defined as the representation of G induced by the identity representation of the subgroup H_F (cf. 13.1).

We define an operator L in the space W_F, carrying a function $f(x)$ into the function defined by

$$(Lf)(x) = \tfrac{1}{4} \sum f(y),$$

where the summation is extended over the four faces y that are adjacent to the face x.

The operator L commutes with the action of G. (Think over this assertion and convince yourself that it is correct.) Therefore we can apply our general theorems on intertwining operators to it. We decompose the representation T into irreducible components. The intertwining number $c(T, T)$ is equal to the number of classes of pairs of faces of the cube (see 13.5). Obviously there are three such classes: two faces that are the same; adjacent faces; and opposite faces. It follows that T is the sum of three pairwise inequivalent representations T_1, T_2, and T_3.

Now it is easy to single out three invariant subspaces in W_F. These are: W_1, the space of constant functions; W_2, the space of even functions the sum of whose values is zero; and the space W_3 of odd functions. (Here evenness or oddness means that the function assumes the same values on opposite faces or changes its sign on opposite faces.) The operator L is a multiple λ_i of the identity operator on each W_i, $i=1, 2, 3$. It is plain that $\lambda_1=1$. For a representative of W_2 we can take a function equal to 1 on one pair of opposite faces, equal to -1 on another pair, and equal to 0 on the third pair. It follows from this that $\lambda_2 = -\frac{1}{2}$. Finally, as a representative of W_3, we take a function equal to 1 on one face, equal to -1 on the opposite face, and equal to 0 on the remaining faces. From this, we obtain $\lambda_3 = 0$.

The operator we are concerned with is L^{30}. It yields multiplication by 1, $(-\frac{1}{2})^{30}$, and 0, on W_1, W_2, and W_3 respectively. Hence we may assert with sufficiently high precision that all of the required numbers are equal to 3.5, since the vector (3.5, 3.5, 3.5, 3.5, 3.5, 3.5) is the projection of the vector $\xi = (1, 2, 3, 4, 5, 6)$ onto the subspace of constant functions. Since the projection of ξ on W_2 has length not greater than

$$\|\xi - \xi_1\| = \sqrt{17.5},$$

the possible error does not exceed

$$\frac{\sqrt{17.5}}{2^{30}} < 4.2 \times 10^{-9}.$$

The reader may if he likes repeat the above arguments, replacing the cube by an octahedron or a dodecahedron. In the first case, the eigenvalues of the operator L are equal to 1, -1, 0. In the second case, they are equal to 1, $-1/5$, $1/\sqrt{5}$, $-1/\sqrt{5}$.

Considering other homogeneous spaces for the group of rotations of the cube, one can construct a complete system of representations of this group. Define the following sets:

X_F is the set of faces of the cube;

X_V is the set of vertices of the cube;

X_D is the set of principal diagonals of the cube;

X_T is the set of regular tetrahedra inscribed in the cube;

X_C is the set of centers of the cube.

We denote the spaces of functions on these sets by W_F, W_V, W_D, W_T, W_C respectively. Their dimensions are 6, 8, 4, 2, 1 respectively. We suggest that the reader verify the following table of intertwining numbers for the corresponding representations T_F, T_V, T_D, T_T, and T_C:

§ 16. Finite Groups

	T_F	T_V	T_D	T_T	T_C
T_F	3	2	1	1	1
T_V	2	4	2	2	1
T_D	1	2	2	1	1
T_T	1	2	1	2	1
T_C	1	1	1	1	1

Let us see what inferences one can draw from the form of this table. First of all, from the form of the diagonal numbers, it follows that T_C is irreducible (this is also clear from the fact that it is one-dimensional). Each of T_D and T_T is the sum of two irreducible components. The representation T_F is the sum of three irreducible components. For the representation T_V, there are two possibilities. It is either the sum of four pairwise inequivalent representations, or it is the sum of two equivalent representations. The second possibility cannot arise, however, since in this case the intertwining number $c(T_V, T_C)$ could not be equal to 1.

It is also obvious from the table that T_C enters in all of the other representations with multiplicity 1. We have seen above that T_F is the sum of representations $T_1 \approx T_C$, T_2, and T_3.

Let us write $T_D = T_1 + T_4$, $T_T = T_1 + T_5$. Since

$$c(T_D, T_F) = c(T_D, T_T) = c(T_F, T_T) = 1,$$

the representations T_1, T_2, T_3, T_4, and T_5 are pairwise inequivalent. It is also obvious from the table that T_V contains the representations T_1, T_4, T_5 and one of the representations T_2 and T_3. Since $\dim T_1 = 1$, $\dim T_2 = 2$, $\dim T_3 = 3$, $\dim T_4 = 3$, $\dim T_5 = 1$, it follows that

$$T_V = T_1 + T_3 + T_4 + T_5.$$

There are five conjugacy classes of elements of the group G: the identity element; rotations around vertices; rotations around the center of an edge; and two types of rotations around the center of a face, by 90° and 180°. Hence the five representations that we have found comprise all of the irreducible representations of G. This is also evident from the equality $|G| = \sum (\dim T_i)^2$, which in our case has the form $24 = 1^2 + 1^2 + 2^2 + 3^2 + 3^2$. We now list the subspaces of functions in which the irreducible representations act.

The representation T_1 obviously acts in the space of constant functions.

The representation T_2 by definition acts in the space of even functions with sum of their values zero on X_F.

The representation T_3 by definition acts in the space of odd functions on X_F, and also in a certain subspace $W' \subset W_V$.

The representation T_4 by definition acts in the space of functions with sum of values zero on X_D, and also in a certain subspace $W'' \subset W_V$.

The representation T_5 by definition acts in the space of functions with sum zero on T_T and also in a certain subspace $W''' \subset W_V$.

We shall describe the spaces W', W'', and W'''.

The space W''' is one-dimensional and consists of functions assuming values of opposite signs on any two adjacent vertices. The space W'' is three-dimensional and consists of even functions with sum zero on X_V. Finally, W' is three-dimensional and consists of the odd functions on X_V that are orthogonal to W'''.

All of these assertions are easily derived by considering the explicit form of the intertwining operators.

16.2. Representations of the Symmetric Group

The group of all automorphisms of a finite set X consisting of n elements is called the *symmetric group of the n-th order*, and is denoted by the symbol S_n. It is clear that the order of the group S_n is equal to the number of all permutations of n symbols, that is, $n!$.

Problem 1. Prove that the group S_n is solvable for $n \leqslant 4$.

Hint. The groups S_1 and S_2 are commutative; S_3 is a semidirect product of S_2 and $\mathbf{Z}_3$; S_4 is a semidirect product of S_3 and $\mathbf{Z}_2 \times \mathbf{Z}_2$.

It is known that for $n \geqslant 5$, the group S_n is not solvable and contains a simple normal subgroup A_n of index 2. This normal subgroup is called the *alternating group* of order n and is defined as the kernel of the homomorphism s (defined below) of the group S_n onto the group of two elements.

Problem 2. Prove that the function

$$s(\sigma) = \prod_{i<j} \operatorname{sgn}(\sigma(i) - \sigma(j)), \quad \sigma \in S_n,$$

is a homomorphism of S_n onto the multiplicative group $\{+1, -1\}$.

Hint. Consider the polynomial in n variables defined by

$$P(x) = \prod_{i<j} (x_i - x_j)$$

and verify that $P(\sigma(x)) = s(\sigma) P(x)$.

Elements of A_n are called *even* permutations, and elements of $S_n \setminus A_n$ are called *odd*.

We can obtain a classification of the irreducible representations of the group S_n by the same method that we used above in considering the group of rotations of the cube. For this purpose, we must construct a sufficiently rich collection of homogeneous spaces on which the group S_n acts.

Let α denote a sequence of natural numbers $n_1, \ldots, n_k$, having the properties

$$n_1 \geqslant n_2 \geqslant \cdots \geqslant n_k, \tag{1}$$

$$n_1 + \cdots + n_k = n. \tag{2}$$

§ 16. Finite Groups

For every sequence α, we denote by X_α the space whose points are all of the partitions of X into subsets $X_1, \ldots, X_k$ such that $|X_i| = n_i$. Plainly the group S_n acts transitively on X_α.

Problem 3. Prove that $|X_\alpha| = \dfrac{n!}{n_1! \ldots n_k!}$.

Hint. Prove that the stationary subgroup S_α of a point $x \in X_\alpha$ in S_n is isomorphic to $S_{n_1} \times \ldots \times S_{n_k}$.

With every space X_α we associate two representations of the group S_n: the representation T_α induced by the trivial one-dimensional representation of the subgroup $S_\alpha = S_{n_1} \times \ldots \times S_{n_k}$; and the representation T'_α, induced by the nontrivial one-dimensional representation s (see problem 2) of the same subgroup.

The classification of all irreducible representations of S_n can be obtained by considering the table of intertwining numbers $c(T_\alpha, T'_\beta)$. We will order the sequences α lexicographically, defining $\alpha = (n_1, \ldots, n_k)$ as greater than $\beta = (m_1, \ldots, m_l)$ if: $n_1 > m_1$; or $n_1 = m_1$ but $n_2 > m_2$; or $n_1 = m_1$, $n_2 = m_2$ but $n_3 > m_3$; and so on.

Furthermore, for every sequence $\alpha = (n_1, \ldots, n_k)$ we define the *adjoint* sequence $\alpha^* = (n_1^*, \ldots, n_l^*)$, where n_i^* denotes the number of the integers $n_1, \ldots, n_k$ that are not less than i.

Problem 4. Prove that the sequence α^*, like α, satisfies conditions (1) and (2). Prove also that $(\alpha^*)^* = \alpha$.

Hint. To every sequence $\alpha = (n_1, \ldots, n_k)$, there corresponds a nonincreasing step function, equal to n_i on the interval from i to $i+1$. Show that the inverse to this step function corresponds to the sequence α^*. (Also see the interpretation of sequences α in terms of Young diagrams, described below.)

Theorem 1 (von Neumann-Weyl).

$$c(T_\alpha, T'_\beta) = \begin{cases} 0 & \text{if } \alpha > \beta^*, \\ 1 & \text{if } \alpha = \beta^*. \end{cases}$$

Proof. Let C be an intertwining operator for the pair T_α, T'_β. As we saw in 13.5, such an operator is defined by a function $c(x, y)$ on $X_\alpha \times X_\beta$ that satisfies the condition

$$c(\sigma(x), \sigma(y)) = s(\sigma) c(x, y). \tag{3}$$

We must show that if $\alpha > \beta^*$, then no such functions exist except for the trivial function $c \equiv 0$. We must also show that if $\alpha = \beta^*$, then there exists only one (up to a constant factor).

Let $x \in X_\alpha$ and $y \in X_\beta$. We recall that x is a partition of X into subsets $X_1, \ldots, X_k$, $|X_i| = n_i$, and that y is a partition of X into subsets $Y_1, \ldots, Y_l$, $|Y_i| = m_i$.

Problem 5. Prove that if $\alpha > \beta^*$, then $|X_i \cap Y_j| \geq 2$ for certain i and j.

Hint. Use the method of contradiction.

Thus, if $\alpha > \beta^*$, there are two elements of the set X that lie in the same subset for both of the partitions α and β. Let σ be the element of the group S_n that permutes these two elements of X and leaves all other elements of X fixed. Then we have $\sigma(x)=x$, $\sigma(y)=y$. Hence

$$c(x, y) = c(\sigma(x), \sigma(y)) = s(\sigma)c(x, y) = -c(x, y).$$

It follows that $c(x, y)=0$, and the first assertion is proved.

Suppose now that $\alpha = \beta^*$. We consider what is called a Young table:

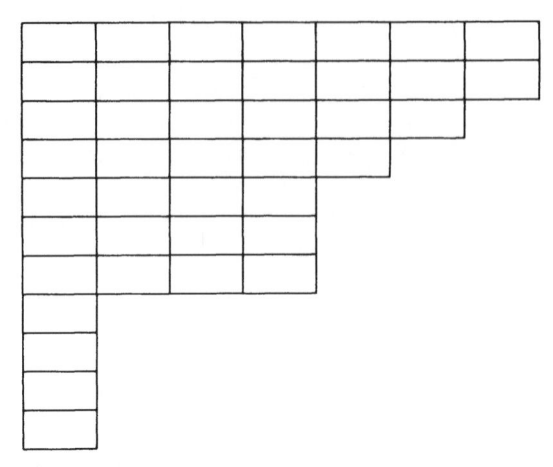

Fig. 4

consisting of k rows of lengths $n_1, \ldots, n_k$ respectively.

Problem 6. Prove that the lengths of the columns of the Young table form the sequence $\beta = (n_1^*, \ldots, n_l^*)$ conjugate to α.

We now suppose that we have placed the elements of our original set X into the boxes of the Young table, in any way we like. (It is convenient to identify X with the set of natural numbers from 1 to n.)

The figure obtained in this way is called a *Young diagram*. To every Young diagram, there corresponds a pair (x, y), $x \in X_\alpha$, $y \in X_\beta$. The point x is the partition of X according to the rows of the table, and the point y is the partition of X according to its columns. It is clear that the pair $(\sigma(x), \sigma(y))$ corresponds to the Young diagram obtained from the original permutation σ. In particular, this shows that all pairs $(\sigma(x), \sigma(y))$ are distinct for distinct σ.

We now fix an arbitrary Young diagram, and write (x_0, y_0) for the pair corresponding to it. We set

$$c_0(x, y) = \begin{cases} s(\sigma) & \text{if } x = \sigma(x_0) \text{ and } y = \sigma(y_0), \\ 0 & \text{in all remaining cases}. \end{cases} \quad (4)$$

It is clear that this function satisfies condition (3) and hence produces an intertwining operator.

§ 16. Finite Groups

Just as in the case $\alpha > \beta^*$, one can convince one's self that the equation (3) has no solutions besides c_0 and multiples of c_0. The theorem is proved.

It follows from theorem 1 that the representations T_α and $T'_{\alpha*}$ have exactly one common irreducible component U_α and that all of the representations U_α are pairwise inequivalent. The classification will be complete if we show that the number of conjugacy classes in S_n is equal to the number of irreducible representations that we have identified.

With every sequence $\alpha = (n_1, \ldots, n_k)$, we associate a conjugacy class in the following way. Let $X_1, \ldots, X_k$ be a partition in X_α and let $\sigma \in S_n$ be a transformation that permutes cyclically the elements of each of the subsets X_i, $i = 1, 2, \ldots, k$.

Problem 7. Prove that the conjugacy class of the element σ depends only on the sequence α, and then every conjugacy class can be obtained in this way.

Hint. For every $\sigma \in S_n$, consider the partition of X into orbits of the subgroup generated by σ.

A closer look at an intertwining operator $C \in \mathscr{C}(T_\alpha, T'_{\alpha*})$ leads to an explicit formula for the character of the representation U_α. This formula is very complicated, and apparently admits no essential simplification. (In this it resembles every "general" function of two sequences of natural number arguments.)

We present here only an expression for the dimension of the representation U_α corresponding to the sequence $\alpha = (n_1, \ldots, n_k)$:

$$\dim U_\alpha = \frac{n! \prod_{i<j} (l_i - l_j)}{l_1! \ldots l_k!}.$$

Here the numbers l_i are defined by $l_i = n_i + k - i$, $i = 1, 2, \ldots, k$.

16.3. Representations of the Group $SL(2, \mathbf{F}_q)$

The family of groups $SL(2, \mathbf{F}_q)$, where $\mathbf{F}_q$ is a finite field (see 3.2) is one of the most studied families of groups. We shall show here how to construct the principal series of representations of these groups.

The group $G = SL(2, \mathbf{F}_q)$ acts on the two-dimensional plane over the field $\mathbf{F}_q$. If we remove the origin of coordinates from this plane, then the group G acts transitively on the remaining set X. The stationary subgroup of the point $(0, 1) \in X$ is the triangular subgroup H, consisting of matrices of the form

$$\begin{pmatrix} 1 & b \\ 0 & 1 \end{pmatrix}, \quad b \in \mathbf{F}_q. \tag{1}$$

It follows in particular from this that

$$|G| = |N| \cdot |X| = q(q^2 - 1). \tag{2}$$

Our goal will be the study of the representation T of the group G in the space V of functions on X, defined by

$$\left[T\begin{pmatrix} a & b \\ c & d \end{pmatrix} f\right](u, v) = f(au+cv, bu+dv). \tag{3}$$

(In other words, T is induced by the trivial representation of N.)

We will find the space $\mathscr{C}(T)$ of intertwining operators for the representation T.

As we know (see 13.5), it is necessary for this purpose to compute the orbits of the group G in the space $X \times X$.

Problem 1. Prove that every G-orbit in $X \times X$ has the form

$$O_\lambda : \{(x, y) \in X \times X : x = \lambda y, \lambda \in F_q^*\}$$

or

$$\tilde{O}_s : \left\{ (x, y) \in X \times X : \begin{vmatrix} x_1 & x_2 \\ y_1 & y_2 \end{vmatrix} = s, s \in F_q^* \right\}.$$

Hint. Every vector in X can be carried into the vector $(0, 1)$ by action of the group G. The vector y either is proportional to x with a coefficient $\lambda \neq 0$ or generates together with x a parallelogram of area $s \neq 0$.

Corollary. $c(T, T) = 2(q-1)$.

Let C_λ and $\tilde{C}_s$ denote the intertwining operators corresponding to the orbits listed above. The structure of the ring $\mathscr{C}(T)$ is defined by the following relations.

Problem 2. Verify the equalities

$$C_{\lambda_1} C_{\lambda_2} = C_{\lambda_1 \lambda_2}, \tag{4}$$

$$C_\lambda \tilde{C}_s = \tilde{C}_{s\lambda^{-1}}, \quad \tilde{C}_s C_\lambda = \tilde{C}_{\lambda s}, \tag{5}$$

$$\tilde{C}_{s_1} \tilde{C}_{s_2} = \sum_{s \neq 0} \tilde{C}_s + q C_{-s_1^{-1} s_2}. \tag{6}$$

Hint. The operators C_λ act in the space V by the formula

$$C_\lambda f(x) = f(\lambda x), \tag{7}$$

and the operators $\tilde{C}_s$ by the formula

$$[\tilde{C}_s f](x) = \sum_{\Delta(x, y) = s} f(y), \tag{8}$$

§ 16. Finite Groups

where $\Delta(x, y) = \begin{vmatrix} x_1 & x_2 \\ y_1 & y_2 \end{vmatrix}$. Use the fact that the system of equations

$$\begin{cases} \Delta(x, z) = s_1, \\ \Delta(z, y) = s_2 \end{cases}$$

in the unknown z has exactly one solution if $\Delta(x, y) \neq 0$ and q solutions if $\Delta(x, y) = 0$.

It is clear that the operators C_λ, $\lambda \in \mathbf{F}_q^*$, form a commutative subring in $\mathscr{C}(T)$ isomorphic to the group ring of the multiplicative group of the field. We consider the corresponding decomposition of the space V (see 8.3):

$$V = \sum_{\chi \in \hat{F}_q^*} V_\chi. \tag{9}$$

The space V_χ consists of the functions on X that satisfy the condition

$$f(\lambda x) = \chi(\lambda) f(x). \tag{10}$$

Problem 3. Prove that the operators $\tilde{C}_s$ carry the space V_χ onto $V_{\chi^{-1}}$.
Hint. Use formulas (5), (7), and (8).

The representations T_χ of the group G arising in the subspaces V_χ are called the representations of the principal series. It is not hard to check that the representation T_χ coincides with $\mathrm{Ind}(G, H, U_\chi)$, where H is the subgroup of triangular matrices of the form

$$\begin{pmatrix} a & b \\ 0 & a^{-1} \end{pmatrix}, \quad a \in \mathbf{F}_q^*, \quad b \in \mathbf{F}_q,$$

and U_χ is the one-dimensional representation of H defined by the formula

$$U_\chi \begin{pmatrix} a & b \\ 0 & a^{-1} \end{pmatrix} = \chi(a).$$

Theorem 1. *The representation T_χ is irreducible if $\chi \neq \chi^{-1}$ and is the sum of two inequivalent irreducible components if $\chi = \chi^{-1}$.*

Proof. Both assertions follow from problem 3. In fact, the most general intertwining operator in $\mathscr{C}(T)$ has the form

$$C = \sum_{\lambda \in \hat{F}_q^*} f(\lambda) C_\lambda + \sum_{s \in \hat{F}_q^*} g(s) \tilde{C}_s.$$

Hence, if $\chi \neq \chi^{-1}$, all of the operators in $\mathscr{C}(T_\chi)$ are scalar. However, if $\chi = \chi^{-1}$, then $\mathscr{C}(T_\chi)$ is generated by the operators $C_1 = 1$ and $\tilde{C}_1$. This means that $c(T_\chi) = 2$ in this case.

We shall study in more detail the exceptional case $\chi = \chi^{-1}$. If q is odd, there are two exceptional characters:

$$\chi_0(x) \equiv 1 \quad \text{and} \quad \chi_\varepsilon(x) = \begin{cases} 1 & \text{if } x = y^2, \\ -1 & \text{if } x \text{ is not a square}. \end{cases}$$

The space V_{χ_0} contains the invariant one-dimensional space of constants and the complementary subspace of functions the sum of whose values is zero.

The space V_{χ_ε} is the sum of two irreducible subspaces of the same dimension $(q+1)/2$.

This result can be obtained without difficulty from the equalities (6) and (10).

If $q = 2^m$, then there is only one exceptional character, $\chi_0 \equiv 1$. The corresponding representation T_{χ_0} is constructed just as in the case of odd q.

It is interesting to note that in the cases both of even and odd q, the sum of the squares of the dimensions of the irreducible representations we have obtained is equal to $q(q+1)(q+3)/2$.

(In the case q odd, this number arises as the sum

$$\frac{q-3}{2}(q+1)^2 + 1 + q^2 + 2\left(\frac{q+1}{2}\right)^2,$$

and in the case q even as the sum

$$\frac{q+2}{2}(q+1)^2 + 1 + q^2 .)$$

This number is roughly half of the order of the group. Hence the group G admits yet another series of irreducible representations, which are not encountered in the decomposition of T. These representations are sometimes called *analytic*, since in the case of the group $SL(2, \mathbf{R})$ they are realized in a space of analytic functions. An analogous realization is possible also for a finite field. It is described in detail in the book [18] (for the case of the field of p-adic numbers). A different realization of this series can be obtained by embedding the group G in the symplectic group $Sp(2, \mathbf{F}_{q^2})$ and considering the spinor representation of the latter group (see 18.2 infra). Depending upon the choice of two possible ways to embed G in $Sp(2, \mathbf{F}_{q^2})$, one can obtain either the principal or the analytic series of representations of G. An interesting generalization of the concept of analytic representation to the case $G = SL(n, \mathbf{F}_q)$ is given in [77].

16.4. Vector Fields on Spheres

One of the most brilliant recent achievements in the topology of manifolds is the solution of the long-outstanding problem of vector fields on a sphere. The statement of the problem is as follows.

Let S^{n-1} be the unit sphere in n-dimensional real Euclidean space $\mathbf{R}^n$.

§ 16. Finite Groups

(A) *What is the maximal number $\rho(n)$ of smooth tangent vector fields on S^{n-1} that are linearly independent at every point?*

Problem 1. Prove that $\rho(n)$ coincides with the maximal number of vector-functions ξ on S^{n-1} with values in $\mathbf{R}^n$ having the property that

$$(\xi_i(x), x) = 0, \quad (\xi_i(x), \xi_j(x)) = \delta_{ij}. \tag{1}$$

Hint. Use the process of orthogonalization.

It is convenient to formulate the solution of problem (A) in the form of three properties of the function $\rho(n)$.

a) The function $\rho(n)$ depends only upon the highest power of 2 that divides n.
b) We have $\rho(16n) = \rho(n) + 8$;
c) For $k = 0, 1, 2, 3$ we have $\rho(2^k) = 2^k - 1$.

Using these properties, it is easy to compute $\rho(n)$ for arbitrary n. For example, we have $\rho(800) = \rho(32) = \rho(2) + 8 = 9$.

We shall show here how to solve an easier problem by using representation theory.

(B) *What is the maximal number $\mu(n)$ of linear vector fields on S^{n-1} that are linearly independent at every point?* (Linear vector fields are those of the form $\xi(x) = Ax$, $A \in \text{End } \mathbf{R}^n$.)

Problem 2. Prove that $\mu(n)$ is equal to the maximal number of operators A in $\mathbf{R}^n$ that have the following properties:

$$A_i' A_i = 1, \quad A_i A_j + A_j A_i = -2\delta_{ij}. \tag{2}$$

Hint. Prove that $(Ax, x) = 0$ for all $x \in S^{n-1}$ if and only if $A + A' = 0$.

It is clear that $\mu(n) \leq \rho(n)$. As we shall see *infra*, after we compute $\mu(n)$, we find that $\mu(n) = \rho(n)$. It would be very interesting to obtain a direct proof of this equality.

One of the possible methods here is to prove that every collection of vector-functions satisfying the conditions (1) can be deformed into a collection of linear vector-functions having the same property.

We shall state problem (B) in the language of representation theory. Let G_k be a group with generators $a_1, \ldots, a_k, \varepsilon$ and with the following relations:

$$a_i^2 = \varepsilon, \quad \varepsilon^2 = 1, \quad a_i a_j = a_j a_i \varepsilon \quad \text{for } i \neq j. \tag{3}$$

Problem 3. Prove that the inequality $k \leq \mu(n)$ is equivalent to the existence of a real n-dimensional representation T of the group G_k under which the element ε goes into -1.

Hint. Consider the representation T of the group G_k which has the form

$$T(a_i) = A_i, \quad T(\varepsilon) = -1$$

on the generators.

We now take up the study of the group G_k and its representations.

Problem 4. Prove the following properties of the group G_k.
a) The group G_k is a central extension of $(\mathbb{Z}_2)^k$ with the aid of $\mathbb{Z}_2$.
b) The center $C(G_k)$ has order 2 for even k and order 4 for odd k.
c) The commutant $[G_k, G_k]$ has order 2 for $k>1$.
d) The set of elements conjugate with the element $g \in G_k$ consists either of g alone (if $g \in C(G_k)$) or of the pair of elements g, εg (if $g \notin C(G_k)$).

Hint. Use the fact that every element $g \in G_k$ can be written uniquely in the form
$$g = a_{i_1} \ldots a_{i_s} \quad \text{or} \quad g = \varepsilon a_{i_1} \ldots a_{i_s}. \tag{4}$$

The results of problem 4 allow us to write down all of the irreducible representations of G_k. The number of these representations is equal to the number of conjugacy classes, that is, $|C(G_k)| + \frac{|G_k| - |C(G_k)|}{2}$. This number is equal to $2^k + 1$ for even k and is equal to $2^k + 2$ for odd k. We note that the group G_k admits exactly 2^k one-dimensional representations, under which ε goes into 1. Furthermore, the sum of the squares of the dimensions of all representations must be equal to $|G_k| = 2^{k+1}$. From this it follows that for even k, there is a single irreducible representation under which ε goes into -1. Its dimension is equal to $2^{k/2}$. For odd k, there are two such representations, each of dimension $2^{(k-1)/2}$.

Let us explain the type of these representations. In agreement with Schur's criterion (see 11.1, theorem 3) we must compute the sign of the quantity $\sum_{g \in G_k} \chi(g^2)$, where χ is the character of the representation under study.

Problem 5. Prove that $g^2 = 1$ or $g^2 = \varepsilon$ for all $g \in G_k$. Furthermore, if g has the form (4), then we have $g^2 = \varepsilon^{s(s+1)/2}$.

Thus the quantity we are interested in is equal to
$$2\chi(e)(1 - C_k^1 - C_k^2 + C_k^3 + C_k^4 - \ldots). \tag{5}$$

Problem 6. Prove that the sign of the quantity (5) is given by the following table:

$k \bmod 8$	0	1	2	3	4	5	6	7
sign	+	0	−	−	−	0	+	+

Hint. The quantity in parentheses in (5) can be written in the form $\operatorname{Re}(1+i)^k - \operatorname{Im}(1+i)^k = \operatorname{Re}(1-i)^{k+1}$.

Let T_k denote a real representation of the group G_k that carries ε into -1 and has the least possible dimension.

Problem 7. Prove that the dimension of T_k is equal to:

$2^{k/2}$ if $k = 8m$ or $8m+6$,
$2^{(k+1)/2}$ if $k = 8m+1$, $8m+3$ or $8m+5$,
$2^{(k+2)/2}$ if $k = 8m+2$ or $8m+4$,
$2^{(k-1)/2}$ if $k = 8m+7$.

Hint. Use the result of problem 6 and Schur's criterion.

To compute $\mu(n)$, it remains to note that the equality $\mu(n)=k$ is equivalent to the condition $\dim T_k = n$. From this consideration and the result of problem 7, it is easy to deduce that the function $\mu(n)$ has the properties a), b), c) stated above, and hence coincides with $\rho(n)$.

We could obtain the same result by studying, not the group G_k, but the algebra over **R** generated by the operators A_i that satisfy the conditions (2). We leave it to the reader to work out completely this method, using the results of 3.5 concerning Clifford algebras over **R**.

§ 17. Compact Groups

17.1. Harmonic Analysis on the Sphere

Consider a geometric body which casts a shadow of constant area when illuminated by parallel rays from an arbitrary direction. Can one infer that the body is a sphere?

It turns out that if we also suppose that the body is convex and possesses central symmetry, then the answer is affirmative. More precisely, we have

Theorem 1. *A convex centrally symmetric body in $\mathbf{R}^n$ is uniquely determined by the areas of its projections on all possible hyperplanes.*

The proof of this theorem uses harmonic analysis on the sphere. Before presenting this proof, we state the theorem in a different way.

Theorem 2. *A convex centrally symmetric body in $\mathbf{R}^n$ is uniquely determined by the areas of its sections by all possible hyperplanes.*

The equivalence of theorems 1 and 2 is inferred as follows. Let K be a centrally symmetric convex body in $\mathbf{R}^n$. Then we can define a norm in $\mathbf{R}^n$ for which the body K is the unit ball. Conversely, if we are given a norm in $\mathbf{R}^n$, its unit ball is a convex centrally symmetric body.

We say that two bodies K and K' are dual, if the corresponding Banach spaces V_K and $V_{K'}$ are duals of each other with respect to the natural scalar product in $\mathbf{R}^n$. In other words, K is dual to K' if the conditions

$$\sup_{y \in K'} (x,y) \leq 1 \quad \text{and} \quad x \in K \tag{1}$$

are equivalent.

In this case the conditions

$$\sup_{x \in K} (x,y) \leq 1 \quad \text{and} \quad y \in K' \tag{2}$$

are also equivalent.

Let $\mathbf{R}^{n-1}$ be a hyperplane in $\mathbf{R}^n$ and p the projection of $\mathbf{R}^n$ onto $\mathbf{R}^{n-1}$.

Problem 1. Prove that if the bodies K and K' are duals of each other in $\mathbf{R}^n$, then $\mathbf{R}^{n-1} \cap K$ and $p(K')$ are dual in $\mathbf{R}^{n-1}$.

Hint. Use either of the conditions (1) or (2).

Thus *the problem of reconstructing a body from its projections is equivalent to the problem of reconstructing the dual body from its sections.*

We proceed to the proof of the theorem. For clarity, we restrict ourselves to the case $n=3$, leaving to the reader the analogous arguments for the general case.

First of all we shall formulate the problem in a language convenient for applying representation theory. We shall describe a convex centrally symmetric body $K \subset \mathbf{R}^3$ with the aid of a function f on the unit sphere $S \subset \mathbf{R}^3$ in the following way. If x is a point of S and l_x is a ray drawn from the origin of coordinates through this point, then we set

$$f(x) = \tfrac{1}{2} r_x^2, \tag{3}$$

where r_x is the length of the interval $l_x \cap K$.

Since the body K is centrally symmetric, the function f on S is even (that is, it assumes the same value at every pair of antipodal points).

The areas of sections of the body K by all possible two-dimensional planes can be very simply expressed in terms of the function f.

Problem 2. Let C be a great circle of the sphere S, the intersection of S with a plane P. Then we have

$$\text{area}(K \cap P) = \int_C f(x) dx. \tag{4}$$

Hint. Use the formula for area in polar coordinates.

Thus our theorem can be stated in the following terms.

An even function on the sphere is uniquely determined by its integrals on all great circles.

We consider the space $L^2(S)$ of all functions on S with summable squares under the usual (area) measure, and the subspace $L_+^2(S)$ of all even square summable functions.

Rotations of the sphere carry these spaces into themselves. Thus we obtain a unitary representation T of the group $SO(3)$ in $L^2(S)$ and a subrepresentation T_+ of T in the subspace $L_+^2(S)$.

Let J be the operator in $L^2(S)$ defined by the formula

$$Jf(x) = \int_{C_x} f(y) dy, \tag{5}$$

where C_x is the great circle with epicenter at the point $x \in S$.

Strictly speaking, the operator J is defined by formula (5) on the subspace $C(S)$ of continuous functions and is extended by continuity to all of $L^2(S)$. One can show, however, that the equality (5) holds almost everywhere on S for an arbitrary $f \in L^2(S)$.

§ 17. Compact Groups

Our problem is that of proving that J has null kernel: that is, if $Jf=0$, then $f=0$.

Problem 3. Prove that the operator J is an intertwining operator for the representations T and T_+.

Hint. Use the invariance of length under rotations of the sphere.

We know from the general theory that the space $L_+^2(S)$ of the representation T_+ is the direct sum of irreducible finite-dimensional spaces in each of which the operator J is scalar.

We shall now compute these subspaces explicitly and also find the eigenvalues of J in these subspaces. All of the eigenvalues are different from zero, and this will prove our assertion.

Let P_n be the space of all functions on S that are restrictions to S of homogeneous polynomials of degree n in R^3.

Problem 4. Prove that $P_n \subset P_{n+2}$ and that $\dim P_n = (n+1)(n+2)/2$.

Hint. The first assertion follows from the equality $x^2+y^2+z^2=1$ on S. The second is easily proved by induction.

Let H_n be the orthogonal complement of P_{n-2} in P_n.

Theorem 3. *The decompositions of the spaces $L^2(S)$ and $L_+^2(S)$ into irreducible subspaces (under the group $SO(3)$) have the forms*

$$L^2(S) = \sum_{n=0}^{\infty} H_n \quad \text{and} \quad L_+^2(S) = \sum_{k=0}^{\infty} H_{2k},$$

respectively.

Proof. First of all, it is clear that the H_n's are invariant subspaces and that the algebraic sum of these spaces coincides with the union of the spaces P_n. By the Stone-Weierstrass theorem (see 4.2) the latter space is dense in $C(S)$ and consequently also in $L^2(S)$. Hence the Hilbert space sum of the H_n's coincides with $L^2(S)$. It is also clear that $H_n \subset L_+^2(S)$ if and only if n is even. To prove the theorem it remains to show that the H_n's are irreducible.

Problem 5. Prove that H_n contains exactly one function (up to a constant factor) that is invariant under the subgroup of rotations about the z-axis.

Hint. Prove that P_n contains $[n/2]+1$ linearly independent functions invariant under the given subgroup, namely, $z^n, z^{n-2}(x^2+y^2),...,z^{n-2[n/2]}(x^2+y^2)^{[n/2]}$.

Problem 6. Prove that every irreducible subspace $V \subset L^2(S)$ contains at least one nonzero function that is invariant under rotations about the z-axis.

Hint. Use the Frobenius reciprocity theorem from 13.5.

The irreducibility of H_n follows at once from the assertions of problems 5 and 6.

It remains for us to find the eigenvalues of the operator J in the spaces H_n. For this, we shall give the explicit form of the function $L_n \in H_n$ that does not change under rotations about the z-axis.

Problem 7. Prove that we can take L_n to be the function

$$L_n(z) = \frac{d^n}{dz^n}[(z^2-1)^n] \tag{6}$$

(the n-th Legendre polynomial).

Hint. Prove that for functions f on S that depend only on the coordinate z, the equality

$$\int_S f(x)dx = \pi \int_{-1}^{1} f(z)dz$$

holds.

Use integration by parts to show that L_n is orthogonal to all polynomials in z of degree less than n, with respect to the following inner product:

$$(f_1, f_2) = \pi \int_{-1}^{1} f_1(z)\overline{f_2(z)}dz.$$

Finally, show that L_n is defined up to a constant factor by these properties.

Let λ_n be the eigenvalue of the operator J in H_n. Substituting L_n for f in formula (5) and taking the point $(0,0,1)$ for x, we obtain

$$\lambda_n L_n(1) = 2\pi L_n(0).$$

The values $L_n(1)$ and $L_n(0)$ are easily computed:

$$L_n(1) = \frac{d^n}{dz^n}[(z-1)^n(z+1)^n]|_{z=1} = n!(z+1)^n|_{z=1} = 2^n \cdot n!,$$

$$L_n(0) = \begin{cases} 0 & \text{if } n \text{ is odd}, \\ (2k)! C_{2k}^k & \text{if } n = 2k. \end{cases}$$

From this we obtain

$$\lambda_n = \begin{cases} 0 & \text{if } n \text{ is odd}, \\ 2\pi \dfrac{(2k-1)!!}{2k!!} & \text{if } n = 2k. \end{cases}$$

17.2. Representations of the Classical Compact Lie Groups

The classical compact Lie groups are the groups of isometries of a finite-dimensional Hilbert space V over **R**, **C**, or **H**. That is, they are the groups $O(n)$, $U(n)$, and $Sp(n)$. (See 5.1, where these groups are denoted by $O(n,0)$, $U(n,0)$, and $Sp(n,0)$ respectively.)

§ 17. Compact Groups

We can study the topological structure of these groups by using a single simple observation.

Problem 1. Let G_n be one of the groups $O(n)$, $U(n)$, $Sp(n)$. Prove that G_n acts transitively on the unit sphere S of the corresponding space V and that the stationary subgroup of a point $s_0 \in S$ is isomorphic to G_{n-1}.

We leave it to the reader to prove for himself the following facts.

Problem 2. Prove that the groups $U(n)$ and $Sp(n)$ are connected, while $O(n)$ consists of two components, one of which is the group $SO(n)$. The fundamental group of $SO(n)$ is isomorphic to $\mathbf{Z}$ for $n=2$ and to $\mathbf{Z}_2$ for $n>2$. The fundamental group of $U(n)$ is isomorphic to $\mathbf{Z}$. The group $Sp(n)$ is simply connected.

We shall show that all irreducible representations of the group G_n are realized in tensors over the space V.

A more precise formulation of this assertion is the following. Let W be a complex linear space, which is either the complexification of V (if V is real), or is V itself (if V is complex) or is obtained from V by restricting the field of scalars (if V is a quaternion space). Let $T^{k,l}(W)$ denote the space of mixed tensors of rank (k,l) over W (see 3.5) and let $\rho^{k,l}$ denote the representation of the group G_n arising in the space $T^{k,l}(W)$.

Theorem 1. *All irreducible representations of the group G_n appear in the irreducible components of the representations $\rho^{k,l}$.*

Proof. Let $A^{k,l}(G)$ be the subspace of $C(G_n)$ generated by all matrix elements of the representation $\rho^{k,l}$ and define $A(G)$ by $A(G) = \sum_{k,l} A^{k,l}(G)$.

Problem 3. Prove that $A(G)$ is a subalgebra of $C(G_n)$.
Hint. Use the fact that $\rho^{k_1 l_1} \otimes \rho^{k_2 l_2} \approx \rho^{k_1+k_2, l_1+l_2}$, and prove that $A^{k_1 l_1} A^{k_2 l_2} \subset A^{k_1+k_2, l_1+l_2}$.

Problem 4. Prove that the algebra A is closed under complex conjugation.
Hint. Use the relation $(\rho^{k,l})^* \approx \rho^{l,k}$ and infer from it that $\overline{A^{k,l}} = A^{l,k}$.

We further note that by the very definition of the space $A^{1,0}$, functions in this space separate points of the group G_n. The Stone-Weierstrass theorem (see 4.2) together with this and the results of problems 3 and 4 imply that the algebra $A(G)$ is everywhere dense in $C(G_n)$. Now let ρ be an arbitrary irreducible representation of the group G_n. If ρ did not appear in the decomposition of the representations $\rho^{k,l}$, then the orthogonality relations of 9.2 would imply that the matrix elements of ρ are orthogonal to all of the functions in $A(G_n)$. But this is impossible, since $A(G_n)$ is dense in $C(G_n)$ and hence also in $L^2(G_n, dg)$. The theorem is proved.

Thus to classify all irreducible representations of the group G_n, it suffices to know the decomposition of the representations $\rho^{k,l}$ into irreducible components. This rather difficult problem was explicitly solved by H. Weyl (see [55]). The main result can be stated in contemporary language as follows.

Theorem 2 (H. Weyl). *Let ρ be a representation of the group G_n in the space $\mathcal{T} = \sum_{k,l} T^{k,l}(W)$. The algebra $\mathscr{C}(\rho, \rho)$ of intertwining operators for this representation is generated by:*

1) *operators of permuting indices (covariant and contravariant indices separately);*
2) *operators of multiplication by invariant tensors;*
3) *operators of convolution (in one covariant and one contravariant index).*

We will not deduce here from theorem 2 a complete classification of the irreducible representations of G_n. We note that this classification can be obtained by many different methods (see [57], [47], and [46]).

We suggest that the reader, as a useful excercise, prove the following special case of the general classification theorem.

Theorem 3. *Let ρ_k be the representation of the group $U(n)$ in skew-symmetric tensors of rank k $(k=1,2,...,n)$. Prove that the Grothendieck ring of the group $U(n)$ (see the definition of this ring in 7.1) is isomorphic to*

$$Z[\rho_1,...,\rho_{n-1},\rho_n,\rho_n^{-1}].$$

(The representation ρ_n is one-dimensional, and so ρ_n^{-1} has a meaning and coincides with ρ_n^*.)

17.3. Spinor Representations of the Orthogonal Group

The groups $SO(n,K)$ for $K=\mathbf{R}$ and $K=\mathbf{C}$ are not simply connected. It is known that their fundamental groups are isomorphic to $\mathbf{Z}_2$ for $n \geqslant 3$. The simply connected covering group of the group $SO(n,K)$ is denoted by $\mathrm{Spin}(n,K)$ and is called a spinor group. Thus we have the exact sequence

$$1 \to \mathbf{Z}_2 \to \mathrm{Spin}(n,K) \to SO(n,K) \to 1. \qquad (1)$$

Every linear representation of the group $SO(n,K)$ generates a representation of $\mathrm{Spin}(n,K)$. However, the group $\mathrm{Spin}(n,K)$ admits also representations not obtained in this fashion. These representations can be considered as two-valued representations of $SO(n,K)$ (that is, projective representations with a multiplier assuming the values ± 1).

The simplest of these representations has dimension $2^{[n/2]}$. It is irreducible if n is odd and is the sum of two inequivalent components of the same dimension if n is even.

We now describe the explicit construction of this representation. Let C_k be the complex Clifford algebra with k generators $e_1,...,e_k$ and with the relations

$$e_i e_j + e_j e_i = 0 \quad \text{for} \quad i \neq j, \quad e_i^2 = 1 \qquad (2)$$

(*cf.* 3.5).

Problem 1. Prove that

$$C_{2k} \approx \mathrm{Mat}_{2^k}(\mathbf{C}), \quad C_{2k+1} \approx \mathrm{Mat}_{2^k}(\mathbf{C}) \oplus \mathrm{Mat}_{2^k}(\mathbf{C}).$$

§ 17. Compact Groups

Hint. Use the method of solving problem 8 of 3.5 and prove that

$$C_{k+2} = C_k \underset{\mathbf{C}}{\otimes} C_2, \quad C_0 = \mathbf{C}, \quad C_1 = \mathbf{C} \otimes \mathbf{C}.$$

Problem 2. Prove that the center of the algebra C_{2k+1} is generated by the elements 1 and $\varepsilon = e_1 e_2 \ldots e_{2k+1}$.

Hint. Consider the commutator of the monomial $e_{i_1} \ldots e_{i_k}$ with the generators $e_1, \ldots, e_n$.

Suppose now that $A \in SO(n, \mathbf{C})$. The transformation

$$e_i \to \sum_j a_{ij} e_j, \quad \|a_{ij}\| = A \tag{3}$$

does not destroy the relations (2) and hence defines an automorphism $\phi(A)$ of the algebra C_n.

Problem 3. Prove that the automorphisms $\phi(A)$, $A \in SO(n, \mathbf{C})$ preserve the elements of the center of C_n.

Hint. Use the result of problem 2 and the fact that $\det A = 1$ for $A \in SO(n, \mathbf{C})$.

The assertion of problem 3 implies that for $n = 2k+1$, the automorphism $\phi(A)$ is given by a certain automorphism of the matrix algebra $\mathrm{Mat}_{2^k}(\mathbf{C})$.

Thus, for an arbitrary n, we can consider ϕ as an automorphism of the complete matrix algebra of order $2^{[n/2]}$ over $\mathbf{C}$.

As we know (cf. 14.1), every automorphism of $\mathrm{Mat}_k(\mathbf{C})$ is inner. Thus we have

$$\phi(A)(X) = \rho(A) X \rho(A)^{-1} \tag{4}$$

for a certain matrix $\rho(A) \in \mathrm{Mat}_{2^{[n/2]}}(\mathbf{C})$.

It is clear that the correspondence $A \mapsto \rho(A)$ is a projective representation of the group $SO(n, \mathbf{C})$. This is the representation that we need. We shall now show that this representation can be made two-valued, that is, we can choose $\rho(A)$ in such a way that the equality

$$\rho(A_1) \rho(A_2) = \pm \rho(A_1 A_2)$$

holds.

For this we use the results of 14.3. Since the group $SO(n, \mathbf{C})$ is semisimple, the representation ρ is obtained from a linear representation $\tilde{\rho}$ of the covering group $\mathrm{Spin}(n, \mathbf{C})$. The corresponding representation $\tilde{\rho}_*(=\rho_*)$ of the Lie algebra is easily computed.

Problem 4. Let $A \in \mathfrak{so}(n, \mathbf{C})$ be a skew-symmetric complex matrix. Then we have

$$\rho_*(A) = \tfrac{1}{2} \sum_{i,j} a_{ij} e_i e_j. \tag{5}$$

Hint. Compare the action of both sides of the equality (5) on the generators $e_1, \ldots, e_n$.

Under the representation ρ_*, the Lie algebra $\mathfrak{so}(n, \mathbf{C})$ thus goes into the subspace $C_n^{(2)}$ of homogeneous elements of degree 2 in C_n. Under an appropriate identification of C_n with a matrix algebra, this subspace is contained in $SO(2^{n/2}, \mathbf{C})$ for even n and in $SL(2^{[n/2]}, \mathbf{C})$ for odd n. The corresponding subgroup in $SO(2^{n/2}, \mathbf{C})$ or $SL(2^{[n/2]}, \mathbf{C})$ is the group $\mathrm{Spin}(n, \mathbf{C})$.

This group can also be characterized as the connected component of the subgroup $\Gamma \subset SO(2^{[n/2]}, \mathbf{C})$ preserving the subspace generated by $e_1,\ldots,e_n$ under the action of inner automorphisms.

From formula (5) it is easy to verify the following. Consider the closed one-parameter subgroup in $SO(n, \mathbf{C})$ corresponding to all rotations in a two-dimensional plane. Under the representation $\tilde{\rho} = \exp\rho_*$, this subgroup goes into the nonclosed curve that connects the points 1 and -1. Hence the representation $\tilde{\rho}$ is two-valued on $SO(n, \mathbf{C})$.

We now investigate the reducibility of the representation ρ.

Problem 5. Prove that the centralizer of the subspace $C_n^{(2)}$ in C_n is generated by the elements 1 and $e_1 e_2 \ldots e_n$.

Thus for odd n, the representation ρ is irreducible, since every operator commuting with the operators in the image of the Lie algebra belongs to the center of C_n and is scalar in each of the two equivalent subspaces of the algebra $C_n = \mathrm{Mat}_{2^{[n/2]}}(\mathbf{C}) \oplus \mathrm{Mat}_{2^{[n/2]}}(\mathbf{C})$.

For even n, the element $e_1 e_2 \ldots e_{2n}$ is a nontrivial intertwining operator for the representation ρ. Hence ρ is the sum of two subrepresentations ρ_+ and ρ_-.

Problem 6. Prove that the representations ρ_+ and ρ_- are inequivalent and have the same dimension.

Hint. Use the relation $c(\rho, \rho) = 2$ and the equality $(e_1 e_2 \ldots e_{2n})^2 = (-1)^n$.

Problem 7. Prove that the restriction of the spinor representation ρ of the group $\mathrm{Spin}(2n+1, \mathbf{C})$ to the subgroup $\mathrm{Spin}(2n, \mathbf{C})$ is equivalent to $\rho_+ + \rho_-$, while the restriction of each of the representations $\rho_\pm$ to $\mathrm{Spin}(2n-1, \mathbf{C})$ is irreducible and equivalent to the spinor representation of this group.

Hint. Compute the corresponding intertwining number and compare the structures of the algebras C_{2n+1}, C_{2n}, and C_{2n-1}.

We can carry out analogous but more complicated computations for the case $K = \mathbf{R}$. The type of the corresponding Clifford algebras depends upon the remainder of the number n modulo 8. The properties of the spinor representations depend also upon this remainder.

We leave it to the reader to prove the following assertions.

Problem 8. The spinor representation ρ of the group $\mathrm{Spin}(2k+1, \mathbf{R})$ is of real type for $k \equiv 0, 3 \pmod 4$ and is of quaternion type for $k \equiv 1, 2 \pmod 4$.

The half-spinor representations ρ_+ and ρ_- of the group $\mathrm{Spin}(2k, \mathbf{R})$ are complex conjugates of each other for $k \equiv 1 \pmod 2$, are real for $k \equiv 0 \pmod 4$ and are of quaternion type for $k \equiv 2 \pmod 4$.

These two representations are duals of each other for odd k and are self-dual for even k.

§ 18. Lie Groups and Lie Algebras

18.1. Representations of a Simple Three-Dimensional Lie Algebra

As we pointed out in 6.2, there are exactly two simple real Lie algebras of dimension 3. These are: the algebra $\mathfrak{g}_1 = \mathfrak{sl}(2, \mathbf{R})$ of real matrices of the second order with zero trace and the algebra $\mathfrak{g}_2 = \mathfrak{so}(3, \mathbf{R})$ of real skew-symmetric matrices of the third order.

In this section, we shall obtain a classification of all finite-dimensional irreducible representations of the algebras $\mathfrak{g}_1$ and $\mathfrak{g}_2$ over the field $\mathbf{C}$.

Every such representation can be extended by linearity to a representation of the complex hull of the corresponding Lie algebra.

Problem 1. Prove that the complex hulls of the algebras $\mathfrak{g}_1$ and $\mathfrak{g}_2$ are isomorphic.

Hint. The required isomorphism of $\mathfrak{sl}(2, \mathbf{C})$ and $\mathfrak{so}(3, \mathbf{C})$ has the form

$$\begin{pmatrix} a & b \\ c & -a \end{pmatrix} \leftrightarrow \begin{pmatrix} 0 & b-c & -i(b+c) \\ c-b & 0 & 2ia \\ i(b+c) & -2ia & 0 \end{pmatrix}.$$

Thus it suffices to study the irreducible representations of the algebra $\mathfrak{g} = \mathfrak{sl}(2, \mathbf{C}) \approx \mathfrak{so}(3, \mathbf{C})$.

In $\mathfrak{g}$, we choose the following basis (over $\mathbf{C}$):

$$X_0 = \begin{pmatrix} 1 & 0 \\ 0 & -1 \end{pmatrix}, \quad X_+ = \begin{pmatrix} 0 & 1 \\ 0 & 0 \end{pmatrix}, \quad X_- = \begin{pmatrix} 0 & 0 \\ 1 & 0 \end{pmatrix}.$$

The commutation relations between the basis vectors are:

$$[X_0, X_\pm] = \pm 2 X_\pm, \quad [X_+, X_-] = X_0. \tag{1}$$

Let T be a finite-dimensional representation of the algebra $\mathfrak{g}$ in a space V. Then the operator $T(X_0)$ has at least one eigenvector ξ in V with eigenvalue λ:

$$T(X_0)\xi = \lambda \xi. \tag{2}$$

We call such a vector ξ a *weight vector* with weight λ.

Problem 2. Prove that if ξ is a vector of weight λ, then $T(X_\pm)\xi$ is either the zero vector or a vector of weight $\lambda \pm 2$.

Hint. Use the relations (1).

Since the space V is finite-dimensional, it contains a weight vector ξ of largest weight λ (we order weights by their real parts; as a matter of fact, as we shall shortly see, all weights of finite-dimensional representations are real). Besides the condition (2), this vector satisfies the condition

$$T(X_+)\xi = 0. \tag{3}$$

We consider the sequence of vectors

$$\xi_0 = \xi, \quad \xi_1 = T(X_-)\xi_0, \quad \xi_2 = T(X_-)\xi_1, \ldots.$$

In view of problem 2, all of the nonzero vectors in this sequence are weight vectors with different weights. Hence they are linearly independent. This means that only a finite number of the terms of the sequence are different from zero.

Let ξ_n be the last nonzero vector in this sequence.

Problem 3. The space generated by the vectors $\xi_0, \ldots, \xi_n$ is invariant under the operators of the representation and is irreducible.

Hint. Using the relations $T(X_-)\xi_k = \xi_{k+1}$, $T(X_0)\xi_k = (\lambda - 2k)\xi_k$, prove by induction that

$$T(X_+)\xi_k = k(\lambda - k + 1)\xi_{k-1}.$$

Problem 4. Prove that the parameter λ coincides with the number n.
Hint. Use the equality $\operatorname{tr} T(X_0) = \operatorname{tr} T(X_+ X_- - X_- X_+) = \operatorname{tr}[T(X_+), T(X_-)] = 0.$
Thus we have proved

Theorem 1. *For every positive integer n, there exists exactly one irreducible representation T_n of the algebra $\mathfrak{g}$ of dimension $n+1$. In a basis of weight vectors, the operators of the representation have the following form:*

$$\left.\begin{aligned} T(X_-)\xi_k &= \xi_{k+1}, \\ T(X_0)\xi_k &= (n - 2k)\xi_k, \\ T(X_+)\xi_k &= k(n - k + 1)\xi_{k-1}. \end{aligned}\right\} \tag{4}$$

We note the following useful

Corollary. *For each irreducible representation T, there is exactly one vector ξ (up to a scalar multiple) annihilated by the operator $T(X_+)$. This vector is a weight vector, and its weight is equal to $\dim T - 1$.*

In view of the results of 6.3 and 6.4, which relate representations of groups and algebras, theorem 1 also yields a classification of the irreducible finite-dimensional representations of the Lie groups $SL(2,\mathbf{R})$ and $\widetilde{SO}(3,R) = SU(2)$, as well as the analytic representations of the complex Lie group $SL(2,\mathbf{C})$.

Problem 5. Let P_n be the space of homogeneous polynomials of degree n in two variables p and q. Define an action of the group $SL(2,\mathbf{C})$ in P_n by the formula

$$\left[T\begin{pmatrix} a & b \\ c & d \end{pmatrix} f\right](p,q) = f(ap + cq, bp + dq).$$

Prove that the corresponding representation T_* of the algebra $\mathfrak{g} = \mathfrak{sl}(2,\mathbf{C})$ is equivalent to T_n.

§ 18. Lie Groups and Lie Algebras

Hint. The operators of the representation of g are given by the formulas

$$T_*(X_+) = p\frac{\partial}{\partial q}, \qquad T^*(X_0) = p\frac{\partial}{\partial p} - q\frac{\partial}{\partial q}, \qquad T_*(X_-) = q\frac{\partial}{\partial p}.$$

The isomorphism between V_n and P_n has the form

$$\xi_k \leftrightarrow n(n-1)\ldots(n-k+1)p^{n-k}q^k,$$

where $\{\xi_k\}$ is the basis in V_n that we defined above.

18.2. The Weyl Algebra and Decomposition of Tensor Products

The *Weyl algebra* $A_n(K)$ is the algebra with unit over a field K generated by the elements

$$p_1, \ldots, p_n, \qquad q_1, \ldots, q_n$$

and governed by the relations

$$p_i p_j = p_j p_i, \qquad q_i q_j = q_j q_i, \qquad p_i q_j - q_j p_i = \delta_{ij} \cdot 1. \tag{1}$$

This algebra is a special case of the algebra $R_{n,k}(K)$ defined in 3.5. In fact, we have $A_n(K) = R_{n,0}(K)$.

We will now show that computations based on the algebra $A_1(K)$ allow us to give explicit formulas for decomposing the tensor product of two irreducible representations of the group $SL(2,K)$ into irreducible components.

Every element of the algebra $A_1(K)$ can be written as a (noncommutative) polynomial in the variables p and q with coefficients in K. These polynomials are not unique. For example, the element pq can be written also in the form $qp+1$ and in the form $(pq)^2 - qp^2q$ and so on. One can give three natural methods of making this form unique:

a) the left form: $f = \sum c_{kl} p^k q^l$;
b) the right form: $f = \sum c_{kl} q^l p^k$;
c) the symmetric form: $f = \sum c_{kl}\sigma(p^k q^l)$. Here σ denotes the *symmetric product*

$$\sigma(x_1, \ldots, x_n) = \frac{1}{n!} \sum_{s \in S_n} x_{s(1)} \ldots x_{s(n)}.$$

Each of these methods identifies the space A_1 (but not the algebra!) with the space $K[p,q]$ of all (commutative) polynomials in p, q with coefficients in K (for a field K of characteristic zero).

Going from one form to another yields an automorphism of the space $K[p,q]$. We now give an explicit form of these automorphisms.

Theorem 1. *Going from the left form to the symmetric form and from the symmetric form to the right form generates the following automorphism in* $K[p,q]$:

$$e^{\frac{1}{2}\frac{\partial^2}{\partial p \partial q}} = \sum_{k=0}^{\infty} \frac{1}{j!2^j} \frac{\partial^{2j}}{\partial p^j \partial q^j}.$$

The proof is by induction, using the identity

$$\sigma(p^k q^l) = \frac{k}{k+l} p\sigma(p^{k-1}q^l) + \frac{l}{k+l} q\sigma(p^k q^{l-1}). \tag{2}$$

It would be interesting to find a shorter and direct proof. (It is useful, for example, to use the realization of the algebra $A_1(R)$ as the ring of differential operators with polynomial coefficients on the real line.)

Suppose that the matrix $\begin{pmatrix} a & b \\ c & d \end{pmatrix}$ belongs to $SL(2,K)$.

Then the transformation

$$\left.\begin{array}{l} p \mapsto ap + bq, \\ q \mapsto cp + dq \end{array}\right\} \tag{3}$$

preserves the relations (1) and consequently can be extended to an automorphism of $A_1(K)$. Furthermore, this transformation can be naturally extended to an automorphism of $K[p,q]$.

Problem 1. Prove that the isomorphism σ between $K[p,q]$ and $A_1(K)$ commutes with the action of the group $SL(2,K)$.

Hint. Use the fact that $K[p,q]$ is generated as a linear space by elements of the form $(p+\lambda q)^k$ and verify the identity

$$\sigma((p+\lambda q)^k) = (p+\lambda q)^k \quad \text{in} \quad A_1(K),$$

by comparing coefficients of individual powers of λ. Compare also problem 8 in 10.4.

As we saw in 18.1, the representation of $SL(2,K)$ in the space $K[p,q]$ decomposes into the direct sum of irreducible finite-dimensional representations. All irreducible finite-dimensional representations of the group $SL(2,K)$ appear just once in this decomposition. The irreducible subspaces $\tilde{V}_n$ consist of homogeneous polynomials of degree n. The dimension of $\tilde{V}_n$ is equal to $n+1$. Let V_n be the image of $\tilde{V}_n$ in $A_1(K)$ under the mapping σ and let T_n be the representation of $SL(2,K)$ in V_n.

Let $V_n \cdot V_m$ denote the subspace of $A_1(K)$ generated by the products xy, $x \in V_n, y \in V_m$.

Theorem 2. *The mapping of the product*

$$x \otimes y \mapsto xy$$

§ 18. Lie Groups and Lie Algebras

can be extended by linearity to an isomorphism of $V_n \otimes V_m$ onto $V_n \cdot V_m$. The space $V_n \cdot V_m$ is the direct sum of subspaces of the form V_{n+m-2k}, $k = 0, 1, \ldots, \min(m,n)$.

Proof. We consider the product of elements $p^n \in V_n$ and $q^m \in V_m$. In view of theorem 1, we have

$$p^n q^m = \sum_j \frac{1}{2^j \cdot j!} \sigma \left(\frac{\partial^{2j}}{\partial p^j \partial q^j} p^n q^m \right)$$

$$= \sigma(p^n q^m) + \frac{nm}{2} \sigma(p^{n-1} q^{m-1}) + \frac{n(n-1)m(m-1)}{8} \sigma(p^{n-2} q^{m-2}) + \ldots .$$

But the element $\sigma(p^{n-k} q^{m-k})$ belongs to V_{n+m-2k}. Hence the space $V_n V_m$ contains all subspaces of the form V_{n+m-2k}, $k = 0, 1, \ldots, \min(m,n)$. The sum of the dimensions of these subspaces is equal to

$$\sum_{k=0}^{\min(m,n)} (n+m-2k+1) = (m+1)(n+1),$$

which is the dimension of $V_n \otimes V_m$. The theorem is proved.

Theorems 1 and 2 allow us to give an explicit expression for the decomposition of the tensor product of two irreducible representations of the group $SL(2,K)$. For this computation, we need to find the symmetric form of the product of two elements both written in the symmetric form.

The method described here can be applied to the study of tensor products of irreducible representations of the symplectic group $Sp(2n,K)$. This group acts naturally as the group of automorphisms of the Weyl algebra $A_n(K)$. Let P_k be the space of homogeneous polynomials of degree k in the variables $p_1, \ldots, p_n$, $q_1, \ldots, q_n$ and let $V_k = \sigma(P_k) \subset A_n(K)$. One can show that the space V_k is irreducible and that an analogue of theorem 2 holds.

Nevertheless, in distinction to the case $k=1$, one obtains in this way far from all of the irreducible representations of the group $Sp(2n,K)$. It would be very interesting so to generalize the method presented here as to obtain the other representations of this group.

18.3. The Structure of the Enveloping Algebra $U(\mathfrak{g})$ for $\mathfrak{g} = \mathfrak{sl}(2,\mathbb{C})$

In this paragraph, we describe the two-sided ideals of the enveloping algebra $U(\mathfrak{g})$ for the Lie algebra $\mathfrak{g} = \mathfrak{sl}(2,\mathbb{C})$.

We find it convenient to use the basis

$$2X = \begin{pmatrix} 0 & 1 \\ -1 & 0 \end{pmatrix}, \quad 2Y = \begin{pmatrix} 0 & -i \\ -i & 0 \end{pmatrix}, \quad 2Z = \begin{pmatrix} -i & 0 \\ 0 & i \end{pmatrix}.$$

This basis is connected with the basis X_+, X_-, X_0 introduced in 18.1 by the relations

$$\tfrac{1}{2} X_+ = \tfrac{1}{2}(X + iY), \quad \tfrac{1}{2} X_- = \tfrac{1}{2}(X - iY), \quad \tfrac{1}{2} X_0 = iZ.$$

The elements X, Y, Z generate $U(\mathfrak{g})$ and satisfy the relations

$$XY - YX = Z, \quad YZ - ZY = X, \quad ZX - XZ = Y. \tag{1}$$

We recall that the symmetric form establishes an isomorphism σ between the space $\mathbf{C}[X, Y, Z]$ of all polynomials in the variables X, Y, Z and the space $U(\mathfrak{g})$, also that this isomorphism commutes with the action of the adjoint group (see problem 8 in 10.4).

In our case the adjoint group is $SO(3, \mathbf{C})$, and the adjoint representation is the well-known homomorphism of $SL(2, \mathbf{C})$ onto $SO(3, \mathbf{C})$ (corresponding to the mapping of $\mathfrak{sl}(2, \mathbf{C})$ onto $\mathfrak{so}(3, \mathbf{C})$ described in 18.1).

Problem 1. Prove that every polynomial $P \in \mathbf{C}[X, Y, Z]$ that is invariant with respect to $SO(3, \mathbf{C})$ has the form $f(R)$, where

$$R = X^2 + Y^2 + Z^2.$$

Hint. The group $SO(3, \mathbf{C})$ acts transitively on the spheres $X^2 + Y^2 + Z^2 =$ const $\neq 0$.

In view of theorem 2 of 10.4, it follows from this that the center $Z(\mathfrak{g})$ of the algebra $U(\mathfrak{g})$ is generated by the element $\sigma(R)$.

We note that the restriction of σ to $\mathbf{C}[R]$ is not an isomorphism of the algebras $\mathbf{C}[R]$ and $Z(\mathfrak{g})$ (although these algebras are in fact isomorphic). For example, one can verify that

$$\sigma(R^2) = [\sigma(R)]^2 - \tfrac{1}{3}\sigma(R).$$

For our purposes we need to know the explicit form of the invariant subspaces for the group $SO(3, \mathbf{C})$ in the space $\mathbf{C}[X, Y, Z]$.

To accomplish this, we introduce the space H_n of homogeneous harmonic polynomials of degree n, that is, polynomials P that satisfy the equation

$$\Delta P = \frac{\partial^2 P}{\partial X^2} + \frac{\partial^2 P}{\partial Y^2} + \frac{\partial^2 P}{\partial Z^2} = 0. \tag{2}$$

Problem 2. Prove that H_n is invariant and irreducible with respect to the group $SO(3, \mathbf{C})$ and that the corresponding representation T of the Lie algebra is equivalent to T_{2n} (in the notation of 18.1).

Hint. Prove that the operator Δ commutes with the action of the group $SO(3, \mathbf{C})$ and that X_+^n is the unique element of H_n (up to a constant factor) annihilated by the operator

$$T(X_+) = -2X_+ \frac{\partial}{\partial X_0} + X_0 \frac{\partial}{\partial X_-}.$$

Problem 3. Prove that every irreducible subspace H in $\mathbf{C}[X, Y, Z]$ has the form

$$H = f(R)H_n.$$

§ 18. Lie Groups and Lie Algebras

Hint. Using the explicit form of the one-parameter subgroup $S_t = e^{tT(X_+)}$, prove that every element $P \in \mathbb{C}[X, Y, Z]$ that is invariant under S_t is a polynomial in X_+ and R. Use also the fact that X_+ has weight 2 and R weight 0.

Problem 4. Prove that every subspace in $U(\mathfrak{g})$ that is irreducible with respect to the adjoint group has the form

$$\sigma(f(R))\sigma(H_n) \qquad (3)$$

for a certain $f \in \mathbb{C}[R]$ and a certain integer $n \geq 0$.

Hint. Use the result of problem 3.

Now let I be a two-sided ideal in $U(\mathfrak{g})$. Then it is invariant under the operation of commutation with elements of $\mathfrak{g}$ and consequently is invariant under the action of the adjoint group. Hence I is the direct sum of subspaces of the form (3) and consequently can be written in the form

$$I = \sum_{n=0}^{\infty} I_n \sigma(H_n), \qquad (4)$$

where I_n is a certain subspace in $Z(\mathfrak{g})$.

Since I is an ideal, it is clear that all of the I_n's are ideals in $Z(\mathfrak{g})$. Now, in the ring of polynomials in one variable, every ideal is principal. Hence there is a certain polynomial f_n, unique up a constant multiple, such that $I_n = f_n(\sigma(R))Z(\mathfrak{g})$.

Thus *every two-sided ideal $I \subset U(\mathfrak{g})$ can be characterized by a collection of polynomials f_n of one variable.*

Problem 5. Prove that $I_n \subset I_{n+1}$ and consequently that f_n is divisible by f_{n+1}.

Hint. Use the invariance of I under multiplication by $\sigma(X_+)$ and the fact that $X_+^n \in H_n$.

We shall now show that the ratio f_n/f_{n+1} is also not arbitrary. For this, we make a closer study of the space $\sigma(H_n)\sigma(H_1)$, which is generated by products of elements in $\sigma(H_n)$ and elements in $\sigma(H_1)$. It follows from the results of 18.2 that a representation of the algebra $\mathfrak{g}$ in $\sigma(H_n)\sigma(H_1)$ can contain only the components T_{2n+2}, T_{2n} and T_{2n-2}.

In view of problem 4, the corresponding representations are realized in the spaces $\sigma(f(R))\sigma(H_{n+1})$, $\sigma(g(R))\sigma(H_n)$ and $\sigma(h(R))\sigma(H_{n-1})$ respectively. Consideration of the highest terms quickly shows that the polynomials f and g have degree zero (that is, are constants) and that the polynomial h is of degree not higher than 1.

Thus we have

$$\sigma(H_n)\sigma(H_1) \subset \sigma(H_{n+1}) + \sigma(H_n) + (\alpha_n \sigma(R) + \beta_n)\sigma(H_{n-1}).$$

We find the explicit form of the decomposition of the element $X_+^n X_- \in \sigma(H_n)\sigma(H_1)$. Since this element has weight $2n-2$, its component in $\sigma(H_{n+1})$ is proportional to $(adX_-)^2 X_+^{n+1}$, its component in $\sigma(H_n)$ is proportional to $adX_- \cdot X_+^n$, and the last component has the form $(\alpha_n \sigma(R) + \beta_n) X_+^{n-1}$.

Using the commutation relations among X_+, X_-, and X_0, we easily compute

$$(adX_-)^2 X_+^{n+1} = -(n+1)[2X_+^n X_- - nX_+^{n-1}(X_0+n)(X_0+n-1)],$$
$$(adX_-)X_+^n = -nX_+^{n-1}(X_0+n-1),$$
$$(\alpha_n\sigma(R)+\beta_n)X_+^{n-1} = 4\alpha_n X_+^n X_- + X_+^{n-1}[\alpha_n X_0(2-X_0)+\beta_n].$$

From this, the coefficients α_n and β_n are found without difficulty. They turn out to be equal to

$$\alpha_n = \frac{n}{4n-2}, \quad \beta_n = \frac{n^3-n}{4n-2}.$$

The result just obtained has an important corollary.

Problem 6. Prove that the ideal I_{n-1} contains $[\sigma(R)+n^2-1]I_n$.

Hint. If $f \in I_n$, then we have $f(\sigma(R))X_+^n X_- \in I$, and so $f(\sigma(R))(\alpha_n\sigma(R)+\beta_n) \in I_{n-1}$.

Thus if $\{F_n\}$ is the collection of polynomials characterizing the ideal I, then the ratio f_{n-1}/f_n is a divisor of the monomial $\sigma(R)+n^2-1$.

Problem 7. Prove that the above condition is not only necessary but also sufficient in order that the sequence $\{f_n\}$ shall correspond to a certain ideal $I \subset U(\mathfrak{g})$.

Hint. Use the fact that the algebra $U(\mathfrak{g})$ is generated by the space $\sigma(H_1)$.

We shall examine in somewhat more detail the collections $\{f_n\}$ that satisfy our condition.

We normalize the polynomials f_n by the requirement that the highest coefficient be equal to 1. It is then clear that the sequence $\{f_n\}$ stabilizes: beginning with a certain N, we shall have $f_N = f_{N+1} = f_{N+2} = \ldots$. In this case, we write f_∞ for f_N. In order to reconstruct the entire sequence $\{f_n\}$ it suffices to know, in addition to f_∞, the subsequence S of the sequence of natural numbers consisting of the numbers n for which $f_n \neq f_{n+1}$. Then we have

$$f_n(x) = f_\infty(x) \prod_{\substack{k \in S \\ k \geq n}} [x+k(k+2)].$$

It is interesting to compare the structure of ideals in $U(\mathfrak{g})$ with the structure of ideals in $Z(\mathfrak{g})$. Every ideal $I \subset U(\mathfrak{g})$ upon being intersected with $Z(\mathfrak{g})$ yields an ideal I_0, which is generated by the polynomial

$$f_0(\sigma(R)) = f_\infty(\sigma(R)) \prod_{k \in S} [\sigma(R)+k(k+2)].$$

If f_0 does not have roots of the form

$$-k(k+2), \quad k = 0, 1, 2, \ldots, \tag{3}$$

then I is uniquely determined by I_0; namely, we have $I = I_0 U(\mathfrak{g})$. If f_0 does have roots of the form (3), there are several ideals $I \subset U(\mathfrak{g})$ having the same intersection

§ 18. Lie Groups and Lie Algebras

I_0 with $Z(\mathfrak{g})$. There are 2^m such ideals, where m is the number of roots of f_0 of the form (3).

In particular, *for prime ideals of $Z(\mathfrak{g})$, for which $f_0(x) = x - \lambda$, there is either one ideal in $U(\mathfrak{g})$, or two.*

It would be very interesting to obtain an analogous description of the structure of the ideals for the enveloping algebras of other Lie algebras.

18.4. Spinor Representations of the Symplectic Group

Let $A_n(\mathbf{R})$ be the Weyl algebra defined in 18.2. Under the operation $[x, y] = xy - yx$, it is an infinite-dimensional real Lie algebra. Let σ be the isomorphism of $\mathbf{R}[p_1, \ldots, p_n, q_1, \ldots, q_n]$ onto $A_n(\mathbf{R})$, given in the symmetric form (see 18.2).

Problem. Prove that if x and y are two polynomials in the variables $p_1 \ldots p_n$, $q_1 \ldots q_n$ of degrees k and l respectively, then $[x, y]$ is a polynomial of degree $\leq k + l - 2$.

It follows from this that the set of polynomials of degree ≤ 2 in the variables $p_1 \ldots p_n, q_1 \ldots q_n$ is a subalgebra of $A_n(\mathbf{R})$. We denote this subalgebra by $\mathfrak{st}(n, \mathbf{R})$. It is also clear that polynomials of degree ≤ 1 form a nilpotent ideal $\mathfrak{n}$ in $\mathfrak{st}(n, \mathbf{R})$, while constants form the center $\mathfrak{z}$ of this ideal. Let $\mathfrak{m}$ denote the subspace generated by the generators p and q, and let $\mathfrak{a}$ denote the subspace of symmetric homogeneous polynomials of the second degree in these generators. Then it is easy to verify that

$$[\mathfrak{a}, \mathfrak{a}] \subset \mathfrak{a}, \quad [\mathfrak{a}, \mathfrak{m}] \subset \mathfrak{m}, \quad [\mathfrak{m}, \mathfrak{m}] \subset \mathfrak{z}.$$

Let $\widetilde{St}(n, \mathbf{R})$ be the connected and simply connected Lie group that corresponds to the algebra $\mathfrak{st}(n, \mathbf{R})$ (see 15.3 just after problem 3). Let A, N, and Z denote the subgroups corresponding to the subalgebras $\mathfrak{a}$, $\mathfrak{n}$, and $\mathfrak{z}$. The group N is usually called the *Heisenberg group*.

Problem 1. Prove that N is isomorphic to the group of matrices of order $n+2$ of the following form:

$$g(a,b,c) = \begin{pmatrix} 1 & a & c \\ 0 & 1_n & b \\ 0 & 0 & 1 \end{pmatrix}.$$

Here a is an n-dimensional row vector, b is an n-dimensional column vector, and 1_n is the unit matrix of order n.

Hint. Show that the corresponding Lie algebras are isomorphic.

Problem 2. Prove that the algebra $\mathfrak{a}$ is isomorphic to the Lie algebra $\mathfrak{sp}(2n, \mathbf{R})$ of the symplectic group $Sp(2n, \mathbf{R})$, and that the group A is isomorphic to the simply connected covering group of $Sp(2n, \mathbf{R})$.

Hint. Consider the adjoint representation of $\mathfrak{a}$ in $\mathfrak{m}$. Prove that $\widetilde{St}(n, \mathbf{R}) = A \cdot N$ (semidirect product).

The group N admits a series of unitary irreducible representations U_λ, $\lambda \in \mathbf{R}^*$. This representation acts in $L^2(\mathbf{R}^n)$ by the formula

$$U_\lambda(g(a,b,c))f(x) = e^{i\lambda(bx+c)}f(x+a). \tag{1}$$

For brevity, we write x instead of $x_1, \ldots, x_n$, bx instead of $\sum_{j=1}^{n} b_j x_j$, and so on. The corresponding representation of the Lie algebra has the form

$$p_j \mapsto \frac{\partial}{\partial x_j}, \quad q_j \mapsto i\lambda x_j, \quad 1 \mapsto i\lambda. \tag{2}$$

We can extend the representation (2) to a representation of $\mathfrak{st}(n, \mathbf{R})$ by the formulas

$$p_j p_k \mapsto \frac{1}{i\lambda} \frac{\partial^2}{\partial x_j \partial x_k}, \quad q_j p_k \mapsto x_j \frac{\partial}{\partial x_k}, \quad q_j q_k \mapsto i\lambda x_j x_k. \tag{3}$$

Problem 3. Prove that the representation obtained in this way corresponds to a certain unitary representation T_λ of the group $\widetilde{St}(n, \mathbf{R})$.

Hint. Use Nelson's criterion given in 10.5.

We denote the restriction of the representation T_λ to the group A by the symbol S_λ and call it the spinor representation.

Problem 4. Prove that the representations S_λ and S_μ are equivalent if and only if $\lambda\mu > 0$.

Hint. Consider the transformation of the space $L^2(\mathbf{R}^n)$ corresponding to a dilation of $\mathbf{R}^n$. The inequivalence of S_λ and S_μ for $\lambda\mu < 0$ follows from considering the spectrum of the operator corresponding to the element q^2.

Problem 5. Prove that the spinor representations S_λ are reducible and decompose into two irreducible components acting in the subspaces of even and odd functions, respectively.

Hint. Prove that the space $\mathscr{C}(S_\lambda)$ is generated by the identity operator and the operator of reflection in the origin.

We consider the case $n=1$ more closely. The group $Sp(2, \mathbf{R})$ is isomorphic to $SL(2, \mathbf{R})$. Every matrix in $SL(2, \mathbf{R})$ can be uniquely written in the form

$$\begin{pmatrix} \cos\phi & \sin\phi \\ -\sin\phi & \cos\phi \end{pmatrix} \exp\begin{pmatrix} a & b \\ b & -a \end{pmatrix}.$$

Hence the topological group $SL(2, \mathbf{R})$ is equivalent to $\mathbf{T} \times \mathbf{R}^2$. Its universal covering space is homeomorphic to $\mathbf{R}^3$ and its fundamental group is isomorphic to $\mathbf{Z}$.

Problem 6. Verify that the isomorphism between $\mathfrak{sl}(2, \mathbf{R})$ and the algebra $\mathfrak{a}$ generated by the elements p^2, q^2, and $pq + qp$ in $A_1(\mathbf{R})$ has the form

$$\begin{pmatrix} a & b \\ c & -a \end{pmatrix} \leftrightarrow \tfrac{1}{2}[a(pq+qp) + bq^2 - cp^2].$$

§ 18. Lie Groups and Lie Algebras

Hint. Use the relations

$$[p^2, q^2] = 2(pq+qp), \quad [pq, p^2] = -2p^2, \quad [pq, q^2] = 2q^2.$$

We shall now show that the spinor representation S_λ is not a single-valued representation of $SL(2, \mathbf{R})$, but becomes single-valued on the two-sheeted covering of this group. For this we consider the element $X = \begin{pmatrix} 0 & 1 \\ -1 & 0 \end{pmatrix}$ of the Lie algebra $\mathfrak{sl}(2, \mathbf{R})$, corresponding to the closed one-dimensional subgroup $SO(2, \mathbf{R}) \subset SL(2, \mathbf{R})$.

The expression $\frac{1}{2}(p^2 + q^2)$ corresponds to this element in the algebra $\mathfrak{a}$, and under the representation $S_{\pm 1}$, it goes into the operator $\mp i(x^2 + d^2/dx^2)$.

To study this operator it is convenient to replace the realization of our representation in $L^2(\mathbf{R})$ by a different realization in the space $H^2(\mathbf{C})$ of holomorphic functions on $\mathbf{C}$ that are square integrable with respect to the measure $e^{-|z|^2} dx dy$ (where $z = x + iy$).

Problem 7. Prove that the operators z and d/dz are conjugate to each other in $H^2(\mathbf{C})$.

Hint. Use the fact that polynomials in z are dense in $H^2(\mathbf{C})$, and also use the formula

$$\iint_\mathbf{C} |z|^{2k} e^{-|z|^2} dx dy = \pi k!.$$

A representation of the algebra $\mathfrak{n}$ in $H^2(\mathbf{C})$ is defined by the formula

$$p \mapsto \frac{i}{\sqrt{2}} \left(z + \frac{d}{dz} \right), \quad q \mapsto \frac{\lambda}{\sqrt{2}} \left(z - \frac{d}{dz} \right), \quad 1 \mapsto i\lambda,$$

and the corresponding representation of $\mathfrak{a}$ has the form

$$p^2 \mapsto \frac{i}{2\lambda} \left(z + \frac{d}{dz} \right)^2, \quad (pq+qp) \mapsto z^2 - \frac{d^2}{dz^2}, \quad q^2 \mapsto -\frac{i\lambda}{2} \left(z - \frac{d}{dz} \right)^2.$$

In particular, for $\lambda = \pm 1$, the element we are concerned with goes into the operator $\pm i \left(z \dfrac{d}{dz} + \dfrac{1}{2} \right)$. We easily compute from this that

$$\left[S_{\pm 1} \begin{pmatrix} \cos \phi & -\sin \phi \\ -\sin \phi & \cos \phi \end{pmatrix} f \right](z) = e^{\pm i\phi/2} f(ze^{\pm i\phi}).$$

Thus we get the identity operator for $\phi = 4k\pi$, $k \in \mathbf{Z}$, which proves our assertion.

18.5. Representations of Triangular Matrix Groups

Let $T(n, \mathbf{R})$ denote the group of all nonsingular real matrices of the n-th order for which all elements lying below the main diagonal are zero, while elements on main diagonal are positive. The corresponding Lie algebra $\mathfrak{t}(n, \mathbf{R})$ consists of all triangular matrices.

Problem 1. Prove that the group $T(n, \mathbf{R})$ is solvable and exponential, that is, the mapping

$$\exp: \mathfrak{t}(n, \mathbf{R}) \to T(n, \mathbf{R})$$

is a homeomorphism.

Thus we can apply to the group $T(n, \mathbf{R})$ the theorem of 15.3 stating that there is a one-to-one correspondence between unitary irreducible representations and orbits in the K-representation.

We shall list here all of the orbits of maximal dimension and shall construct the corresponding representations.

We use the method pointed out in 15.1. We realize the space $\mathfrak{t}(n, \mathbf{R})^*$ in the form of lower triangular matrices, and we give the K-representation by the formula

$$K(g): F \mapsto (gFg^{-1})_{\text{lower}} \tag{1}$$

where the index "lower" means that we are interested only in the lower triangular part of the matrix so obtained, and that all elements lying above the diagonal are to be replaced by zeros.

The structure of orbits in the K-representation is somewhat different for even and odd n.

For specificity, we consider the case $n = 2k$. It is convenient to write the matrix $g \in T(2k, \mathbf{R})$ in the form

$$g = \begin{pmatrix} A & C \\ 0 & B \end{pmatrix}, \tag{2}$$

where $A, B \in T(k, \mathbf{R})$, $C \in \text{Mat}_k(\mathbf{R})$. We write the element $F \in \mathfrak{t}(2k, \mathbf{R})^*$ in the form

$$F = \begin{pmatrix} X & 0 \\ Z & Y \end{pmatrix}, \tag{3}$$

where $X, Y \in \mathfrak{t}(k, \mathbf{R})^*$, $Z \in \text{Mat}_k(\mathbf{R})$.

Using this notation, we find that the K-representation looks like this:

$$K(A,B,C): (X,Y,Z)$$
$$\mapsto ([(AXA^{-1} + CZA^{-1})]_{\text{lower}}, [B(Y - ZA^{-1}C)B^{-1}]_{\text{lower}}, BZA^{-1}). \tag{4}$$

We first examine what can be done with the matrix Z by transformations of the form (4).

§ 18. Lie Groups and Lie Algebras

Let J be the matrix with ones on the secondary diagonal and with zeros elsewhere, and let J_ε be the matrix obtained from J by replacing some of the ones by -1.

Lemma. *Almost every matrix $Z \in \mathrm{Mat}_k(\mathbf{R})$ (except for a manifold of lower dimension) can be written in the form $Z = BJ_\varepsilon A^{-1}$, where $A, B \in T(k, \mathbf{R})$.*

To prove this assertion, one can consider the linear matrix equation $ZA = BJ_\varepsilon$ in the unknowns A and B, and study its consistency.

One can also use the following geometric considerations. We consider what is called the *manifold of oriented flags*.

A point of this space is a collection of oriented spaces $V_1 \subset V_2 \subset \ldots \subset V_{k-1} \subset V_k = \mathbf{R}^k$ such that $V_k = k$.

The name "flag manifold" comes from the fact that in the case $k = 3$, it is convenient to picture the pair $V_1 \subset V_2$ to one's self as a flagpole with a flag hanging from it.

Problem 2. Verify that the group $GL(k, \mathbf{R})$ acts transitively on X and that the stationary subgroup of a certain point is the group $T(k, \mathbf{R})$.

Hint. Let $x_0 \in X$ be the flag $\mathbf{R}_1^1 \subset \mathbf{R}_1^2 \subset \ldots \subset \mathbf{R}_k$ with standard orientations. If $V_0 \subset V_1 \subset V_2 \subset \ldots \subset V_k$ is any other flag x, then one can choose a basis $e_1, \ldots, e_k$ in $\mathbf{R}_k$ such that $e_1, \ldots, e_k$ is a positively oriented basis in V_j for all j from 1 to k. Then the transformation from $e_1, \ldots, e_k$ to the standard basis carries x into x_0.

The space $Y = T(k, \mathbf{R}) \backslash GL(k, \mathbf{R}) / T(k, \mathbf{R})$, as we already know from 2.2, can be identified with a set of orbits in the space of pairs of flags. The assertion above is equivalent to the fact that this space contains an everywhere dense open set consisting of 2^k points, corresponding to the matrices J_ε.

Problem 3. Prove that the space Y is finite and consists of $2^k \cdot k!$ points.

Hint. As representatives of the double cosets, one can take matrices that permute basis elements and matrices obtained from these by changing some of the ones to -1.

We turn now to a study of the orbits. Choosing the transformation (4) in a suitable way, we can bring the matrix Z to the form J_ε. We will choose our subsequent transformations in such a way that this form is preserved.

Problem 4. Prove that if $A, B \in T(k, \mathbf{R})$ and $AJ_\varepsilon B^{-1} = J_\varepsilon$, then A and B are diagonal matrices, which are obtained from each other by reflection in the secondary diagonal.

Hint. If A is an upper triangular matrix, then $J_\varepsilon A J_\varepsilon^{-1}$ is a lower triangular matrix.

Further, if we consider the transformation (4) for $A = B = 1_k$ and $Z = J_\varepsilon$, we easily see that we can make Y zero and X a diagonal matrix by properly choosing C.

Thus we have brought the element F into canonical form:

$$\begin{pmatrix} D & 0 \\ J_\varepsilon & 0 \end{pmatrix}. \tag{5}$$

Problem 5. Prove that different elements of the form (5) belong to different orbits.

Hint. Prove that a transformation of the form (4) that carries an element in canonical form to another element in canonical form has the properties that $C=0$ and that A and B are diagonal.

Thus the orbits of elements in canonical form are identified by k real numbers and the index ε.

One can show (see the hint for problem 4) that the stationary subgroups of elements in canonical form have the form $g(A, JAJ^{-1}, 0)$ where A is a diagonal matrix.

Therefore the corresponding orbits have dimension $k(2k+1)-k=2k^2$. Although we could, we shall not prove here that we obtain in this way all orbits of dimension $2k^2$ and that all other orbits have smaller dimension.

As an admissible subalgebra for the functional F of the form (5), we can take the algebra $\mathfrak{h}$ of matrices of the form $g(A, JAJ^{-1}, C)$, where A is a diagonal matrix. The corresponding subgroup consists of matrices of the same form with positive elements on the diagonal. We set

$$\chi_{D,\varepsilon}(g(A, JAJ^{-1}, C)) = e^{2\pi i \operatorname{tr} J} \varepsilon^C \sum_{j=1}^{k} a_j^{2\pi i d_j},$$

and

$$T_{D,\varepsilon} = \operatorname{Ind}(T(n, \mathbf{R}), H, \chi_{D,\varepsilon}).$$

This is the principal series of representations of the group $T(n, \mathbf{R})$.

There is an interesting problem here, that of defining the parameters D and ε directly from the representation T that corresponds to these parameters.

It turns out that the elements d of the matrix D can be expressed in terms of the generalized infinitesimal character of the representation T (see 11.3). It would appear that the parameter ε can be found as the spectrum of certain operators corresponding to the enveloping algebra of the Lie algebra of the group $T(2k, \mathbf{R})$.

It is a useful exercise to compute explicitly the characters of the representations $T_{D,\varepsilon}$ as generalized functions on the group. For the subgroup $T_0(n, \mathbf{R})$ of triangular matrices with ones on the main diagonal, this computation has been carried out in [99].

§ 19. Examples of Wild Lie Groups

We can construct a very simple example of a Lie group not belonging to type I as follows. Let α be an irrational real number. Consider the group G of matrices of the form

$$\begin{pmatrix} e^{it} & 0 & z \\ 0 & e^{i\alpha t} & w \\ 0 & 0 & 1 \end{pmatrix}, \quad t \in \mathbf{R}, \quad z, w \in \mathbf{C}. \tag{1}$$

§ 19. Examples of Wild Lie Groups

It is clear that G is a Lie group. As a topological space, G is homeomorphic to $\mathbf{R}^5$. As a group, it is a semidirect product of the one-dimensional subgroup $z=w=0$ and the four-dimensional commutative normal subgroup $t=0$.

We consider the K-representation of this group. Identifying the space $\mathfrak{g}^*$ with the space of matrices of the form

$$\begin{pmatrix} i\tau & 0 & 0 \\ 0 & i\alpha\tau & 0 \\ a & b & 0 \end{pmatrix}, \qquad \tau \in \mathbf{R}, \quad a, b \in \mathbf{C}, \tag{2}$$

we obtain the following explicit form of the action of G in $\mathfrak{g}^*$:

$$K(t, z, w) : (a, b, \tau) \mapsto (a e^{-it}, b e^{-it}, b e^{-i\alpha t}, \tau + \mathrm{Im}(az + bw)).$$

Thus G-orbits in $\mathfrak{g}^*$ are two-dimensional cylinders, the generators of which are the axes τ, and the bases of which are curves defined by the parametric equations

$$a = a_0 e^{it}, \quad b = b_0 e^{i\alpha t}, \quad t \in \mathbf{R}. \tag{3}$$

It is clear that the curve defined by (3) is everywhere dense on the surface of the torus

$$|a| = |a_0|, \quad |b| = |b_0|.$$

The partition of $\mathfrak{g}^*$ into G-orbits does not satisfy the condition of 9.1, and the corresponding factor space $\mathcal{O}(G)$ is not a T_0 space.

Problem 1. Prove that G admits a two-sided invariant measure which in the parameters $t, x = z + iy, w = u + iv$ has the form

$$dg = dt\, dx\, dy\, du\, dv.$$

We consider the regular representation T of the group G. It acts in $L^2(G, dg)$ by right translations, and in the coordinates t, z, w can be described by the formula

$$[T(\tau, \xi, \eta) f](t, z, w) = f(t + \tau, z + e^{it}\xi, w + e^{i\alpha t}\eta). \tag{4}$$

We shall now show that this representation can be decomposed into irreducible components in two different ways.

First method. We carry out the Fourier transform in the variables z and w:

$$\tilde{f}(t, a, b) = \int\int\int\int f(t, z, w) e^{i\mathrm{Re}(a\bar{z} + b\bar{w})} dx\, dy\, du\, dv. \tag{5}$$

Problem 2. Verify that under the action of the Fourier transform, T goes into the representation T_1 defined by the equality

$$[T_1(\tau, \xi, \eta) \tilde{f}](t, a, b) = e^{i\mathrm{Re}(e^{it} a\bar{\xi} + e^{i\alpha t} b\bar{\eta})} \tilde{f}(t + \tau, a, b). \tag{6}$$

Formula (6) shows that T_1 is the continuous sum of representations $U_{a,b}$, $a, b \in \mathbf{C}$, acting in $L^2(\mathbf{R}, dt)$ by the formula

$$[U_{a,b}(\tau, \xi, \eta)\phi](t) = e^{i\operatorname{Re}(e^{it}a\bar{\xi} + e^{i\alpha t}b\bar{\eta})}\phi(t+\tau). \tag{7}$$

Problem 3. Prove that all of the representations $U_{a,b}$ for $a \neq 0$, $b \neq 0$ are irreducible.

Hint. Verify that every operator commuting with the operators $U_{a,b}(0, \xi, \eta)$ is an operator of multiplication by a function.

Problem 4. Prove that representations $U_{a,b}$ and U_{a_1, b_1} are equivalent if and only if there is a real number τ such that $a_1 = ae^{i\tau}$, $b_1 = be^{i\alpha\tau}$.

Hint. Verify that the intertwining operator for U_{a_1, b_1} and $U_{a,b}$ can be only the operator of translation by τ.

Second method. We make a change of variables:

$$f(t, z, w) = \phi(t, e^{-it}z, e^{-i\alpha t}w)$$

and then take the Fourier transform:

$$\tilde{\phi}(s, a, b) = \int\int\int\int\int \phi(t, z, w)e^{i\operatorname{Re}(ts + a\bar{z} + b\bar{w})} dt\, dx\, dy\, du\, dv. \tag{8}$$

Problem 5. Verify that after the transformation (8), the representation T goes into a representation T_2 defined by the equality

$$[T_2(\tau, \xi, \eta)\tilde{\phi}](s, a, b) = e^{i\operatorname{Re}(\tau s + e^{i\tau}a\bar{\xi} + e^{i\alpha\tau}b\bar{\eta})}\tilde{\phi}(s, e^{i\tau}a, e^{i\alpha\tau}b). \tag{9}$$

Hint. It is useful to examine separately the transformations $T_2(0, \xi, \eta)$ and $T_2(\tau, 0, 0)$.

Let $X_{r,\rho}$ be the two-dimensional surface in $\mathbf{C}^2$ given by the equations $|a| = r$, $|b| = \rho$, where r and ρ are nonnegative real numbers.

The representation T_2 decomposes naturally into a continuous sum of representations $V_{r,\rho,s}$ acting in $L^2(X_{r,\rho})$ by the formula

$$[V_{r,\rho,s}(\tau, \xi, \eta)\phi](a, b) = e^{i\operatorname{Re}(\tau s + e^{i\tau}a\bar{\xi} + e^{i\alpha\tau}b\bar{\eta})}\phi(e^{i\tau}a, e^{i\alpha\tau}b). \tag{10}$$

Problem 6. Prove that the representations $V_{r,\rho,s}$ are irreducible for all r, ρ, and s.

Hint. Use the fact that every measurable function on $X_{r,\rho}$ that is invariant under $V_{r,\rho,s}(\tau, 0, 0)$ is a constant.

Problem 7. Prove that representations $V_{r,\rho,s}$ and $V_{r,\rho,s'}$ are equivalent if and only if the difference $s - s'$ can be written in the form $m + \alpha n$, $m, n \in \mathbf{Z}$.

Hint. An intertwining operator is an operator of multiplication by a function of the form $a^m b^n$.

Problem 8. Prove that no representation $U_{a,b}$ is equivalent to a representation $V_{r,\rho,s}$.

§ 19. Examples of Wild Lie Groups

Hint. Consider the operator that restricts these representations to the subgroup $t=0$.

Thus the two decompositions obtained above are essentially different. This shows that the group G does not belong to type I.

We can construct a more refined example of a Lie group not of type I, in which the second condition of Kostant-Auslander fails (see 15.3, theorem 3), as follows.

Let $\mathfrak{g}$ be the Lie algebra of dimension 7 over $\mathbf{R}$ with basis X, Y, U, V, S, T, R and the following nonzero commutators:

$$[S, X] = Y, \quad [S, Y] = -X,$$
$$[T, U] = V, \quad [T, V] = -U, \quad [S, T] = R.$$

We can realize this algebra by the blockwise diagonal matrices of the 6th order with diagonal blocks

$$\begin{pmatrix} is & 0 & z \\ 0 & it & w \\ 0 & 0 & 0 \end{pmatrix} \quad \text{and} \quad \begin{pmatrix} 0 & s & r \\ 0 & 0 & t \\ 0 & 0 & 0 \end{pmatrix},$$

where $z = x + iy$, $w = u + iv$ and x, y, s, t, u, v, r are real parameters (coordinates in the basis X, Y, S, T, U, V, R).

Let G be the connected and simply connected Lie group that corresponds to the algebra $\mathfrak{g}$.

Problem 9. Prove that orbits Ω of maximal dimension in $\mathfrak{g}^*$ are topologically equivalent to $\mathbf{T}^2 \times \mathbf{R}^2$ and that the cohomology class defined by the canonical form B_Ω is different from zero.

Hint. Considering X, Y, S, T, U, V, R as linear functions on $\mathfrak{g}^*$, prove that the equations

$$X^2 + Y^2 = r_1^2, \quad U^2 + V^2 = r_2^2, \quad R = r_3$$

give an orbit for $r_1 \neq 0, r_2 \neq 0, r_3 \neq 0$. Prove that we have

$$B_\Omega = d\phi \wedge dS + d\psi \wedge dT + R d\phi \wedge d\psi$$

on this orbit. Here ϕ and ψ are parameters on Ω defined by the equalities $X = r_1 \cos\phi$, $Y = r_1 \sin\phi$, $U = r_2 \cos\psi$, $V = r_2 \sin\psi$.

The fact that the group G does not belong to type I can be established in the following way. There is a commutative normal subgroup of G whose Lie algebra is spanned by the vectors X, Y, U, V. Applying the reasoning of 13.3 and 14.1 to this case, we easily show that a study of the representations of the group G can be reduced to a study of the representations of the group G_0 of matrices of the form

$$\begin{pmatrix} 1 & m & r \\ 0 & 1 & n \\ 0 & 0 & 1 \end{pmatrix}, \quad m, n \in \mathbf{Z}, \quad r \in \mathbf{R}.$$

Problem 10. Prove that the group G_0 does not belong to type I.

Hint. Let H be the subgroup of G_0 defined by the condition $m=0$. For arbitrary real numbers λ and ϕ, we define a character $\chi_{\lambda,\phi}$ of the group H by the following equality:

$$\chi_{\lambda,\phi}(n,r) = e^{i(\lambda r + n\phi)}.$$

Prove that the representations $T_{\lambda,\phi} = \mathrm{Ind}(G_0, H, \chi_{\lambda,\phi})$ are irreducible and that $T_{\lambda,\phi} \sim T_{\lambda_1,\phi_1}$ if and only if $\lambda = \lambda_1$ and also $\phi - \phi_1$ has the form $2\pi k + \lambda l$, where k and l are integers.

Infer from this that if λ is not a rational multiple of π, then the equivalence classes of the representations $T_{\lambda,\phi}$ and T_{λ,ϕ_1} cannot be separated by any open set in $\hat{G}$.

It would be very interesting to apply the method of orbits to the study of representations of wild Lie groups. It seems extremely likely that the rôle of orbits will be taken over by ergodic G-invariant measures in $\mathfrak{g}^*$, while the formula (Φ) expresses the relative trace of an operator of the representation in the sense of von Neumann. It is also possible that one can express in terms of orbits the decomposition of the regular representation into factors and give an explicit formula for the generalized Plancherel measure in the sense of I. Segal (see 12.4)[1].

---

[1] After this passage was written, the author received a preprint of an interesting article by L. Pukanszky, in which a part of this program is realized.

A Short Historical Sketch and a Guide to the Literature

The theory of group representations has existed as an independent domain of study for about 80 years.

The first period of its development (approximately 1890 to 1920) is connected with the names of G. Frobenius and Schur, W. Burnside and F. E. Molin. Only finite groups and finite-dimensional representations were considered in the works of this period.

The original impetus for the development of the theory of representations was the invention by Frobenius of the generalization of the notion of character. This was based on a suggestion of Dedekind. In present-day terminology, a character in the sense of Frobenius is a multiplicative functional on the center of a group ring. It was soon discovered that this character can also be defined as the trace of a matrix representation. Thus the theory of Frobenius and the theory of characters of commutative groups, which had been developed earlier, were united in a new single theory of representations.

A striking achievement of this epoch was the systematic use by I. Schur of the process of averaging over a group and his famous lemma on intertwining operators for irreducible representations. The orthogonality relations were obtained in general form in the works of Burnside, along with an elucidation of the structure of irreducible matrix algebras. This result, as well as the theorem of F. E. Molin on the semisimplicity of a group algebra, connected the theory of representations of finite groups with the theory of finite-dimensional algebras. The theory of projective representations of finite groups was constructed in one of the papers of Schur.

Besides the general results outlined above, the first period saw the accumulation of many concrete facts on representations of specific groups and of certain special classes of groups.

The second period is marked by the creation of the theory of representations of compact topological groups. The most important general results of this period are the theorem of Haar-von Neumann on the existence of a finite invariant measure and the theorem of F. Peter-H. Weyl on the completeness of the system of finite-dimensional representations. During the same period, H. Weyl and E. Cartan evolved the theory of finite-dimensional representations of Lie groups. These results were not only astonishing for their beauty, but they also found wide applications in diverse fields of mathematics and physics (the theory of symmetric spaces and the theory of moments in quantum mechanics). The theory of group representations has become an applied science and its popularity is increasing rapidly.

The necessity of considering noncompact groups and their infinite-dimensional representations became clear at an early stage. For example, the well-known theorem of Stone-von Neumann on the uniqueness of the Schrödinger operator is in effect equivalent to the classification of the infinite-dimensional unitary representations of the simplest nilpotent group (called the Heisenberg group). E. Wigner in [137] made the first attempt to construct a theory of elementary particles on the basis of the theory of infinite-dimensional representations.

A systematic study of infinite-dimensional group representations, which is the principal theme of the third period, began in the 1940's. It is natural to mark the beginning of this period with the 1943 paper of I. M. Gel'fand and D. A. Raĭkov on the completeness of the system of unitary irreducible representations for locally compact groups. During this period, Murray and von Neumann completed their fundamental study of operator algebras [123]. The theory of von Neumann algebras was united with the theory of group representations in the works of G. M. Adel'son-Vel'skiĭ [1], F. Mautner [116], [117], and R. Godement [86], [87].

The first classification theorems for infinite-dimensional representations were obtained in 1947 by I. M. Gel'fand and M. A. Naĭmark [84], [83], and V. Bargmann [60]. The monograph [21] of I. M. Gel'fand and M. A. Naĭmark appeared in 1950, in which the infinite-dimensional representations of the groups $SL(n, \mathbf{C})$, $SO(n, \mathbf{C})$, and $Sp(n, \mathbf{C})$ were obtained for arbitrary n.

This work achieved wide recognition, and since its appearance the flow of investigations of infinite-dimensional representations has increased continuously.

We shall trace here the course of several streams in this great flow.

General theory. One of the main achievements is the recognition of the class of groups which we call *tame* in the present book. The equivalence of various definitions of this class has been proved in the past 10 to 15 years. The main results belong to G. Mackey, J. Fell, J. Dixmier, and J. Glimm. Their work is recapitulated in the monograph [15] of J. Dixmier.

Harish-Chandra proved in [90] that all semisimple groups are tame. A simpler and more elegant proof was found by R. Godement [88]. O. Takenouchi [134] established that all exponential groups are tame. The first example of a wild Lie group was apparently presented by F. Mautner (unpublished). Later this example was repeatedly rediscovered. L. Auslander and B. Kostant [59] found criteria for a solvable Lie group to be tame.

Up to this time, such criteria for general Lie groups have not been found.

Induced representations. The Frobenius reciprocity theorem was extended to compact groups by A. Weil in the book [54].

Even in the works of E. Wigner [137] and I. M. Gel'fand and M. A. Naĭmark [84], [83], mention is made of the fact that their construction is a generalization of the concept of induced representation. A systematic study of this construction for locally compact groups was undertaken by G. Mackey. One of his basic results is the criterion for inducibility, given in [113].

Using this criterion, Mackey pointed out a convenient algorithm for constructing and studying unitary representations of group extensions [114].

The concept of a holomorphically induced representation was implicitly used by Bargmann in [60] and explicitly introduced by I. M. Gel'fand and M. I. Graev in [79]. R. Blattner in [65] undertook a systematic study of induced and holomorphically induced representations. He also introduced the notion of a partially holomorphic representation and gave a simpler proof of Mackey's criterion.

In [66], R. Bott constructed a realization of the (finite-dimensional) representations of a semisimple compact Lie group G in cohomology spaces of a sheaf of germs of holomorphic sections of certain homogeneous line bundles. R. P. Langlands in [112] suggested that an analogous construction (with ordinary cohomology replaced by the so-called L^2-cohomology) could be applied to construct representations of noncompact semisimple groups. For a more detailed description of this construction, see for example [104]. G. Mackey has proposed still another generalization of the notion of induced representation. It arose from an attempt to extend the criterion of inducibility to nonhomogeneous ergodic G-spaces. More details about this generalization and the concept of virtual subgroup, which arises here, may be found in the survey [115] of Mackey. Some further results were obtained in [94].

Much work has been devoted to various generalizations of Frobenius reciprocity to noncompact groups (see for example [115]). The most convenient case for such a generalization is that of a compact factor space. An important theorem of this genre was obtained by I. M. Gel'fand and I. I. Pjateckiĭ-Šapiro for a semisimple group G and a discrete subgroup Γ. Interest in this case is heightened by the fact that the reciprocity theorem for the pair (G, Γ) and the trace formula arising from it allow the application of representation theory to certain difficult problems of modern number theory. More details about this circle of questions can be found in the monograph [18].

A general reciprocity theorem for the case of a compact factor space was obtained by G. I. Ol'šanskiĭ in [124]. In [118], C. Moore made an interesting attempt to preserve Frobenius's theory while sacrificing the condition that the representation be unitary.

Representations of semisimple Lie groups. The principal outlines of the theory of infinite-dimensional representations of complex semisimple groups were elucidated in the monograph [21]. It turned out in particular that what are called parabolic subgroups play a basic rôle in this theory. (A parabolic subgroup P of a complex semisimple group G can be defined as a connected complex subgroup having one of the two following equivalent properties. 1) The factor space G/P is compact. 2) The subgroup P contains a maximal solvable subgroup in G.) It was shown in papers of M. A. Naĭmark and D. P. Želobenko [120], [139] that all irreducible (also nonunitary) representations of a complex semisimple group G are contained in its elementary representations. (These are representations induced by a one-dimensional representation of a parabolic subgroup.) The study of the category of elementary representations is essentially simplified by the fact that the spaces of double cosets $P_1 \backslash G / P_2$ are finite for an arbitrary pair of parabolic subgroups P_1 and P_2. In particular, when P_1 and P_2 are both the Borel subgroup B (that is, the maximal solvable subgroup in G), the points of this space are identified with the elements of the Weyl group W of the group G.[1] In the monograph [21], characters of elementary representations are computed and Plancherel's formula is obtained for the complex classical groups.

Harish-Chandra has extended these results to general complex semisimple groups [90], [91].

F. A. Berezin has obtained in [62] a classification of irreducible representations in Banach spaces, using his own investigation of Laplace operators on semisimple Lie groups. We note that the identification of the unitary representations among those that he constructed turned out to be a very complicated problem technically. The simpler question of the existence of a G-invariant Hermitian form has a simple answer in terms of the inducing character of a parabolic subgroup. However, the problem of determining when this form is positive-definite has not been completely settled up to the present (cf. [89], [132]).

The theory of representations of real semisimple groups presents additional difficulties. In this case, the elementary representations are insufficient for the construction of a complete system, even if we use the concept of a holomorphically induced representation. I. M. Gel'fand and M. I. Graev obtained the first series of representations that are not elementary, for the group $SL(2, \mathbf{R})$. They called this the "strange" series. Only after the enunciation of Langlands' conjecture did it become clear that this series is naturally realized not in functions (that is, in zero-dimensional cohomology spaces) but in cohomology spaces of higher dimension.

Harish-Chandra has obtained very strong results in the theory of representations of real semisimple Lie groups. His work [93], which completed a long cycle of investigations, contains a complete classification of the so-called discrete series (that is, representations with square integrable matrix elements).

In particular, it has been shown that the real semisimple Lie groups admitting discrete series of representations are exactly those containing compact Cartan subgroups.

Nevertheless, in Harish-Chandra's work, the discrete series are described only in an indirect fashion. Namely, the restriction of the character of a representation to the regular part of a compact Cartan subgroup is pointed out, and it is proved that there exists only one representation (up to equivalence) having a character with this property.

A very interesting unsolved problem is that of finding an explicit realization of representations of the discrete series. One of the most promising approaches to this problem is the construction of representations in L^2-cohomology spaces, proposed by Langlands, and already cited above.

The most recent steps on the road to establishing Langlands' conjecture belong to W. Schmid [129], K. Okamoto and M. Narasimhan [121], and R. Parthasarathy [126].

---

[1] This important result was established by Gel'fand and Naĭmark for the complex classical groups, and by Harish-Chandra and Chevalley for the general case. In the literature it has gone by the name of "Bruhat's lemma" since the appearance of the paper [67], in which this lemma is used for studying intertwining operators.

The theory of infinite-dimensional representations of semisimple groups has led to new and essential progress in classical questions of the finite-dimensional theory.

I. M. Gel'fand and M. L. Cetlin succeeded in obtaining explicit formulas for finite-dimensional representations of the linear and orthogonal groups (see the appendix to the book [20]).

The results of D. P. Želobenko in [138] lead one to hope that analogous formulas can be written also for the symplectic group.

The methods of the theory of infinite-dimensional representations have also been useful in the theory of representations of algebraic groups over finite fields (see [133], [77]).

The method of orbits. The method of orbits first appeared in the author's article [99], devoted to representations of nilpotent Lie groups. The possibility of generalizing this method to other classes of groups was pointed out also in this article. Further developments were obtained in the papers of P. Bernat [64], B. Kostant [109], [110], L. Auslander and B. Kostant [59], L. Pukanszky [127], [128], M. Duflo [72], [74], and the author [100], [102], [103]. The method of orbits has also proved to be useful in certain other questions in the theory of representations of Lie algebras (see [71]), in studying the center of the enveloping algebra (cf. [78] and [74]).

B. Kostant in [109] pointed out the connection of the method of orbits with mechanics. A classification of homogeneous symplectic manifolds was obtained independently about five years ago by B. Kostant, J. M. Souriau, and the author. Expositions of this classification have appeared in [110], [50], and [105].

At the present time, the third period of the development of representation theory is conceptually complete (although many difficult concrete problems are still awaiting solution). Nevertheless, it would be wrong to consider the theory of representations as a completed discipline. Even now one can point out directions which will perhaps form the basis of the next, fourth, period.

In the first instance, we have the theory of infinite-dimensional representations of groups that arise in the modern theory of numbers and algebraic geometry: algebraic groups over local fields and rings of adèles. As an introduction to this field, one may study the monograph [18] and also Godement's exposition of the well-known work of Jacquet and Langlands [33] on automorphic forms on $GL(2)$.

Second, we have the theory of representations of infinite-dimensional Lie groups. One can point out at least three types of such groups, the study of which has special interest and is connected with many applications.

1) Groups of invertible operators in infinite-dimensional linear spaces (for example, the group $U(H)$ of all unitary operators in a Hilbert space H or its subgroup $U_0(H)$ consisting of the operators comparable with the identity operator modulo compact operators).

2) Groups of diffeomorphisms of smooth manifolds (for example, groups of canonical transformations of symplectic manifolds).

3) Continuous products of finite-dimensional Lie groups, that is, groups $C^\infty(M, G)$ of smooth functions on a manifold M with values in a Lie group G.

It would appear that the method of orbits will occupy an essential place in these new divisions of representation theory.

Bibliography[1]

a) Textbooks and monographs

1. Adel'son-Vel'skiĭ, G. M. [Адельсон-Вельский, Г. М.]: Spectral analysis of a ring of bounded linear operators [Спектральный анализ кольца ограниченных линейных операторов]. Dissertation, Moscow State University 1948. See also Doklady Akad. Nauk SSSR (N. S.) **67** (1949), 957—959 (MR 11, p. 115).
2. Atiyah, Michael F.: Lectures on K-theory. New York, N. Y.: Benjamin 1967 (MR 36, #7130). Russian translation, Moscow: Izdat. "Mir" 1969.
3. Bogoljubov, N. N. [Боголюбов, Н. Н.]: Lectures on the theory of symmetry of elementary particles [Лекции по теории симметрии элементарных частиц]. Moscow: Izdat. MGU 1966.
4. Borel, Armand, and George D. Mostow, editors: Algebraic groups and discontinuous subgroups. Proc. of Sympos. Pure Math., Boulder, Colo. 1965, Vol. IX. American Mathematical Society: Providence, R. I. 1966.
5. Borevič, Z. I., and I. R. Šafarevič [Боревич, З. И. и И. Р. Шафаревич]: Number theory [Теория чисел]. Moscow: Izdat. "Nauka" 1964 (MR 30, #1080). English translation, New York, N. Y.: Academic Press 1966 (MR 33, #4001).
6. Bourbaki, Nicolas: Éléments de mathématique. Algèbre. Chapitres 1 à 3. Paris: Hermann & Cie. 1970 (MR 43, #2). English translation, Reading, Mass.: Addison-Wesley 1974. Russian translation of an earlier edition, Moscow: Izdat. "Nauka" 1962, 1965, 1966.
7. Bourbaki, Nicolas: Éléments de mathématique, Fasc. II. Première partie. Livre III. Topologie générale, édition entièrement refondue. Actualités Sci. et Indust., No. 1142. Paris: Hermann & Cie. 1961 (MR 25, #4480). English translation, Reading, Mass.: Addison-Wesley 1966 (MR 34, #5044a). Russian translation, Moscow: Izdat. "Fizmatgiz" 1958.
8. Bourbaki, Nicolas: Éléments de mathématique, Fasc. XIII. Intégration, Chapitres 1—4. Deuxième édition revue et augmentée. Actualités Sci. et Indust., No. 1175. Paris: Hermann & Cie. 1965 (MR 36, #2763). Russian translation, Moscow: Izdat. "Nauka" 1967, 1970.
9. Bourbaki, Nicolas: Éléments de mathématique, Fasc. XXXII. Théories spectrales. Actualités Sci. et Indust., No. 1332. Paris: Hermann & Cie. 1967 (MR 35, #4725). Russian translation, Moscow: Izdat. "Mir" 1972.
10. Cartan, Henri, and Samuel Eilenberg: Homological algebra. Princeton, N. J.: Princeton University Press 1956 (MR 17, p. 1040). Russian translation, Moscow: Izdat. IL 1960.
11. Chevalley, Claude: Theory of Lie groups, I. Princeton, N. J.: Princeton University Press 1946 (MR 7, p. 412). Russian translation, Moscow: Izdat. IL 1948.

---

[1] Translator's note. A reference to the review in *Mathematical Reviews* has been supplied for every monograph and article listed for which such a review could be found. These are indicated by "MR", followed by the volume number and either the page or review number.

12. Cohen, Paul J.: Set theory and the continuum hypothesis. New York, N. Y.: Benjamin 1966 (MR 38, #999). Russian translation, Moscow: Izdat. "Mir" 1969.

13. Curtis, Charles W., and Irving Reiner: Representation theory of finite groups and associative algebras. New York, N. Y.: Interscience 1962 (MR 26, #2519).

14. Day, Mahlon M.: Normed linear spaces. Second printing corrected. Ergebnisse der Math. u. Grenzgebiete, N. F., Heft 21. Berlin-Göttingen-Heidelberg: Springer-Verlag 1962 (MR 26, #2847). Russian translation of first printing, Moscow: Izdat. IL 1961.

15. Dixmier, Jacques: Les C*-algèbres et leurs représentations. Deuxième edition. Paris: Gauthier-Villars 1969. Cahiers Scientifiques, Fasc. XXIX (MR 39, #7442). Russian translation in preparation.

16. Dunford, Nelson, and Jacob T. Schwartz: Linear operators. Part I: General theory. New York, N. Y.: Interscience Publishers, Inc. 1958 (MR 22, #8302). Russian translation, Moscow: Izdat. IL 1962.

17. Fraenkel, A. A., Y. Bar-Hillel, and A. Levy: Foundations of set theory. Second rev. edition. Amsterdam: North-Holland 1973. Russian translation of earlier edition, Moscow: Izdat. "Mir" 1966.

18. Gel'fand, I. M., M. I. Graev, and I. I. Pjateckiĭ-Šapiro [Гельфанд, И. М., М. И. Граев и И. И. Пятецкий-Шапиро]: The theory of representations and automorphic functions. Generalized functions, Fasc. 6. [Теория представлений и автоморфные функции (Обобщенные функции, вып. 6)]. Moscow: Izdat. "Nauka" 1966. English translation, Philadelphia, Penna.: Saunders 1968 (MR 38, #2093).

19. Gel'fand, I. M., M. I. Graev, and N. Ja. Vilenkin [Гельфанд, И. М., М. И. Граев и Н. Я. Виленкин]: Integral geometry and questions of the theory of representations connected with it. Generalized functions, Fasc. 5 [Интегральная геометрия и связанные с ней вопросы теории представлений. (Обобщенные функции, вып. 5)]. Moscow: Izdat. "Fizmatgiz" 1962 (MR 28, #3324). English translation, New York, N. Y.: Academic Press 1966 (MR 34, #7726).

20. Gel'fand, I. M., R. A. Minlos, and Z. Ja. Šapiro [Гельфанд, И. М., Р. А. Минлос и З. Я. Шапиро]: Representations of the rotation group and of the Lorentz group [Представления группы вращений и группы Лоренца]. Moscow: Fizmatgiz 1958 (MR 22, #5694). English translation, New York, N. Y.: Pergamon Press 1963.

21. Gel'fand, I. M., and M. A. Naĭmark [Гельфанд, И. М. и М. А. Наймарк]: Unitary representations of the classical groups [Унитарные представления классических групп]. Trudy Mat. Inst. Steklov, v. 36. Moscow: Izdat. Akad. Nauk SSSR 1950 (MR 13, p. 722).

22. Gel'fand, I. M., D. A. Raĭkov, and G. E. Šilov [Гельфанд, И. М., Д. А. Райков и Г. Е. Шилов]: Commutative normed rings [Коммутативные нормированные кольца]. Moscow: Fizmatgiz 1960. English translation, New York, N. Y.: Chelsea Publishing Co. 1964 (MR 34, #4940).

23. Gel'fand, I. M., and N. Ja. Vilenkin [Гельфанд, И. М. и Н. Я. Виленкин]: Certain applications of harmonic analysis. Accoutered Hilbert spaces. (Generalized functions, fasc. 4). [Некоторые применения гармонического анализа. Оснащенные гильбертовы пространства. (Обобщенные функции, вып. 4)]. Moscow: Fizmatgiz 1961 (MR 26, #4173). English translation, New York, N. Y.: Academic Press 1964 (MR 30, #4152).

24. Godement, Roger: Topologie algébrique et théorie des faisceaux. Actualités Sci. et Indust. 1252. Paris: Hermann & Cie. 1958 (MR 21, #1583). Russian translation, Moscow: IL 1961 (MR 24, #A544).

25. Grothendieck, Alexandre: Sur quelques points d'algèbre homologique. Tôhoku Math. J. (2) 9, 119—221 (1957). (MR 21, #1328). Russian translation, Moscow: IL 1961.

26. Gunning, Robert C., and Hugo Rossi: Analytic functions of several complex variables. Englewood Cliffs, N. J.: Prentice-Hall 1965 (MR 31, #4927). Russian translation, Moscow: Izdat. "Mir" 1969.

27. Halmos, Paul R.: Measure theory. New York, N. Y.: Van Nostrand Co. 1950 (MR 11, p. 504). Russian translation, Moscow: IL 1953.

28. Hamermesh, M.: Group theory and its application to physical problems. Reading,

Mass.: Addison-Wesley 1962 (MR 25, #132). Russian translation, Moscow: Izdat. "Mir" 1966.

29. Helgason, Sigurdur: Differential geometry and symmetric spaces. New York, N. Y.: Academic Press 1962 (MR 26, #2986). Russian translation, Moscow: Izdat. "Mir" 1964.

30. Hewitt, Edwin, and Kenneth A. Ross: Abstract harmonic analysis, I, II. Berlin-Heidelberg-New York: Springer-Verlag 1963, 1970 (MR 28, #158, 41, #7378). Russian translation, Moscow: Vol. I, Izdat. "Nauka" 1976; Vol. II, Izdat. "Mir" 1975.

31. Hu, Sze-tsen: Homotopy theory. New York, N. Y.: Academic Press 1959 (MR 21, #5186). Russian translation, Moscow: Izdat. "Mir" 1964.

32. Husemoller, Dale H.: Fiber bundles. New York, N. Y.: McGraw-Hill 1966 (MR 37, #3280). Russian translation under title Расслоенные пространства, Moscow: Izdat. "Mir" 1970.

33. Jacquet, H., and R. P. Langlands: Automorphic forms on GL(2). Berlin-Heidelberg-New York: Springer-Verlag 1970. Lecture Notes in Mathematics, #114.

34. Jacobson, Nathan: Lie algebras. New York, N. Y.: Interscience 1962 (MR 26, #1345). Russian translation, Moscow: Izdat. "Mir" 1964.

35. Kaplansky, Irving: An introduction to differential algebra. Actualités Sci. et Indust., No. 1251. Paris: Hermann & Cie. 1957 (MR 20, #177). Russian translation, Moscow: IL 1959.

36. Kelley, John L.: General topology. New York, N. Y.: D. Van Nostrand Co. 1955 (MR 16, p. 1136). Russian translation, Moscow: Izdat. "Nauka" 1968.

37. Kokkedee, J.: The quark model. New York, N. Y.: Benjamin 1969. Russian translation, Moscow: Izdat. "Mir" 1971.

38. Kolmogorov, A. N., and S. V. Fomin [Колмогоров, А. Н. и С. В. Фомин]: Elements of the theory of functions and functional analysis, I and II [Элементы теории функций и функционального анализа, I и II]. Revised edition, Moscow: Izdat. "Nauka" 1972. English translation of first edition, Rochester, N. Y.: Graylock Press 1957, 1961 (MR 19, p. 44, 22, #9566a).

39. Lang, Serge: Algebra. Reading, Mass.: Addison-Wesley 1965 (MR 33, #5416). Russian translation, Moscow: Izdat. "Mir" 1968.

40. Ljubarskiĭ, G. Ja. [Любарский, Г. Я.]: The theory of groups and its application to physics [Теория групп и ее применение к физике]. Moscow: Gostehizdat 1957 (MR 21, #5441). English translation, New York, N. Y.: Pergamon Press 1960 (MR 22, #7709).

41. Mac Lane, Saunders: Homology. Berlin-Heidelberg-New York: Springer-Verlag 1963 (MR 28, #122). Russian translation, Moscow: Izdat. "Mir" 1966.

42. Naĭmark, M. A. [Наймарк, М. А.]: Normed rings [Нормированные кольца]. Second edition, revised. Moscow: Izdat. "Nauka" 1968. Revised German translation under title Normierte Algebren. Berlin: VEB Deutscher Verlag der Wiss. 1959 (MR 22, #1824). English translation under title Normed algebras, 3rd edition. Groningen: Wolters-Noordhoff 1972.

43. Phelps, Robert R.: Lectures on Choquet's theorem. Princeton, N. J.: D. Van Nostrand Co. 1966 (MR 33, #1690). Russian translation, Moscow: Izdat. "Mir" 1968.

44. Pontrjagin, L. S. [Понтрягин, Л. С.]: Continuous groups [Непрерывные группы]. Third edition, corrected. Moscow: Izdat. "Nauka" 1973. German translation of second edition under title Topologische Gruppen, I, II. Leipzig: B. G. Teubner 1957, 1958 (MR 19, p. 152, 20, #3925). English translation of second edition under title Topological groups, New York, N. Y.: Gordon & Breach 1966 (MR 34, #1439).

45. Schaefer, Helmut H.: Topological vector spaces. Third printing, corrected. Berlin-Heidelberg-New York: Springer-Verlag 1971 (MR 33, #1689). Russian translation, Moscow: Izdat. "Mir" 1971 (MR 43, #2461).

46. Séminaire "Sophus Lie" de l'École Normale Supérieure, 1954/1955. Théorie des algèbres de Lie. Topologie des groupes de Lie. Paris: Secrétariat Mathématique 1955 (MR 17, p. 384). Russian translation, Moscow: IL 1962.

47. Serre, J.-P.: Lie algebras and Lie groups. New York, N. Y.: Benjamin 1965 (MR 36, #1582). Russian translation, Moscow: Izdat. "Mir" 1969 (MR 40, #5795).

48. Serre, J.-P.: Représentations linéaires des groupes finis. 2ᵉ édition. Paris: Hermann & Cie. 1971 (MR 38, #1190). Russian translation of first edition, Moscow: Izdat. "Mir" 1970.

49. Šilov, G. E. [Шилов, Г. Е.]: Mathematical analysis. Special course [Математический анализ. Специальный курс]. Moscow: Fizmatgiz 1961 (MR 24, #A1180). English translation, New York, N. Y.: Pergamon Press 1965 (MR 32, #2519).

50. Souriau, J.-M.: Structure des systèmes dynamiques. Paris: Dunod 1970 (MR 41, #4866).

51. Spivak, Michael: Calculus on manifolds. New York, N. Y.: Benjamin 1965 (MR 35, #309). Russian translation, Moscow: Izdat. "Mir" 1968.

52. Steenrod, N. E., and S. Eilenberg: Foundations of algebraic topology. Princeton, N. J.: Princeton University Press 1952 (MR 14, p. 398). Russian translation, Moscow: IL 1956.

53. Vilenkin, N. Ja. [Виленкин, Н. Я.]: Special functions and the theory of group representations [Специальные функции и теория представлений групп]. Moscow: Izdat. "Nauka" 1965 (MR 35, #420). English translation, Providence R. I.: American Mathematical Society 1968.

54. Weil, André: L'intégration dans les groupes topologiques et ses applications. Actualités Sci. et Indust. 869, 1145. Paris: Hermann & Cie. 1940, 1951 (MR 3, p. 198). Russian translation, Moscow: IL 1950.

55. Weyl, Hermann: The classical groups, their invariants and representations. Princeton, N. J.: Princeton University Press 1939 (MR 1, p. 42). Russian translation, Moscow: IL 1947.

56. Whitney, Hassler: Geometric integration theory. Princeton, N. J.: Princeton University Press 1957 (MR 19, p. 309). Russian translation, Moscow: IL 1960.

57. Želobenko, D. P. [Желобенко, Д. П.]: Compact Lie groups and their representations [Компактные группы Ли и их представления]. Moscow: Izdat. "Nauka" 1970. English translation, Providence, R. I.: American Mathematical Society 1973.

b) Journal articles

58. Atiyah, M. F., and Raoul Bott: A Lefschetz fixed point theorem for elliptic complexes, I. Ann. of Math. (2) **86**, 374—407 (1967) (MR 35, #3701).

59. Auslander, Louis, and Bertram Kostant: Quantization and representations of solvable Lie groups. Bull. Amer. Math. Soc. 73, 692—695 (1967) (MR 39, #2910).

60. Bargmann, Valentine: Irreducible unitary representations of the Lorentz group. Ann. of Math. (2) **48**, 568—640 (1947) (MR 9, p. 133).

61. Bargmann, Valentine: On unitary ray representations of continuous groups. Ann. of Math. (2) **59**, 1—46 (1954) (MR 15, p. 397).

62. Berezin, F. A. [Березин, Ф. А.]: Laplace operators on semisimple Lie groups [Операторы Лапласа на полупростых группах Ли]. Trudy Moskov. Mat. Obšč. **6**, 371—463 (1957) (MR 19, p. 867). Letter to the editor [Письмо в редакцию] same Trudy **12**, 453—466 (1963) (MR 28, #3335). English translation: Amer. Math. Soc. Transl. (2) **21**, 239—339 (1962) (MR 24, #B280).

63. Berezin, F. A., and I. M. Gel'fand [Березин, Ф. А. и И. М. Гельфанд]: Some remarks on the theory of spherical functions on symmetric Riemannian manifolds [Несколько замечаний к теории сферических функций на симметрических римановых многообразиях]. Trudy Moskov. Mat. Obšč. **5**, 311—351 (1956) (MR 19, p. 152). English translation: Amer. Math. Soc. Transl. (2) **21**, 193—238 (1962) (MR 27, #1910).

64. Bernat, P.: Sur les représentations unitaires des groupes de Lie résolubles. Ann. Sci. École Norm. Sup. (3) **82**, 37—99 (1965) (MR 33, #2763).

65. Blattner, Robert J.: On induced representations, I and II. Amer. J. Math. **83**, 79—98, 499—512 (1961) (MR 23, #A2757, 26, #2885).

66. Bott, Raoul: Homogeneous vector bundles. Ann. of Math. (2) **66**, 203—248 (1957) (MR 19, p. 681).

67. Bruhat, F.: Sur les représentations induites des groupes de Lie. Bull. Soc. Math. France **84**, 97—205 (1956) (MR 18, p. 907).

68. Bruhat, F.: Distributions sur un groupe localement compact et applications à l'étude des représentations des groupes p-adiques. Bull. Soc. Math. France **89**, 43—75 (1961) (MR 25, #4354).

69. Delaroche, C., and A. A. Kirillov: Sur les relations entre l'espace dual d'un groupe et la structure des ses sous-groupes fermés. Séminaire Bourbaki, 1967/1968, Exp. 343, 1—22.

70. Dixmier, Jacques: Sur les représentations unitaires des groupes de Lie algébriques. Ann. Inst. Fourier, Grenoble, **7**, 315—328 (1957) (MR 20, #5820).

71. Dixmier, Jacques: Sur les représentations induites des algèbres de Lie. J. Math. Pures Appl. **50**, 1—24 (1971).

72. Duflo, Michel: Caractères des groupes et des algèbres de Lie résolubles. Ann. Sci. École Norm. Sup. (4) **3**, 23—74 (1970) (MR 42, #4672).

73. Duflo, Michel: Representations of the principal series of a semisimple Lie group [Представления основной серии полупростой группы Ли], Funkcional. Anal. i Priložen. **4**, 38—42 (1970). English translation under title Fundamental-series representations of a semisimple Lie group. Functional Anal. Appl. **4**, 122—126 (1970).

74. Duflo, Michel: Représentations induites d'algèbres de Lie. C. R. Acad. Sci. Paris Sér. A-B **272**, A1157—A1158 (1971) (MR 43, #3311).

75. Ernest, John A.: Hopf-von Neumann algebras. Proc. Conf. Functional Analysis, Irvine, Calif., 1966. London: Academic Press 1967, pp. 195—215 (MR 36, #6956).

76. Fell, J. M. G.: The structure of algebras of operator fields. Acta Math. **106**, 233—280 (1961) (MR 29, #1547).

77. Gel'fand, S. I. [Гельфанд, С. И.]: Representations of the full linear group over a finite field [Представления полной линейной группы над конечным полем]. Mat. Sb. **83 (125)**, 15—41 (1970) (MR 42, #7797).

78. Gel'fand, I. M. [Гельфанд, И. М.]: The center of an infinitesimal group ring [Центр инфинитезимального группого кольца]. Mat. Sb. **26 (68)**, 103—112 (1950) (MR 11, p. 498).

79. Gel'fand, I. M., and M. I. Graev [Гельфанд, И. М. и М. И. Граев]: Unitary representations of the real unimodular group (principal nondegenerate series) [Унитарные представления вещественной унимодулярной группы (основные невырожденные серии)]. Izvestija Akad. Nauk SSSR, Ser. Mat. **17**, 189—248 (1953) (MR 15, p. 199).

80. Gel'fand, I. M., and M. I. Graev [Гельфанд, И. М. и М. И. Граев]: The geometry of homogeneous spaces, representations of groups in homogeneous spaces, and related questions of integral geometry, I [Геометрия однородных пространств, представления групп в однородных пространствах и связанные с ними вопросы интегральной геометрии, I]. Trudy Moskov. Mat. Obšč. **8**, 321—390 (1959), addendum **9**, 562 (1960) (MR 23, #A4013).

81. Gel'fand, I. M., and A. A. Kirillov [Гельфанд, И. М. и А. А. Кириллов]: Sur les corps liés aux algèbres enveloppantes des algèbres de Lie. Publ. Math. I.H.E.S. No. 31, 5—19 (1966) (MR 34, #7731).

82. Gel'fand, I. M., and A. A. Kirillov [Гельфанд, И. М. и А. А. Кириллов]: Structure of the Lie skew-field connected with a semisimple decomposable Lie algebra [Структура тела Ли, связанного с полупростой расщепимой алгеброй Ли]. Funkcional. Anal. i Priložen. **3**, no. 1, 7—26 (1969) (MR 39, #2827).

83. Gel'fand, I. M., and M. A. Naĭmark [Гельфанд, И. М. и М. А. Наймарк]: Unitary representations of the group of linear transformations of the line [Унитарные представления группы линейных преобразований прямой]. Doklady Akad. Nauk SSSR **55**, 567—570 (1947) (MR 8, p. 563).

84. Gel'fand, I. M., and M. A. Naĭmark [Гельфанд, И. М. и М. А. Наймарк]: Unitary representations of the Lorentz group [Унитарные представления группы Лоренца]. Izvestija Akad. Nauk SSSR, Ser. Mat. **11**, 411—504 (1947) (MR 9, p. 495).

85. Geronimus, A. Ju. [Геронимус, А. Ю.]: The Grothendieck topology and the theory of representations [Топология Гротендика и теория представлений]. Funkcional. Anal. i Priložen. **5**, no. 3, 22—31 (1971) (MR 44, #2888).

86. Godement, Roger: Les fonctions de type positif et la théorie des groupes. Trans. Amer. Math. Soc. **63**, 1—84 (1948) (MR 9, p. 327).

87. Godement, Roger: Theorie des caractères. I, Algèbres unitaires. II, Définition et propriétés générales des caractères. Ann. of Math. (2) **59**, 47—62, 63—85 (1954) (MR 14, p. 620, 15, p. 441).

88. Godement, Roger: Theory of spherical functions, I. Trans. Amer. Math. Soc. **73**, 496—556 (1952) (MR 14, p. 620).

89. Gross, K. I.: The dual of a parabolic subgroup and a degenerate principal series of Sp(n, c). Amer. J. Math. **93**, 398—428 (1971) (MR 46, #3693).

90. Harish-Chandra: Representations of a semisimple Lie group on a Banach space, I. Trans. Amer. Math. Soc. **75**, 185—243 (1953) (MR 15, p. 100).

91. Harish-Chandra: The Plancherel formula for complex semisimple Lie groups. Trans. Amer. Math. Soc. **76**, 485—528 (1954) (MR 16, p. 111).

92. Harish-Chandra: Invariant eigendistributions on a semisimple Lie group. Trans. Amer. Math. Soc. **119**, 457—508 (1965) (MR 31, #4862d).

93. Harish-Chandra: Discrete series for semisimple Lie groups, I and II. Acta Math. **113**, 241—318 (1965), **116**, 1—111 (1966) (MR 36, #2744, #2745).

94. Ismagilov, R. S. [Исмагилов, Р. С.]: On the irreducible cycles connected with a dynamical system [О неприводимых циклах, связанных с динамической системой]. Funkcional. Anal. i Priložen. **3**, no. 3, 92—93 (1969) (MR 40, #2816).

95. Kac, G. I. [Кац, Г. И.]: Generalized functions on a locally compact group and decompositions of unitary representations [Обобщенные функции на локально компактной группе и разложения унитарных представлений]. Trudy Moskov. Mat. Obšč. **10**, 3—40 (1961) (MR 27, #5863).

96. Kac, G. I. [Кац, Г. И.]: Ring groups and the principle of duality, I and II [Кольцевые группы и принцип двойственности, I и II]. Trudy Moskov. Mat. Obšč. **12**, 259—301 (1963), **13**, 84—113 (1965) (MR 28, #164, 31, #4857).

97. Kac, G. I., and V. G. Paljutkin [Кац, Г. И. и В. Г. Палюткин]: Finite ring groups [Конечные кольцевые группы]. Trudy Moskov. Mat. Obšč. **15**, 224—261 (1966) (MR 34, #8211).

98. Kac, G. I. [Кац, Г. И.]: Extensions of groups that are ring groups [Расширения групп, являющиеся кольцевыми группами]. Mat. Sb. **76** (**118**) 473—496 (1968) (MR 37, #4639).

99. Kirillov, A. A. [Кириллов, А. А.]: Unitary representations of nilpotent Lie groups [Унитарные представления нильпотентных групп Ли]. Uspehi Mat. Nauk **17**, vyp. 4, 57—110 (1962) (MR 25, #5396).

100. Kirillov, A. A. [Кириллов, А. А.]: On the Plancherel measure for nilpotent Lie groups [О мере Планчереля для нильпотентных групп Ли]. Funkcional. Anal. i Priložen. **1**, no. 4, 84—85 (1967) (MR 37, #347).

101. Kirillov, A. A. [Кириллов, А. А.]: The method of orbits in the theory of unitary representations of Lie groups [Метод орбит в теории унитарных представлений групп Ли]. Funkcional. Anal. i Priložen. **2**, no. 2, 96—98 (1968) (MR 38, #2251).

102. Kirillov, A. A. [Кириллов, А. А.]: Characters of unitary representations of Lie groups [Характеры унитарных представлений групп Ли]. Funkcional. Anal. i Priložen. **2**, no. 2, 40—55 (1968) (MR 38, #4615).

103. Kirillov, A. A. [Кириллов, А. А.]: Characters of unitary representations of Lie groups. Reduction theorems [Характеры унитарных представлений групп Ли. Редукционные теоремы]. Funkcional. Anal. i Priložen. **3**, no. 1, 36—47 (1969) (MR 40, #1538).

104. Kirillov, A. A. [Кириллов, А. А.]: Constructions of unitary irreducible representations of Lie groups [Конструкции унитарных неприводимых представлений групп Ли]. Vestnik Moskov. Univ. Ser. I, Meh. **25**, no. 2, 41—51 (1970) (MR 43, #418).

105. Kirillov, A. A. [Кириллов, А. А.]: Lectures on the theory of group representations, IV. (Representations of groups and mechanics) [Лекции по теории представлений групп, IV. (Представления групп и механика)]. Moscow: Izdat. MGU 1971.

106. Kirillov, A. A. [Кириллов, А. А.]: Representations of infinite-dimensional unitary groups [Представления бесконечномерных унитарных групп]. Doklady Akad. Nauk SSSR **212**, 288—290 (1973).

107. Kirillov, A. A. [Кириллов, А. А.]: Representations of certain infinite-dimensional

Lie groups [Представления некоторых бесконечномерных групп Ли]. Vestnik Moskov. Univ. Ser. Mat. no. 1, 75—83 (1974).

108. Kirillov, A. A. [Кириллов, А. А.]: Unitary representations of the group of diffeomorphisms and of certain of its subgroups [Унитарные представления группы диффеоморфизмов и некоторых ее подгрупп]. Preprint no. 82 of Institute of Applied Mathematics Akad. Nauk SSSR 1974. 40 pages.

109. Kostant, Bertram: Orbits, symplectic structures and representation theory. Proc. U. S.-Japan Seminar in Differential Geometry (Kyoto, 1965). Tokyo: Nippon Hyoronsha 1966, p. 71 (MR 35, #4340).

110. Kostant, Bertram: Quantization and unitary representations. Part I. Prequantization. Lectures in Modern Analysis and Applications, III. Berlin-Heidelberg-New York: Springer-Verlag, Lecture Notes in Mathematics, Vol. 170, 1970, pp. 87—208 (MR 45, #3638).

111. Kreĭn, S. G., and A. M. Šihvatov [Крейн, С. Г. и А. М. Шихватов]: Linear differential equations on a Lie group [Линейные дифференциальные уравнения на группе Ли]. Funkcional. Anal. i Priložen. 4, no. 1, 52—61 (1970). English translation: Functional Anal. Appl. 4, 46—54 (1970).

112. Langlands, R. P.: Dimension of spaces of automorphic forms. Algebraic Groups and Discontinuous Subgroups. Proc. Sympos. Pure Math., Boulder, Colo. 1965. Providence, R. I.: American Mathematical Society 1966, pp. 253—257 (MR 35, #3010).

113. Mackey, George W.: Imprimitivity for representations of locally compact groups, I. Proc. Nat. Acad. Sci. U.S.A. 35, 537—545 (1949) (MR 11, p. 158).

114. Mackey, George W.: Unitary representations of group extensions, I. Acta Math. 99, 265—311 (1958) (MR 20, #4789).

115. Mackey, George W.: Infinite dimensional group representations. Bull. Amer. Math. Soc. 69, 628—686 (1963) (MR 27, #3745).

116. Mautner, F. I.: Unitary representations of locally compact groups, I and II. Ann. of Math. (2) 51, 1—25 (1950), 52, 528—556 (1950) (MR 11, p. 324, 12, p. 157).

117. Mautner, F. I.: Note on the Fourier inversion formula on groups. Trans. Amer. Math. Soc. 78, 371—384 (1955) (MR 16, p. 692).

118. Moore, Calvin C.: On the Frobenius reciprocity theorem for locally compact groups. Pacific J. Math. 12, 359—365 (1962) (MR 25, #5134).

119. Moore, Calvin C.: Extensions and low dimensional cohomology theory of locally compact groups, I and II. Trans. Amer. Math. Soc. 113, 40—63, 64—86 (1964) (MR 30, #2106).

120. Naĭmark, M. A. [Наймарк, М. А.]: On the description of all unitary representations of the complex classical groups, I and II [Об описании всех унитарных представлений комплексных классических групп]. Mat. Sb. N. S. 35 (77), 317—356 (1954), 37 (79), 121—140 (1955) (MR 16, p. 567, 17, p. 61).

121. Narasimhan, M. S., and K. Okamoto: An analogue of the Borel-Weil-Bott theorem for hermitian symmetric pairs of non-compact type. Ann. of Math. (2) 91, 486—511 (1970) (MR 43, #419).

122. Nelson, E.: Analytic vectors. Ann. of Math. (2) 70, 572—615 (1959) (MR 21, #5901).

123. von Neumann, John: Collected Works, Vol. III: Rings of operators. New York, N. Y.: Pergamon Press 1961 (MR 28, #1103).

124. Ol'šanskiĭ, G. I. [Ольшанский, Г. И.]: On the Frobenius reciprocity theorem [О теореме двойственности Фробениуса]. Funkcional. Anal. i Priložen. 3, no. 4, 49—58 (1969) (MR 43, #421).

125. Parthasarathy, R.: A note on the vanishing of certain 'L^2-cohomologies'. J. Math. Soc. Japan 23, 676—691 (1971).

126. Parthasarathy, R.: Dirac operator and the discrete series. Ann. of Math. (2) 96, 1—30 (1972) (MR 47, #6945).

127. Pukanszky, L.: On the unitary representations of exponential groups. J. Functional Analysis 2, 73—113 (1968) (MR 37, #4205).

128. Pukanszky, L.: Characters of algebraic solvable groups. J. Functional Analysis 3, 435—494 (1969) (MR 40, #1539).

129. Schmid, W.: Homogeneous complex manifolds and representations of semi-simple Lie groups. Proc. Nat. Acad. Sci. U.S.A. **59**, 56—59 (1968) (MR **37**, #1520).

130. Segal, Irving E.: An extension of Plancherel's formula to separable unimodular groups. Ann. of Math. (2) **52**, 272—292 (1950).

131. Smorodinskiĭ, Ja. A. [Смородинский, Я. А.]: The unitary symmetry of elementary particles [Унитарная симметрия элементарных частиц]. Uspehi Fiz. Nauk **84**, 3—36 (1964). English translation: Soviet Physics Uspekhi 7, 637—655 (1965) (MR **31**, #4467).

132. Stein, Elias M.: Analysis in matrix spaces and some new representations of SL(N, c). Ann. of Math. (2) **86**, 461—490 (1967) (MR **36**, #2749).

133. Steinberg, Robert: A geometric approach to the representations of the full linear groups over a Galois field. Trans. Amer. Math. Soc. **71**, 274—282 (1951) (MR **13**, p. 317).

134. Takenouchi, O.: Sur la facteur-représentation d'un groupe de Lie résoluble de type (E). Math. J. Okayama Univ. **7**, 151—161 (1957) (MR **20**, #3933).

135. Takesaki, Masamichi: Duality and von Neumann algebras. Bull. Amer. Math. Soc. **77**, 553—557 (1971) (MR **46**, #4225).

136. Thoma, Elmar: Eine Charakterisierung diskreter Gruppen vom Typ I. Invent. Math. **6**, 190—196 (1968) (MR **40**, #1540).

137. Wigner, Eugene P.: On unitary representations of the inhomogeneous Lorentz group. Ann. of Math. (2) **40**, 149—204 (1939).

138. Želobenko, D. P. [Желобенко, Д. П.]: The classical groups. Spectral analysis of finite-dimensional representations [Классические группы. Спектральный анализ конечномерных представлений]. Uspehi Mat. Nauk **17**, vyp. 1, 27—120 (1962) (MR **25**, #129).

139. Želobenko, D. P., and M. A. Naĭmark [Желобенко, Д. П. и М. А. Наймарк]: A description of the completely irreducible representations of a semisimple complex Lie group [Описание вполне неприводимых представлений полупростой комплексной группы Ли]. Izvestija Akad. Nauk SSSR, Ser. Mat. **34**, 57—82 (1970) (MR **42**, #1949).

Subject Index

abelian group 12
absolutely continuous measure 40
— simple Lie algebra 92
abstract group 12
action (of a group) on the left 13
— — on the right 14
adjoining a unit 25
adjoint representation of a Lie algebra 97
— — of a Lie group 97
admissible chart 63
— subalgebra 236
Ado's theorem 88
Aleksandrov's line 63
algebra of type I 60
— over K 32
algebraic integer 157
— manifold 68
— number field 28
algebraically irreducible representation 110
almost everywhere 39
alternating group 262
amenable group 129
analytic vector 155
antirepresentation 14
associativity 12
atlas 62
automorphism 11, 13

Banach algebra 45
— space 37
—-Steinhaus theorem 42
base 78
— for a topology 6
basis 31
— measure 54
bicommutant 58
bifunctor 6
bigebra 178
binary relation 2
Borel set 6
—-Weil-Bott theorem 140
boundary of a manifold 77

bounded set 41
Burnside's theorem 140

C*-algebra 49
— of a group 143
Campbell-Hausdorff-Dynkin formula 108
closed set 6
canonical embedding 4
— mapping 2
— matrix 92
— parameter 85
— projection 4
Cartan's criterion 92
— subalgebra 93
category 3
central extension 20
centralizer 17
change of scalars 111
character group 167
— of Banach algebra 46
— of a finite dimensional representation 156
— of a infinite dimensional representation 161
chart 62
Choquet's theorem 42
Clifford algebra 34
class C^k 63
classical simple algebra 92
closed form 76
— operator 114
— set 6
co-adjoint representation 226
coboundary 10, 21
— operator 10
cochain 10, 21
cocycle 10, 21
cohomology group 21
coinduced representation 187
commutant 58
commutation 86
commutative algebra 87
— group 12
commutator of vector fields 72

Subject Index

compact 7
— operator 42
complete measure 39
completely continuous operator 42
— irreducible representation 112
— reducible representation 110
completion 9
complex manifold 68
complexification 111
component of a section 79
conjugate representation 109
— subgroup 15
connected set 7
continuous sum of spaces 58
— product of algebras 60
continuously diagonal operator 58
contragredient representation 109
contravariant functor 5
— tensor 33
convolution 140
— of generalized functions 148
cotangent bundle 79
countably additive measure 38
— normable space 37
Courant's priniciple 56
covariant functor 5
— tensor 33
cross norm 42
— section 79
cyclic group 16
— representation 53
— vector 53

decomposable extension 20
— operator 58
— representation 110
decreasing filtration 26
derivative along a vector 69
— of a family of mappings 71
— of a mapping 70
— of a measure 40
determining (family of prenorms) 36
diagonal operator 58
diffeomorphic manifolds 64
diffeomorphism 64
differentiability along a vector field 73
differential 74
— of a function 70
— form 74
— — of type $(0, r)$ 82
dimension of a minifold 62
— of a vector space 31
direct sum 31
— — of bundles 79
directed set 3
direction 3

disjoint measures 40
— representations 109
distance 8
division algebra 27
Dixmier-Glimm-Sakai theorem 162
Dolbeault's theorem, generalized 82
double coset 16
dual category 4
— group 167
— module 30
— object of a group 113
Dynkin diagramm 94

Engel's theorem 91
envelopping algebra 149
epimorphism 13
equivalent atlases 67
— extensions 19
— measures 40
— representations 48, 109
even permutation 262
exact form 74
exponential Lie group 103
— mapping 103
extension (of a group) 18
exterior algebra 34
— product 34
— differentiation 74

factor 62
— group 15
— Lie algebra 33
— module 29
— representation 110
— ring 25
— set 2
— space 7
faithful representation 13, 48
fiber 78
field 27
filtering to the right 3
filtration 26
finite measure 39
— representation 115
flag manifold 291
formal dimension 139
Fourier transform 166
free algebra 33
— K-module 29
functor 6
fundamental direction 9
— group 9

Gårding space 153
Gel'fand theorem 49, 151
— transform 46
—-Kirillov conjecture 152

Gel'fand-Mazur theorem 45
—-Naĭmark theorem 83
—-Raĭkov theorem 146
generalized infinitesimal character 165
— — generating function 232
— — Gleason-Montgomery-Zippin theorem 83
G-orbit 14
graded 26
graph of an operator 121
Grassmann algebra 34
— manifold 65
Grothendieck group 110
group 12
— algebra 140
— ring 139

Hahn-Banach theorem 42
Hamiltonian vector field 231
Hausdorff space 7
Heisenberg group 237, 287
Hermitian element 48
Hilbert space 37
— tensor product 44
—-Schmidt operator 44
holomorphic bundle 80
— cross section 80
holomorphicaly induced representation 204
homogeneous algebra 60
— G-space 15
— symplectic manifold 233
homomorphism 12
homotopic mappings 9
homotopy class 9

I-adic topology 27
ideal of a Lie algebra 89
idempotent 122
identity 12
image 13
increasing filtration 26
induced representation 182
— — in the sense of Mackey 188
inductive limit 5
infinitesimal character 163
— Lie group 86
inner automorphism 13, 97
integrable function 39
integral 39
integral of a form 76
— orbit 239
intertwining number 109
— operator 109
invariant mean 129
— subgroup 15
inverse morphism 4

— relation 2
isomorphic objects 4
— Lie groups 99
isomorphism 4, 13

Jacobson topology 147
joint spectrum 47

k-irreducible 113
K-representation 226
Každan's theorem 181
kernel 13
Killing form 91
Kostant-Auslander theorem 241
Kreĭn-Mil'man theorem 42

Lagrange distribution 246
law of a composition 3
left coset 15
— fraction 152
— G-space 13
— Haar measure 130
— ideal 25
— K-module 28
Leray's theorem 11
Lie algebra 86
— — homomorphism 87
Lie group 83
— operator 73
— skew field 152
Lie's theorem 90
linear mapping 29
— operator 31
— representation 108
— — of a Lie group 87
— space 31
linearization of a projective representation 220
linearly ordered set 2
local coordinates 62
locally convex space 36
— finite measure 39
— isomorphic Lie groups 99

Mackey's criterion of inducibility 196
— theorem 197
manifold 62
— with boundary 77
matrix element 135
Maschke's theorem 140
measure 38
measurable function 39
— set 38
— vector function 57
metric space 8
Minkowski functional 36
mixed tensor 33

Subject Index

modulus of a group   131
monomial group   198
— representation   182
monomorphism   13
morphism   3
multiplicity function   54

neighborhood   6
Nelson's theorem   145
net   3
Neumann-Weyl theorem   263
Neumann algebra   91
nilpotent group of class k 17
— Lie algebra   91
nondegenerate representation   134
nonsingular root system   93
norm   36
normable space   37
normal element   48
— subgroup   15
normalizer   16
nuclear space   43
— operator   43

object   3
odd permutation   262
Ol'šanskiĭ's theorem   199
one-parameter subroup   101
open set   6
operator representation   48
— irreducible representation   114
ordered set   2
orientable manifold   67
orientation   67
orthogonal idempotent   122
— matrix   66

p-adic integer   27
— number   28
partially holomorphic bundle   81
— — cross section   81
— ordered set   2
Pfaffian   253
Pontrjagin's theorem   168
Poisson bracket   232
positive measure   38
— functional   50
positively connected charts   66
prenorm   36
presheaf   67
prime ideal   147
primary representation   122
product   4
— of relations   2
produced module   187
profinite group   24
projection (of a bundle)   78
— measure   56

projective representation   108
— tensor product   43
— unitary representation   218
pseudo-orthogonal matrix   67
pseudo-symplectic matrix   67
pseudo-unitary matrix   67
Pukanszky's condition   236

quadratic form   34
quantization   242
quasiinvariant measure   131

radical   25
— ring   26
rank of a Lie algebra   94
— of nilpotency   90
— of solvability   90
real form   92
realification   111
realization   13
reduced root system   93
reducible representation   109
reflexive relation   2
regular element   94
— measure   40
— representation   88
relatively compact set   43
representable functor   6
representation of a category   175
— of a group   13
— of a Lie algebra   87
— with a simple spectrum   127
— with a homogeneous spectrum   127
resolvent   45
de Rham cohomology ring   76
right coset   15
— fraction   152
— G-space   14
— Haar measure   130
— ideal   25
— K-module   29
ring   25
— group   178
ringed space   67
R-near   8

Schauder-Tihonov theorem   42
Schwartz's theorem   149
Schur's criterion   156
— lemma   119
— theorem   220
semisimple Lie algebra   91
— ring   26
semidirect product   20
semiseparated space   7
seminorm   36
set   1

simple module 29
— function 39
— group 17
— Lie algebra 91
simply connected covering group 85
— — topological space 9
skew field 27
— tensor 34
smooth manifold 63
— curve 68
— tensor field 74
— vector field 71
smoothly connected charts 62
solvable group of class k 17
— Lie algebra 90
source 53
space of a bundle 56
spatialy isomorphic 52
spectral radius 45
spectrum 45
spinor representation 276, 287
square integrable representation 130
stabilizer 15
stationary subgroup 15
structure 3
— constants 86
— sheaf 67
Stone-Weierstrass theorem 42
strictily Hamiltonian vector field 234
— homogeneous symplectic manifold 233
strong topology 41
subspace 7
subring 25
submodule 29
subordinate subalgebra 236
subalgebra (of a Lie algebra) 89
subcategory 3
sum 4
support of a measure 41
— of a generalized function 148
symmetric group 62
— product 281
— relation 2
— representation 48
— tensor 34
— algebra 34, 52
— maximal ideal 52
symplectic manifold 230
— matrix 66
system of roots 93

tame group 127
tangent bundle 79
— vector 69, 70
Tannaka's theorem 176
tensor algebra 33
— bundle 79

— field 74
— product of algebras 32
— — of bundles 79
— — of modules 29, 30
— — of operators 31
— — of representations 110
tensorialy irreducible representation 115
topological conjugate space 41
— factor representation 111
— group 22
— G-space 23
— linear space 35
— ring 27
— skew field 27
— space 6
— subrepresentation 111
topologically decomposable 111
— finite 111
— irreducible 111
topology 6
— of a dual object 113
tower of sets 3
transitive action 15
transition function 78
trivial bundle 78
T_0-space 7
two-sided ideal 25
— module 29

uniform convergence 41
— space 8
uniformly continuous 8
unimodular group 131
unit 25
unitary element 48
— matrix 64
— representation of a group 112
— — of a right G-space 191
universal formula for characters 251

variation of a measure 38
vector 31
— bundle 77
— field 71
— space 31
Vergne's theorem 238
virtual subgroup 191

weak convergence 41
— containment 147
— tensor product 43
— topology 41
weakly bounded set 41
— integrable function 131
weight vector 279
wild group 127

Subject Index

Weyl algebra 281
— chamber 93
— group 93
Weyl's theorem 275

Young diagram 264

Zermelo's axiom 2
Zorn's lemma 2

Die Grundlehren der mathematischen Wissenschaften in Einzeldarstellungen mit besonderer Berücksichtigung der Anwendungsgebiete

Eine Auswahl

23. Pasch: Vorlesungen über neuere Geometrie
41. Steinitz: Vorlesungen über die Theorie der Polyeder
45. Alexandroff/Hopf: Topologie. Band 1
46. Nevanlinna: Eindeutige analytische Funktionen
63. Eichler: Quadratische Formen und orthogonale Gruppen
102. Nevanlinna/Nevanlinna: Absolute Analysis
114. Mac Lane: Homology
123. Yosida: Functional Analysis
127. Hermes: Enumerability, Decidability, Computability
131. Hirzebruch: Topological Methods in Algebraic Geometry
135. Handbook for Automatic Computation. Vol. 1/Part a: Rutishauser: Description of ALGOL 60
136. Greub: Multilinear Algebra
137. Handbook for Automatic Computation. Vol. 1/Part b: Grau/Hill/Langmaack: Translation of ALGOL 60
138. Hahn: Stability of Motion
139. Mathematische Hilfsmittel des Ingenieurs. 1. Teil
140. Mathematische Hilfsmittel des Ingenieurs. 2. Teil
141. Mathematische Hilfsmittel des Ingenieurs. 3. Teil
142. Mathematische Hilfsmittel des Ingenieurs. 4. Teil
143. Schur/Grunsky: Vorlesungen über Invariantentheorie
144. Weil: Basic Number Theory
145. Butzer/Berens: Semi-Groups of Operators and Approximation
146. Treves: Locally Convex Spaces and Linear Partial Differential Equations
147. Lamotke: Semisimpliziale algebraische Topologie
148. Chandrasekharan: Introduction to Analytic Number Theory
149. Sario/Oikawa: Capacity Functions
150. Iosifescu/Theodorescu: Random Processes and Learning
151. Mandl: Analytical Treatment of One-dimensional Markov Processes
152. Hewitt/Ross: Abstract Harmonic Analysis. Vol. 2: Structure and Analysis for Compact Groups. Analysis on Locally Compact Abelian Groups
153. Federer: Geometric Measure Theory
154. Singer: Bases in Banach Spaces I
155. Müller: Foundations of the Mathematical Theory of Electromagnetic Waves
156. van der Waerden: Mathematical Statistics
157. Prohorov/Rozanov: Probability Theory. Basic Concepts. Limit Theorems. Random Processes
158. Constantinescu/Cornea: Potential Theory on Harmonic Spaces
159. Köthe: Topological Vector Spaces I
160. Agrest/Maksimov: Theory of Incomplete Cylindrical Functions and their Applications
161. Bhatia/Szegö: Stability Theory of Dynamical Systems
162. Nevanlinna: Analytic Functions
163. Stoer/Witzgall: Convexity and Optimization in Finite Dimensions I
164. Sario/Nakai: Classification Theory of Riemann Surfaces
165. Mitrinović/Vasić: Analytic Inequalities
166. Grothendieck/Dieudonné: Eléments de Géométrie Algébrique I
167. Chandrasekharan: Arithmetical Functions
168. Palamodov: Linear Differential Operators with Constant Coefficients
169. Rademacher: Topics in Analytic Number Theory
170. Lions: Optimal Control of Systems Governed by Partial Differential Equations
171. Singer: Best Approximation in Normed Linear Spaces by Elements of Linear Subspaces

172. Bühlmann: Mathematical Methods in Risk Theory
173. Maeda/Maeda: Theory of Symmetric Lattices
174. Stiefel/Scheifele: Linear and Regular Celestial Mechanics. Perturbed Two-body Motion—Numerical Methods—Canonical Theory
175. Larsen: An Introduction to the Theory of Multipliers
176. Grauert/Remmert: Analytische Stellenalgebren
177. Flügge: Practical Quantum Mechanics I
178. Flügge: Practical Quantum Mechanics II
179. Giraud: Cohomologie non abélienne
180. Landkof: Foundations of Modern Potential Theory
181. Lions/Magenes: Non-Homogeneous Boundary Value Problems and Applications I
182. Lions/Magenes: Non-Homogeneous Boundary Value Problems and Applications II
183. Lions/Magenes: Non-Homogeneous Boundary Value Problems and Applications III
184. Rosenblatt: Markov Processes. Structure and Asymptotic Behavior
185. Rubinowicz: Sommerfeldsche Polynommethode
186. Handbook for Automatic Computation. Vol. 2. Wilkinson/Reinsch: Linear Algebra
187. Siegel/Moser: Lectures on Celestial Mechanics
188. Warner: Harmonic Analysis on Semi-Simple Lie Groups I
189. Warner: Harmonic Analysis on Semi-Simple Lie Groups II
190. Faith: Algebra: Rings, Modules, and Categories I
191. Faith: Algebra II: Ring Theory
192. Mal'cev: Algebraic Systems
193. Pólya/Szegö: Problems and Theorems in Analysis I
194. Igusa: Theta Functions
195. Berberian: Baer*-Rings
196. Athreya/Ney: Branching Processes
197. Benz: Vorlesungen über Geometrie der Algebren
198. Gaal: Linear Analysis and Representation Theory
199. Nitsche: Vorlesungen über Minimalflächen
200. Dold: Lectures on Algebraic Topology
201. Beck: Continuous Flows in the Plane
202. Schmetterer: Introduction to Mathematical Statistics
203. Schoeneberg: Elliptic Modular Functions
204. Popov: Hyperstability of Control Systems
205. Nikol'skii: Approximation of Functions of Several Variables and Imbedding Theorems
206. André: Homologie des Algèbres Commutatives
207. Donoghue: Monotone Matrix Functions and Analytic Continuation
208. Lacey: The Isometric Theory of Classical Banach Spaces
209. Ringel: Map Color Theorem
210. Gihman/Skorohod: The Theory of Stochastic Processes I
211. Comfort/Negrepontis: The Theory of Ultrafilters
212. Switzer: Algebraic Topology—Homotopy and Homology
213. Shafarevich: Basic Algebraic Geometry
214. van der Waerden: Group Theory and Quatum Mechanics
215. Schaefer: Banach Lattices and Positive Operators
216. Pólya/Szegö: Problems and Theorems in Analysis II
217. Stenström: Rings of Quotients
218. Gihman/Skorohod: The Theory of Stochastic Processes II
219. Duvaut/Lions: Inequalities in Mechanics and Physics
220. Kirillov: Elements of the Theory of Representations
221. Mumford: Algebraic Geometry I: Complex Projective Varieties